PHYSICAL ENVIRONMENTS

A case-study approach to AS and A2 Geography

JACK and MEG GILLETT

Hodder & Stoughton

A MEMBER OF THE HODDER HEADLINE GROUP

Dedication

For Jean and Jenny

Orders: please contact Bookpoint Ltd, 130 Milton Park, Abingdon, Oxon OX14 4SB. Telephone: (44) 01235 827720. Fax: (44) 01235 400454. Lines are open from 9.00–6.00, Monday to Saturday, with a 24 hour message answering service. You can also order through our website www.hodderheadline.co.uk

British Library Cataloguing in Publication Data
A catalogue record for this title is available from the British Library

ISBN 0 340 782072

First Published 2003
Impression number 10 9 8 7 6 5 4 3 2 1
Year 2009 2008 2007 2006 2005 2004 2003

Typeset by Fakenham Photosetting, Fakenham, Norfolk.
Printed in Italy for Hodder & Stoughton Educational, a division of Hodder Headline Plc, 338 Euston Road, London NW1 3BH.

Acknowledgements

The publishers would like to thank the following individuals, institutions and companies for permission to reproduce copyright illustrations in this book: © B & C Alexander, 4.16; © Paul Almasy/Corbis, 4.24c; AP Photo/Sizuo Kanbayashi, 5.38h; AP Photo/Enric Marti, 5.24; AP Photo/Karel Prinsloo, 1.48; AP Photo/Murad Sezer, 5.28; AP Photo/Str, 5.25; © Yann Arthus-Bertrand/Corbis, 4.28; Mr Bagshawe, White Scar Cave, 3.40 and 3.43;© Jonathan Blair/Corbis, 5.8c; Cambridge University Collection of Air Photographs, 4.24b, 4.24d; CEGB, 4.41; © Lloyd Cluff/Corbis, 1.51, 152 top right, 5.20; John Connor Press Associates, Lewes, 2.8a; Digital image © 1996 Corbis, original image courtesy of NASA/Corbis, 1.50; © Alan Curtis/Leslie Garland Picture Library, 3.36; © Ecoscene/Corbis, 4.40; © Mark Ferguson/Life File, 5.14b;G.S.F. Picture Library, 2.24, 3.31, 3.41, 4.12, 4.27, 5.8a, 5.8b, 5.14a, 5.14c, 5.14d, 5.14e; Jack and Meg Gillett, 1.23, 1.24, 1.55, 1.56, 1.58, 2.5, 2.19, 2.20, 2.23a, 2.23b, 3.38, 3.42, 4.35, 4.39, 4.43, 4.44, 5.14f; © George Hall/Corbis, 5.21; © Robert Holmes/Corbis, 1.53; Hulton Archive/Fox Photos, 3.26; Hulton Archive/Keystone Collection, 1.42; Keystone, 3.24; Kyodo News, Tokyo, 5.38c, 5.38d, 5.38e, 5.38g; © Andrew Lambert/Leslie Garland Picture Library, 3.30; © Vincent Lowe/Leslie Garland Picture Library, 3.46; P & A Macdonald Aerographica, 5.41, 5.42; © Kenneth Marthinsen/Leslie Garland Picture Library, 4.34; PA Photos/Paul Barker, 1.40; Panos Pictures/© Jim Holmes 1992, 2.37; Popperfoto/Reuters, 5.26; © PRPA, 2.15, 5.40; © Colin Raw/Leslie Garland Picture Library, 3.47; © Peter Reynolds/Leslie Garland Picture Library, 2.32; © Paul Ridsdale/Leslie Garland Picture Library, 3.45; © Skyscan Photolibrary, 4.38; © Skyscan Photolibrary/Malcolm Bradbury, 2.4; © Skyscan Photolibrary/William Cross, 4.36; © Skyscan Photolibrary/Bob Croxford, 3.33; © Skyscan Photolibrary/Bob Evans, 2.16; © Skyscan Photolibrary/R&R Photography, 2.9; © Hubert Stadler/Corbis, 4.23; © Liba Taylor/ActionAid, 2.36; © *The Times* 21 July 2000, photo by Andrew Hasson, 2.8b; © Topham Picturepoint, 5.38f; © David and Peter Turnley/Corbis, 1.31, 1.32; © University of Dundee, 5.43; USGS Photo Archive, 5.11, 5.12; © Michael S. Yamashita/Corbis, 5.38a, 5.38b.

The publishers would also like to thank the following for permission to reproduce material in this book: Copyright © Guinness World Records 2001. Use by permission. For the statistics about the largest levées; © Longman Group UK Limited 1990, reprinted by permission of Pearson Education Limited; Nelson Thornes for extracts from *Environmental Science* by K. Byrne, Nelson, and *Geography: An Integrated Approach* by D. Waugh, Nelson; this product includes mapping data licensed from Ordnance Survey® with the permission of the Controller of Her Majesty's Stationery Office. © Crown copyright. All rights reserved. Licence No. 100019872; © Arthur N. Strahler. Used by permission for altitudinal zoning of life zones and climates in Africa.

Every effort has been made to trace and acknowledge ownership of copyright. The publishers will be glad to make suitable arrangements with any copyright holders whom it has not been possible to contact.

Contents

Unit 4: Natural environments 2

Unit 5: Plate tectonics, volcanoes and earthquakes

Book unit number and title	Main themes/topics in unit	AQA 'A'		AQA 'B'		Edexcel 'A'		Edexcel 'B'		OCR A		OCR B	
		AS	A2	AS	A2	AS	A2	AS	A2	AS	A2	AS	A2
	Systems, ecosystems	10.3			13.1		4.3		4	5.1.2			
1 Rivers	Hydrological cycle; drainage basins/patterns; river valley features; river flooding/ management; water supply issues	10.1		10.1	1.2	4.1		Unit 1	5.2	5.1.1	5.4.2	5.1.1 5.1.2	5.5.1
2 Coasts	Coastal features/formation processes; coastline exploitation/ management/protection; coastal flooding/management		13.1	11.2		1.3		Unit 1	4.5c 4.6	5.1.2	5.4.1	5.1.3	5.5.1
3 Natural Environments 1	Biomes British Isles climate Granite/limestone landscapes	10.3 10.2	13.2	10.1	13.1	1.13	4.3 4.1		Unit 4 Unit 4	5.1.4	5.4	5.1.1 5.1.1	5.5.1
4 Natural Environments 2	Ice/glacial feature formation; ecology/economies of high altitude/latitude zones; wilderness area protection		13.2 14.1	11.1	13.1		4.2	Unit 4	4.5c 4.6 5.2		5.4.3		5.5.1 5.5.3
5 Plate tectonics, volcanoes and earthquakes	Plate tectonics; tectonic hazard effects/precautions/ prediction		13.3		13.2	1.1			5.2	5.1.4			5.5.1

Specification links table

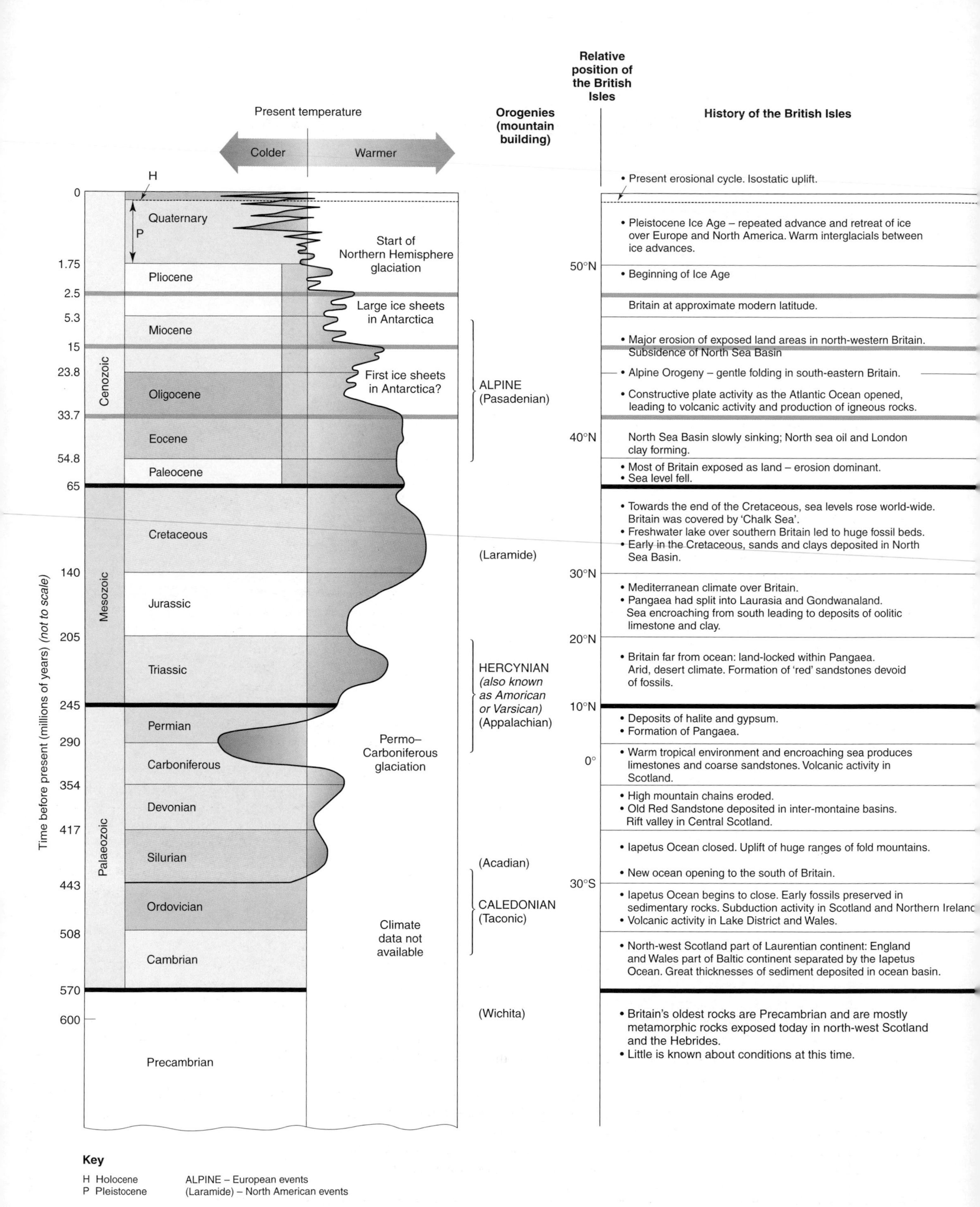

The Stratigraphical Column

INTRODUCTION

The newspaper article in Figure 1 appeared in the very short-lived *The Planet on Sunday* newspaper on 12 December 1999. This article was written to increase public awareness of the effects of natural disasters on the loss of human life. Only ten days later, 30 000 people were killed and 250 000 homes destroyed by mudflows triggered by exceptionally heavy tropical downpours in the Andes. In the last week of 1999, further disasters causing fatalities included floods on the south coast of England, storms over Paris and avalaches in the Alps.

Such disasters have major long term economic implications for the regions concerned, but they also remind us of the inescapable link between natural processes and the 'quality of life' of both the human and the wildlife populations of this planet. If the effects of such events are to be minimised in the future, it is vital that detailed studies are made of past events and the processes which led to them.

The chief aim of *Physical Environments* is to provide its readers with broadly-based, yet sufficiently detailed subject matter to meet the requirements of the AS and A2 geography specifications introduced in September 2000. It has been written with the knowledge that GCSE syllabuses tend to be somewhat 'light' on physical geography topics, and that Year 9 option restrictions may have denied some students the opportunity of studying the subject to GCSE level. Its authors have, therefore, taken care to introduce topics in ways which will allow *all* students to acquire a sound knowledge of physical concepts, processes and features, as well as their implications for the world's populations. The sequence of teaching units has been ordered so that AS-dominant topics are introduced first (in Units 1 and 2); whereas Units 4 and 5 examine themes which are predominantly A2 material.

Much of the content required by the specifications is delivered through case studies rich in location-specific material. These studies are supported, where appropriate, by theory units containing key terms and fundamental concepts presented in easily revisable formats. A number of investigation suggestions are incorporated into activities within Units 1–5; Unit 2 also includes a role-play exercise. The final section of the book provides a number of revision questions ranging in difficulty from an awareness of basic key terms to synoptic essays requiring detailed knowledge of a broad range of topics.

Jack and Meg Gillett

Natural disasters claim 52,000 lives worldwide

Insurance firms pay out £13.75bn

DISASTERS have claimed 52, 000 lives worldwide so far in 1999 and will cost insurance companies £13.75 billion, according to a recent report.

Swiss Reinsurance Company said it was the fourth most expensive year in the history of the industry. And the cost to insurers would have been far higher but for very low insurance coverage in Turkey and Taiwan, both hit by devastating earthquakes.

Insurance losses peaked in 1992 as a result of the devastating Hurricane Andrew in the United States, with costs estimated at £20.63 billion in today's prices.

"The trend toward high losses from natural catastrophes and man-made disasters seems to be continuing unabated," the study said.

Swiss Re said increased building in endangered regions, the world's expanding population, and losses of expensive new technology are contributing to the trend.

It estimated total losses directly due to major disasters this year would total more than £40.6 billion.

Damage in Turkey's August earthquake totalled £12.5 billion, but only £1.25 billion of that was insured.

In Taiwan, losses totalled £8.75 billion but just £625 million was insured.

Nearly 25,000 people lost their lives in earthquakes and more than 16,000 as a result of storms, Swiss Re said. More than four-fifths of all the deaths were caused by natural disasters.

Swiss Re is the world's second largest re-insurer, meaning it insures insurance companies against major losses.

Its periodic reports are seen as authoritative guides to the global insurance industry.

Planet on Sunday, 12 December 1999

FIGURE 1 Natural disasters in the news

Systems

The natural world is very complex. Until the 1960s, geography was chiefly concerned with describing 'how things were'. Then, in an attempt to unravel and understand the complex inter-relationships between apparently separate processes on the Earth's surface, geographers introduced the use of **models.** Models are simplified representations of reality – in much the same way as a model train is a scaled-down version of the real thing. One type of model is a **system** i.e. a method of identifying a single aspect of reality (a **unit**), analysing the relationships between its component parts and then, possibly, discovering how the unit interacts with both its immediate and its wider environment. A systems model is usually displayed as a **flow diagram** which, in its most simple form, can be shown as a **black box system** (see Figure 2).

FIGURE 2 A black box system

Such models allow us to identify **inputs** to a system (i.e. the entries of matter and/or energy) and **outputs** from a system (i.e. the mass, energy or change of state that is produced when an input passes through a system). However, the 'black box' (i.e. something we cannot see into) does not allow us to examine the **processes** which operate within the system. A more sophisticated model would incorporate information about what is happening internally and might look more like that in Figure 3, where:

- A **store** is a part of the system that can hold energy or matter
- A **transfer** is part of the system that redistributes energy or matter from one point to another.
- A **flow** is any movement within the system.

All systems can be classified in one of three ways:

1 Isolated – where there is no input or output of either energy or matter. Whilst it might be that our own universe is a system of this type, this is not a useful model for geographers, whose chief concern is to understand inputs, outputs and transfers at a variety of scales.

2 Closed – where there are inputs and outputs of energy but not of matter (or mass). The Earth is generally thought of as being a 'closed' system, as it receives energy from the Sun but only negligible amounts of matter (in the form of cosmic dust and occasional meteorites) from Outer Space. Whilst Earth losses energy back into Space in the form of heat, there is no mass or matter transfer back into the universe. The **global hydrological cycle** is a system of this type.

3 Open – where there are inputs and outputs of both energy and matter. This means that the system routinely interacts with other, co-existing systems as well as the environment generally.

Systems modelling can be a very useful tool in promoting our understanding within both physical and human geography and it also helps us to focus on the inter-relationships between the physical and human 'worlds'. In physical geography, we usually model real, tangible elements of the world, for example:

- landforms and their associated processes
- interactions between plants, animals and soils
- aspects of hydrology.

In human geography, systems modelling often incorporates concepts, perceptions and the somewhat less tangible aspects of reality such as regions, organisations and the socio-political aspects of human activity.

There are several types of system, but three are particular useful in the study of physical geography:

1 Morphological systems are the simplest type. In such systems, the component materials are identified together with the structural links which exist between them. Often, it is these linkages which are the focus of attention and morphological systems diagrams therefore sometimes indicate the strength of relationships between components. Such a method of modelling applied to a river system is shown in Figure 1.10 on page 17.

2 Cascading systems are more complex. They seek to represent how energy or matter flows (or **cascades**) through the system and how this eventually transforms inputs into outputs. The whole system is seen as a related set of **sub-systems** through which the energy or matter passes. A sub-system is a small unit with identifiable inputs and outputs which forms part of a larger unit or system.

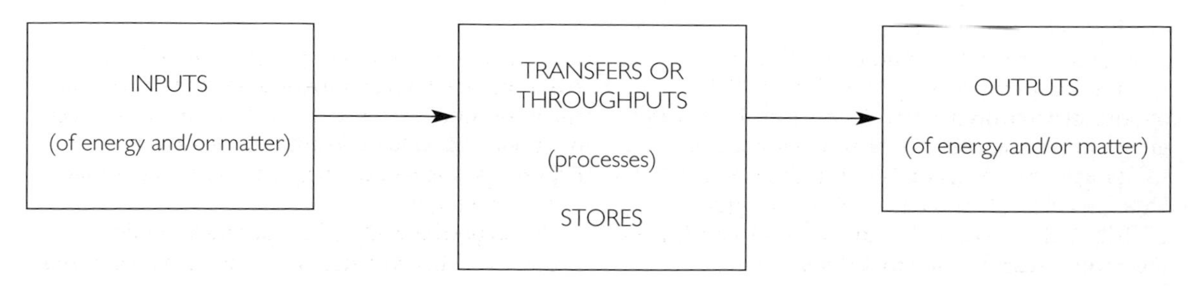

FIGURE 3 A typical systems diagram

An example of a cascading system can also be seen in an erosional drainage basin (i.e. a system which focuses upon the **erosional** processes within a basin rather than its hydrology). Here, debris and water are transferred from the sloping valley sides (one sub-system) into the stream channel (another sub-system). Such an example shows how the focus of interest can vary – even when considering the same unit. We can create very different systems diagrams to represent co-existing aspects of the same unit. **Ecosystems** (discussed later in this unit) are a specialised type of systems representation which can be viewed as being a cascading system i.e. having both energy and mass passing through a succession of **trophic levels** (sub-systems) of the unit.

3 **Process-response systems** are created by amalgamating morphological and cascade systems in order to demonstrate how energy/matter flows can affect various components as they move through the system. They are used to stress the relationship between **system form** and **system process**.

A fourth, highly-specialised and much more complex type of systems representation also exists, but is more often utilised in human geography. This is the **control system**. Such models are used when the key components in the phenomenon being modelled are directly controlled by human activity. Occasionally physical geographers may encounter these when, for example, studying a drainage basin in which river flow is entirely dependent upon the release of water from a dam/reservoir complex.

Finally, geographers may also refer to **hierarchical systems**. These are not true systems models but are representations of the world as a mesh of sub-systems which are ordered hierarchically in such a way that a system at one level (or of one order) may become a component part of the next higher order system, and so on throughout the hierarchy (Figure 1.4 on page 13).

When a system's inputs and outputs are balanced, it is said to be in a state of **dynamic equilibrium**. If one element in a system changes due to an outside influence, it then upsets the balance and affects other components in the system. This process is called **feedback** and may be either positive or negative. **Positive feedback** occurs when the effect of the change is increased and consequently moves the whole system away from equilibrium. **Negative feedback** occurs when the entire system acts to lessen the effect of the initial change. The processes within the system then work to restore the balance or equilibrium of the entire system (Figure 4).

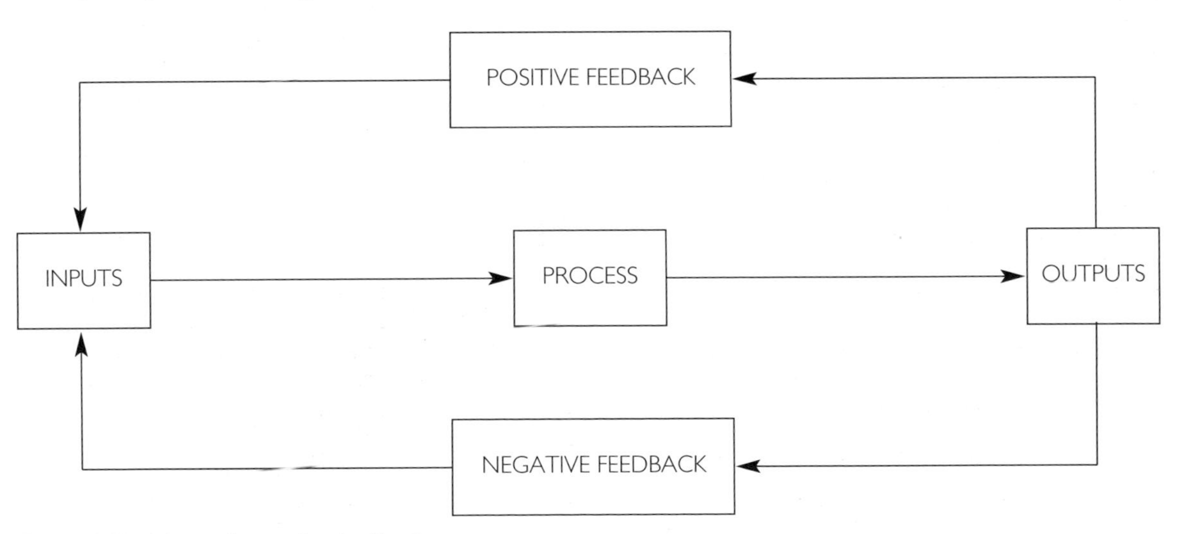

FIGURE 4 Positive and negative feedback

Ecosystems

A systems approach to studies of the living environment was developed within the field of biological/environmental sciences long before any geographers incorporated it into their way of thinking. 'Eco' has its roots in the Greek word '*oikos*', meaning 'a place to live'. The unique contribution of ecosystems to our understanding of the natural world is due to its focus on the *interactions* between living organisms and the environment. Some eminent scholars believe that the importance of these interactions has often been overlooked and that we are now beginning to experience some of the first serious adverse effects of this oversight.

Ecology (the study of the processes which operate within ecosystems) has its own specialised vocabulary, and geographers have borrowed many of its key terms. Ecology is also the study of **organisms**. An organism is any individual form of life; this means that you, the reader, are one form of organism. Groups of organisms which resemble each other in terms of appearance, genetics, chemistry and behaviour are known as **species**. Humans belong to the species *Homo Sapiens*, or Humankind. The total number of members of one species inhabiting a given area at a particular time is known as its **population**. The physical space, region or area in which a population lives is called a **habitat** and all the species which share a population's habitat are known as a **community**. An ecosystem is created when we consider a community of species interacting with each other and the non-living (**abiotic**) environment of energy and matter. All of the Earth's ecosystems comprise that which geographers call the **biosphere** – although other scientists prefer the term **ecosphere**.

Not all ecosystems are natural; reservoirs, agricultural fields and garden ponds are all examples of ecosystems which have been created by human intervention within the natural world. The size of any ecosystem is determined entirely by the requirements of the researchers studying it. For example, the ecology of river basins may be studied at a national scale so as to include all drainage networks within Britain, at a regional scale to focus on one particular river network or at a local scale involving a study of a limited stretch of a minor tributary stream.

Ecosystems have both inputs and outputs (just like any other open system), and the outputs of one often form the inputs to a co-existing ecosystem. Ecosystems are not only linked in this way, but often illustrate very well the concept of hierarchical ordering. This is because their component organisms, species, populations, etc. form sub-units, each having its own highly individualised set of processes, inputs and outputs, yet each sub-unit is an integral part of the larger ecosystem.

There are few clear-cut boundaries between ecosystems. Even the apparently clear interface between land and sea is far from being a distinct boundary; this is partly because animals have adapted to function equally well both in the sea and on land, but is also the result of the development of major ecosystems such as salt marshes which form **transition zones (ecotones)** between land and sea. Ecotones are often richly populated habitats, supporting communities from both adjacent ecosystems as well as habitat-specific populations.

Many ecosystem links are extremely fragile and it is very important that we have a clear understanding of them in order to maintain the stability of natural environments, and also in order to predict the likely outcomes of human initiatives having an impact on the natural world.

Climate patterns are the most important factor in determining which organisms can survive in a particular type of habitat. Certain animal species (most notably humans) have been able to **adapt** remarkably well to contrasting climatic conditions, whereas other species (particularly plants) have adapted slowly, over long periods of geological time, to live in very specific habitats and are now restricted to that particular climate and/or soil type. This tendency for plant species to be environmentally sensitive has led biologists to divide the Earth's land surface (the ecosphere) into large regions typified by distinctive climate and life forms (especially flora); such regions are called **biomes** (see Unit 3). Within every biome are many ecosystems, reflecting plant adaptations to local soils, drainage, relief and micro-climates (i.e. local variations in temperature, moisture and light conditions).

The **hydrosphere** is sub-divided into **aquatic life zones** instead of biomes, the key factor being water **salinity** instead of climate. Lakes and streams make up the **freshwater life zone**, whilst estuaries, coasts, seas and deep oceans comprise the **marine life zone**. Each life zone incorporates and supports a range of ecosystems appropriate to its individual environmental conditions.

In order for any terrestrial or marine ecosystem to be **sustainable** over time, it must possess both the energy and the nutrients necessary to support its resident organisms as well as the resources to dispose of and recycle their waste products. The living part of an ecosystem is known as its **biotic** component and the individual organisms which reside there are often referred to as **biota**. The non-living parts of an ecosystem (e.g. solar energy, water, air and nutrients) form the abiotic environment. All ecosystems must comprise both

ACTIVITIES

1 How does the ecosystem approach contribute to our understanding of the natural world?

2 Write a short, clear definition of each of the following important terms:

- abiotic
- community
- ecosphere
- ecotone
- habitat
- organism
- population
- species

3 a) Sketch a hierarchical representation of yourself as an individual organism living within your own family group and its neighbourhood community; take care to name the various components in your hierarchy diagram.

b) Which aspects of your community's abiotic environment are you able to identify? Mark these on your diagram.

c) Which other species can you identify as sharing your habitat:

- closely
- more distantly?

biotic and abiotic components. The biotic organisms within ecosystems are categorised according to the way in which they obtain their food. **Autotrophs** (self-feeders) are green plants which have the ability to produce sugars and other food compounds directly from abiotic nutrients via **photosynthesis**. During this process, chlorophyll converts solar energy, carbon dioxide and water into oxygen and chemical energy such as glucose. The oxygen by-product of photosynthesis is essential for maintaining most animal life on Earth. In the hydrosphere, plants and algae are the main autotrophs in both fresh water and coastal environments. However, in deep ocean waters, **phytoplankton** are the dominant autotrophs. Autotrophs are often referred to as **primary producers**; all other organisms are known as **consumers** as such organisms must consume other organisms in order to gain energy. Consumers are more correctly known as **heterotrophs**. Heterotrophs may be sub-divided into:

- **Herbivores** (also known as **primary consumers**) which eat only primary producers i.e. plants.
- **Carnivores** (meat eaters) who may be secondary or tertiary consumers. **Secondary consumers** eat herbivores (i.e. primary consumers) whilst **tertiary consumers** eat other carnivores.
- **Omnivores** (meat and plant eaters); with the exception of humankind, most omnivores are hunters.

There are also several other groups of consumers which, whilst fitting into the above categories, are recognised separately because they fulfil quite different roles within the recycling process:

- **Scavengers**, which feed on organisms killed by others or which have died of natural causes.
- **Detritivores**, which live off the waste products of other living organisms.
- **Decomposers** (mainly consumers such as bacteria and fungi), which complete the recycling of organic materials by breaking down such products and releasing the resultant inorganic compounds back into the soil and water – where they become available once again as nutrients for the primary producers.

ACTIVITIES

4 What is the chief difference between an ecosystem and a biome?
5 There are many biomes, but only two major aquatic life zones. Suggest reasons why this should be the case.
6 What do you understand by the term 'sustainable ecosystem'?
7 Explain why it is crucial for humankind to protect and conserve Earth's vegetation cover.

Energy flows through an ecosystem via **food chains** and **food webs**. The rate of energy flow is directly linked to the size of the resident populations and/or local human management initiatives. Any organism is potentially a source of food for another organism and the hierarchy of organisms within an ecosystem is called a food chain.

In every ecosystem, the origin of all its energy is sunlight, which is converted into energy by plants through the process of photosynthesis. As energy moves upwards through a food chain, much of it is lost – often as heat given out to the atmosphere. The remaining energy passes to the animals which eat the plants, and then to other animals which consume the plant-eaters. In its most simple form, a plant-to-animal food chain might look like the representation in Figure 5.

Biologists refer to each level within a food chain as a feeding level or **trophic level**. In the example given in Figure 5, there are three trophic levels.

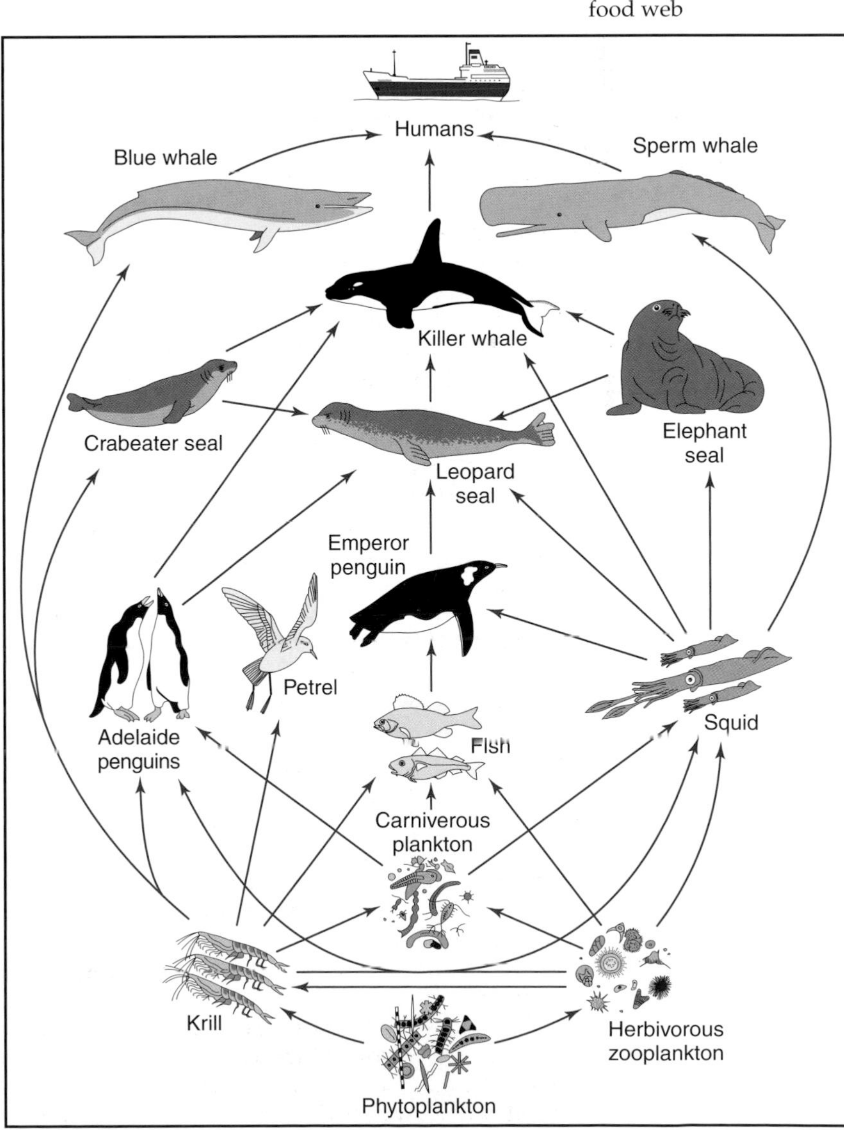

FIGURE 6 A simplified food web

SUNLIGHT ⟶ **GRASS** Producer *1st trophic level* ⟶ **COW** Primary Consumer *2nd trophic level* ⟶ **HUMAN** Secondary Consumer *3rd trophic level*

FIGURE 5 A trophic level food chain

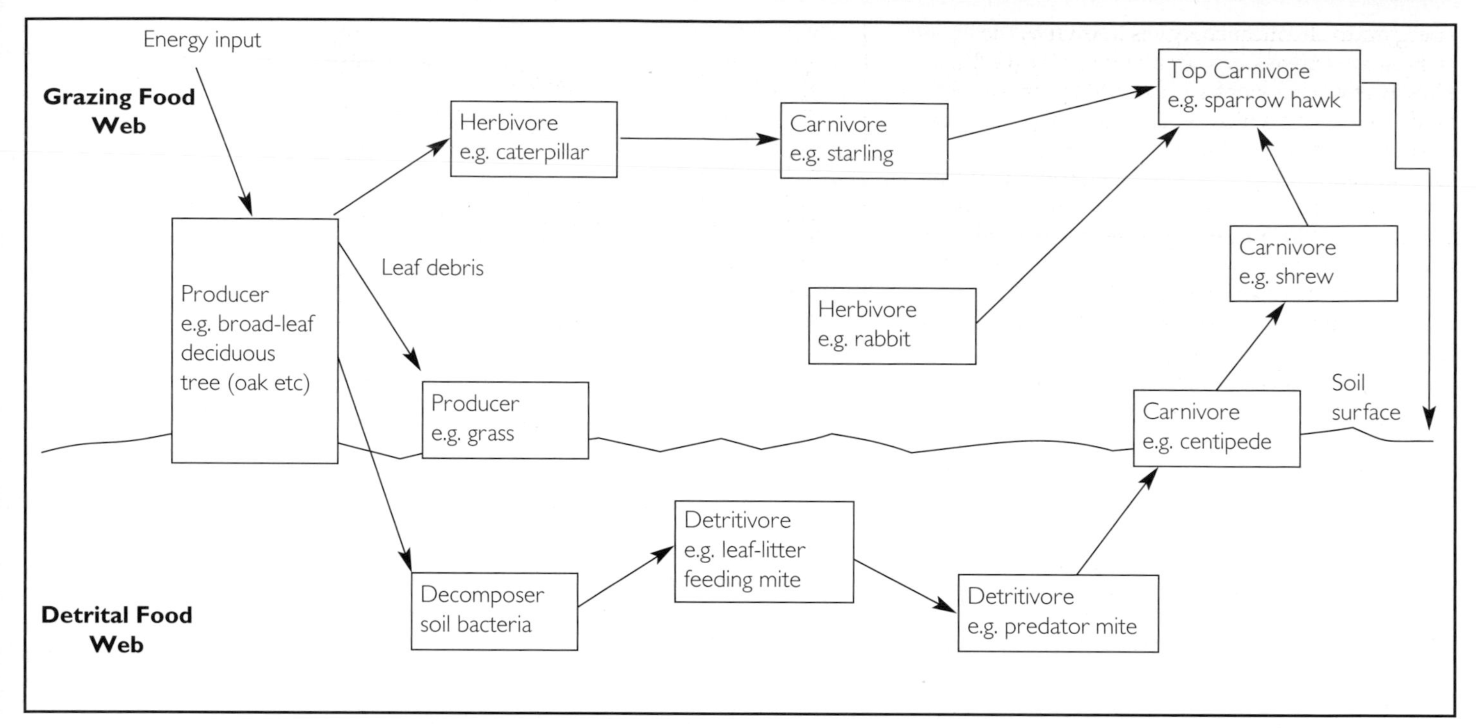

FIGURE 7 Grazing and detrital food webs

In most ecosystems, the daily flows of energy are far more complex than such simple food chains suggest; many animals are components in a large number of food chains and some can be consumers at up to three levels. Such a network of interconnected feeding levels and chains may be displayed as a food web (Figure 6). Still greater accuracy in the consideration of energy flow actually reflects that the flow of energy is a two-way process – up the food web and back down it – as energy is recycled for the 'next round' of the energy cycle. This recycling is usually shown as two food webs – but remember that they are interconnected, not separate:

- The **grazing food web**, through which energy travels to the **top carnivores** (Figure 7a).
- The **detrital food web**, which represents the recycling of energy through organic waste materials. (also Figure 7b).

Much energy is lost as it flows through a food chain (or web), most of it to the environment as heat. In addition, some residual energy actually passes straight through the animal unutilised and is excreted as waste. It is quite usual for as little as 10 per cent of the total energy intake to be utilised by the consumer. This extracted energy is digested and converted into the organism's bodily material. The amount of energy actually available at each trophic level declines as the energy travels up the food web. Significantly, the energy loss is often as great as 90 per cent at each trophic level. This means that if green plants capture 1000 units of energy, only 100 units are likely to be available to the herbivore(s) consuming the plants, and a mere 1 per cent of the original energy reaches the carnivore at the next level in the food web. Such transfers and losses are often represented diagrammatically as a **trophic pyramid** or an **energy flow pyramid** (Figure 8). As a result of such a high rate of energy loss, it is rare for food webs to incorporate more than five trophic levels; four levels is usually the maximum. Beyond this level, there is very little residual energy left and so top carnivores find it almost impossible to support their own existence beyond trophic level five.

FIGURE 8 A trophic pyramid

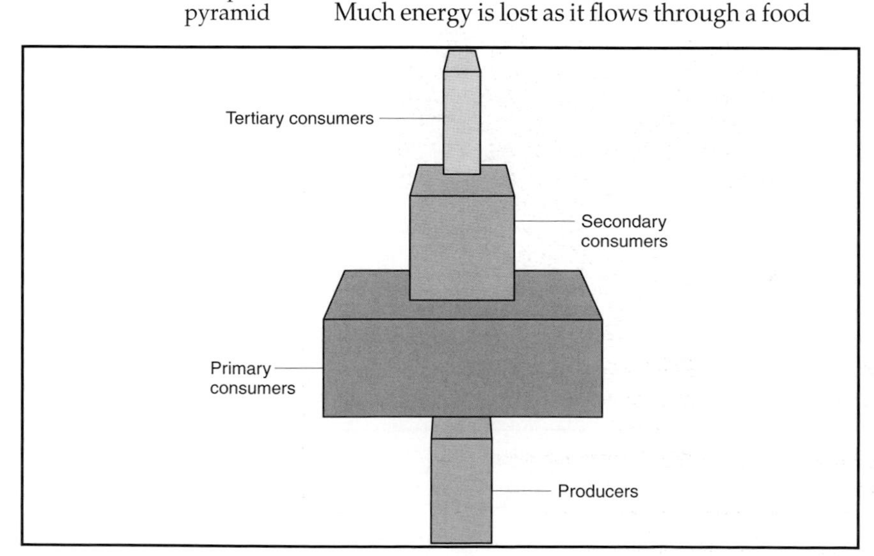

The total weight of all the dry organic matter at any single trophic level in a pyramid is known as its **biomass**. Scientists calculate **dry weight** simply because the water which all organisms contain is of no use as a nutrient source or an energy source. Biomass is used by scientists as a means of representing the chemical energy stored at each trophic level and can also be represented in pyramid form (although biomass pyramids are not always triangular in shape). For example, the biomass pyramid for a typical ocean has a small base – representing its primary producers – but a large mass for primary consumers (Figure 9).

The rate at which any ecosystem converts solar

energy into chemical energy is known as the ecosystem's **gross primary productivity** (GPP). However, such a measure is of little practical use, as all of the energy which is produced by plants can never be available to the primary consumers (because some of the energy which every plant produces must be utilised to sustain the producer). The residue available to consumers is the **net primary productivity** (NPP), which may be obtained by using the formula shown in Figure 10.

NPP is usually measured in units of energy (e.g. kilocalories) or mass (e.g. grams) per unit of area per year; such measurements will, therefore, have units such as kcals/m^2/yr or g/m^2/yr. Where large quantities of biomass are involved, the unit of measurement may be expressed as kilograms or tonnes (i.e. kg/m^2/yr or t/m^2/yr). There are considerable variations in NPP between different types of ecosystem/biome (Figure 11). Tropical rain forests, estuaries and marshes are especially productive – but their primary productivity is not directly accessible to humans as food; estuarine and marsh vegetation is inedible by humans and most rainforest nutrients are stored within the vegetation rather than the soil, which is the main reason why deforested areas used as arable land become infertile within only two or three seasons. Not surprisingly, tundra regions, deserts and deep oceans have particularly low overall NPPs per unit area; the impressive total global productivity of oceanic areas is due to the fact that oceans cover 71 per cent of the Earth's surface rather than a high unit/area output. In direct contrast is the considerable annual contribution of the rainforests – which cover only 2 per cent of Earth's surface.

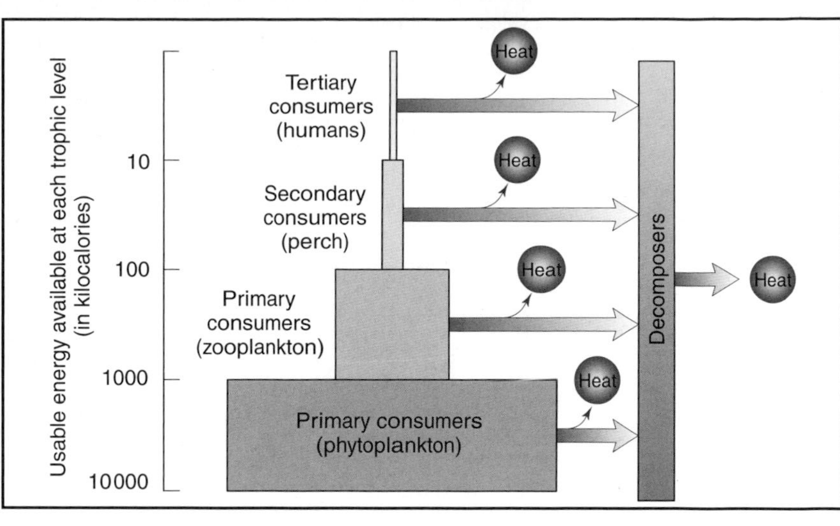

FIGURE 9 A typical oceanic biomass pyramid

$$\text{NPP} = \text{rate at which plants store chemical energy} - \text{rate at which plants use chemical energy}$$

FIGURE 10 Net primary productivity (NPP)

ACTIVITIES

8 Explain what you understand by the terms:
- biome
- food web
- photosynthesis
- detrital food web
- trophic pyramid
- food chain
- detritivore
- grazing food web
- GPP
- NPP

9 Re arrange these organisms so as to create an appropriate food chain:
Cabbage Human Sunlight Chicken Caterpillar

10 a) Explain how energy is lost as it passes up a trophic pyramid.
b) Trophic pyramids are often used to support the theory that the world's food shortages could be reduced if most people became vegetarians. Outline the reasoning behind this thinking.

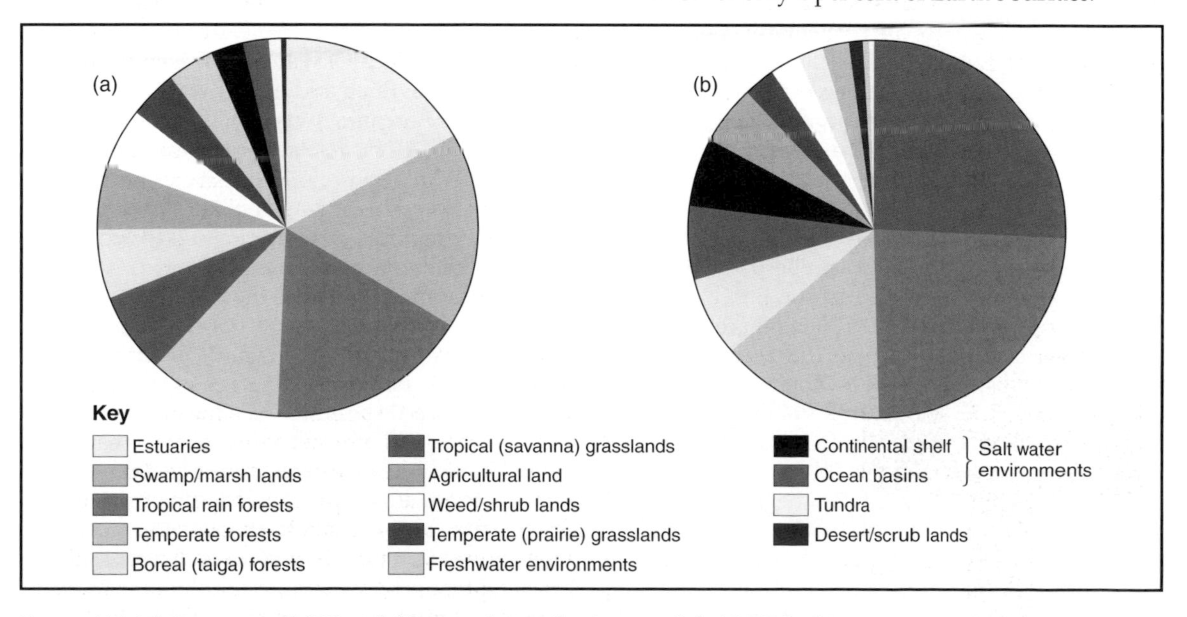

FIGURE 11 (a) Mean annual NPP and (b) annual contributions to global NPP by biome

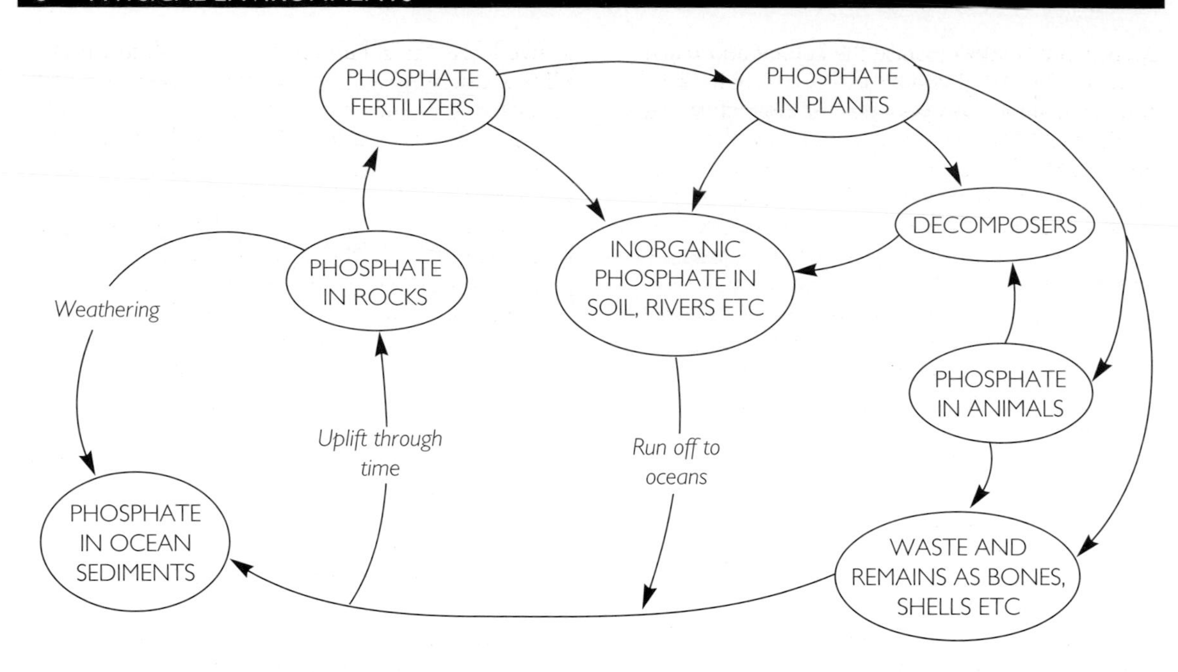

FIGURE 12 The phosphorous cycle

In addition to energy flow, **flows of nutrients** also take place within ecosystems. Nutrients, which are a fundamental requirement of all organisms, can flow through water and/or air. The **carbon cycle** and the **nitrogen cycle** are both **atmospheric cycles**. Carbon (as CO_2) and nitrogen are available within the atmosphere and may occur in cycles at local, regional or global scales. The recycling of both phosphorous and sulphur are also important aspects of ecosystems. Phosphorous is an essential nutrient for plants and animals as it is a critical part of DNA; it is also important for respiration and is present in bones, teeth and shells. Unlike nitrogen and carbon, phosphorous does not circulate in the atmosphere, because it cannot exist on Earth's surface in gaseous form; the phosphorus cycle is, therefore, a **sedimentary cycle** (Figure 12). Sedimentary cycles tend to be slow and occur only at local and regional scales. The recycling of phosphorous in soil can be particularly slow, and resultant phosphorous deficiencies can seriously inhibit plant growth over a long period of time.

One of the most notable features of any community or ecosystem is that, whilst its natural systems' response is to move toward balance or equilibrium, it exists in a state not of stability but of constant change. Such change is gradual and is the inevitable result of the ecosystem perpetually seeking balance. Environmental conditions are constantly subject to modification and ecosystems reflect these changes over time. Individual organisms and species have to adapt to such change – or risk extinction. Therefore, a stable ecosystem is not a static one; it continually adapts and rebalances itself in ways which maintain it in a constant state of flux. Plant **adaptation** is often the first discernable indicator of such change – unless the ecosystem is responding to some sudden, catastrophic event. Adaptation occurs when any organism develops beneficial mutations which allow the individual to cope more effectively in changing environmental conditions and to produce offspring with the same adaptive traits. In plants, this often enhances the individual's ability to compete for light, space and nutrients. Vegetation adaptations usually have a knock-on effect to the animal communities sharing their particular habitat. This gradual adjustment process is called **ecological succession** (or **community development).**

Ecological succession is of two kinds. **Primary succession** involves the development of biotic communities in areas which lack soil (or bottom sediment in water). Such environments may include recently-cooled lava flows, freshly-cut quarry faces and new garden ponds. On land, plants cannot survive without soil; therefore **pioneer species** which colonise newly-exposed surfaces are always soil-forming species such as mosses and lichens. Such species are able to take the nutrients which they need directly from the rock surface. As soil begins to form, the area will be colonised by species such as bacteria, fungi and insects which, when they die, add to the organic content of the embryonic soil. Eventually short grasses, herbs and ferns are also able to move in. Such colonising species are known collectively as **early-successional species**. It may take several hundred years for the soil to become deep and fertile enough to support **mid-successional species** such as the taller grasses and shrubs. Eventually, however, trees will begin to colonise the area and grow towards maturity. As the range of plant species increases within an area, so too do the animal species co-existing alongside them; animals can only move in after their sources of food have become secure, and early successional animal species are always small herbivores. Large grazers often arrive much later, long after small predator carnivores have become well established. Provided that primary succession is not interrupted by natural or human interventions, **late-successional**

species ultimately colonise the habitat and reach maturity. The originally barren area will eventually support a stable, complex and mature community.

Today, it is increasingly rare to find undisturbed primary succession communities; even in our most remote 'wilderness' areas, past interventions by human activity have disturbed the ecosystem in some way. **Secondary succession** is far more common. This occurs where the original natural vegetation has been disturbed or wholly destroyed but, crucially, where the soil has remained *in situ*. It is unusual for secondary succession to fully replicate the primary community which it replaces. This may be because there have been subtle or marginal changes in the environment which mean that slightly different species will now colonise the area. More likely, such changes will occur simply because the whole process is not starting again from scratch. As soil is already in place, the initial colonisation process does not have to be repeated; this will create a different evolutionary process which will mean that an identical succession cannot redevelop in this habitat. If, on the other hand, the soil cover is damaged (e.g. by erosion) it is even less likely that the primary community will be able to re-establish itself. Changes in the soil's composition, depth and fertility will facilitate the development of a new and different secondary vegetation succession – with a major knock-on effect to the other animal species which move into the habitat as vegetation cover develops.

Traditionally, ecological succession has been viewed as a natural and orderly growth towards a **climax community.** Climax communities used to be regarded as the ultimate goal of the developmental process. They were typified by a stable, predictable community based upon one or more long-lived plant species having the ability to sustain and replace themselves well into the future. However, more recent research suggests that both primary and secondary successions are highly unpredictable processes. Certainly, recent studies of secondary succession show that the same mixture of plant species rarely re-grows in any habitat following clearance. Many of today's ecologists hold the view that even single, random or chance events can create major changes to ecological succession within habitats. Such ecologists believe that there is no 'ecological plan' by which stable, climax communities are produced in any area of the world. Instead, they believe that succession is the product of unique struggles between competing species for light, air, water and essential nutrients. They tend to use the term **relatively stable community** (or **mature community**) instead of the traditional concept of climax community. Even so, implicit within the concept of a mature community's development is an assumption that the succession has remained uninterrupted by human interference. Today, such communities are very rare indeed. Primary or secondary successions which are halted or modified by human interference (be it intentional or not) are known as **plagioclimax successions**.

The sequence of events across time which produces a mature community from a barren

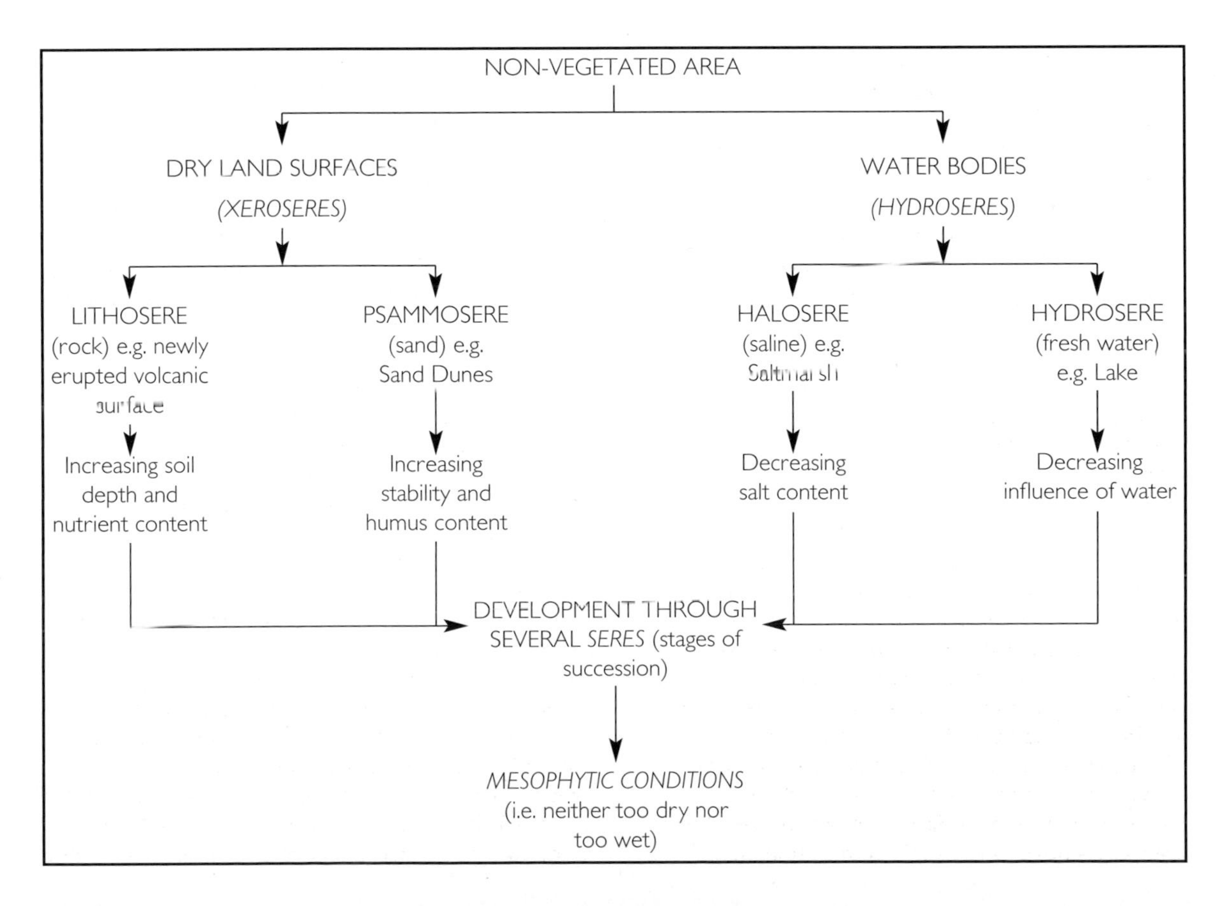

FIGURE 13 Prisere development

expanse of land or sea is known as a **prisere** and each individual phase of development towards maturity is referred to as a **sere**. Communities which develop upon dry land are classified as **xeroseres**, whilst those adapted to water are known as **hydroseres**. Specialised xerosere communities which develop in the extreme dry conditions found in sandy areas (e.g. upon sand dunes) are sub-classified as **psammoseres** whilst those communities which develop on land areas possessing moisture-retaining properties are sub-classified as **lithoseres**. In aquatic environments, the established sub-classification can be a little confusing because all water adaptations are known as hydroseres, but this term is also the sub-classification for all fresh water ecosystems. Communities which adapt to saline conditions are sub-classified as **haloseres**. Many hydroseres are, in fact, transitional seres in the development of mature land-based communities. For example, saltmarsh communities inevitably dry out as the sea retreats, eventually becoming dry land. Salt traces are eventually washed from the soil and land-based plants are then able to colonise the 'reclaimed' land – ultimately generating fully mature xerosere communities. Figure 13 traces all the above developments separately within dry and wet environments.

ACTIVITIES

11 Copy and compete the table below, which describes the various factors having the potential to disturb or even destroy an ecosystem. Do this by brainstorming with a partner, then recording a wide selection of the factors which you have discussed together.

	Catastrophic events	Gradual events
Natural events		
Human activities		

12 Write notes to show that you understand the similarities and the differences between primary and secondary succession. Give named, located examples from both your personal knowledge and your research.

13 a) Working with a partner or within a small group, identify the 'authentic primary succession' of Britain. You may wish to concentrate your attention initially on one particular part of Britain (probably that in which you live) – but take care not to ignore the wider, whole-region perspective.

b) Repeat the exercise in part a) above – but focus your research upon one very different biome overseas.

c) Present your findings for parts a) and b) both orally and in map/diagrammatic form – to either a small group or the whole class.

14 a) Write a short paragraph (not notes or bullet points!) which explains your understanding of the concept of 'climax community'.

b) Now add a further paragraph to your piece of writing to show that you understand fully why ecologists currently prefer the concept of 'maturity' to that of climax community.

Unit 1
RIVERS

The hydrological cycle

The hydrological cycle (also known as the **water cycle**). The movement of water between air, land and sea (the atmosphere, the lithosphere and the hydrosphere). Acts as a recycling system, providing the biosphere with a source of pure water.

Evaporation Changing of a liquid into a gas (water into water vapour) at a temperature less than the boiling point of the liquid. Energy is needed for this change. The rate of evaporation depends upon the existing moisture content of the atmosphere, air temperature, wind speed and surface heat.

Transpiration Biological process in which plants lose water vapour through the pores (stomata) in their leaves. Transpiration is affected by the time of year, type and amount of vegetation, the length of the growing season, plus the factors affecting evaporation (above).

Condensation Changing of vapour or gas into a liquid (water vapour into water or ice). Although generally considered to be the result of air cooling, condensation may be caused by a variety of events.

Precipitation Deposition of moisture from the atmosphere onto the Earth's surface. Can be in the form of rain, hail, sleet, snow, frost or fog. Precipitation varies over time and space. As a general rule, the greater the intensity, the shorter the duration of the precipitation. Precipitation may be the result of convection, relief or the passage of a frontal system.

Infiltration Movement of water into soil or rocks. Infiltration is controlled by a range of factors including cracks, cultivation, freezing, the intensity and type of precipitation and soil **porosity.** Infiltration may not occur at all if the speed of the water is too great (see run-off) or if the rock/soil is either **saturated** or **impermeable**.

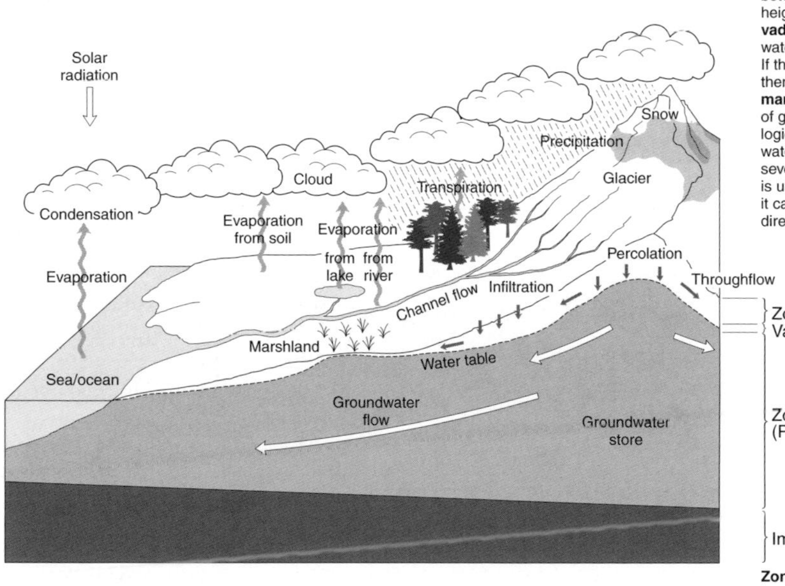

Groundwater At some point in its underground journey, water may reach an impermeable rock and its downward journey then stops. It collects at this point and gradually fills up all the pore spaces – creating a **zone of saturation** (also known as the **phreatic zone**). The top of this zone (the boundary with non-saturated rock above) is called the **water table**. The height of the water table varies as the saturation zone increases and decreases. This area between the maximum and minimum heights of the water table is called the **vadose zone**. The near-stationary body of water below the water table is groundwater. If the water table reaches the land surface then this will saturate. If the land is flat, **marshland** will develop. The deepest layers of groundwater are not active in the hydrological cycle. Below 1000 m, much of the water is saline, fossil water which has taken several millennia to collect. This **deep storage** is unlikely to discharge into a river channel; it can only seep into other basins or **outflow** directly into the oceans.

Zone of aeration The underground zone between the water table and the soil capillary zone. After prolonged/intensive rainfall it may reduce as the water table rises to the surface.

Percolation Filtration of water downwards through the soil and permeable rocks to groundwater storage areas (**aquifers**). Water travels along bedding planes/joints or through pores in porous rocks.

Run-off All of the water entering a stream or river which then flows out of the drainage basin. It is the surface discharge of water generated as overland flow, **throughflow** and groundwater flow.

Channel flow Refers to all of the water present in streams/rivers and which subsequently leaves the drainage basin via river flow. Although some precipitation falls directly into channels, most water reaches the channel through run-off.

Groundwater flow (groundflow) Although groundwater cannot move downward, it may move slowly sideways (laterally); such underground lateral movement is known as groundwater flow. It can account for leakage from one basin to another.

FIGURE 1.1 The global hydrological cycle

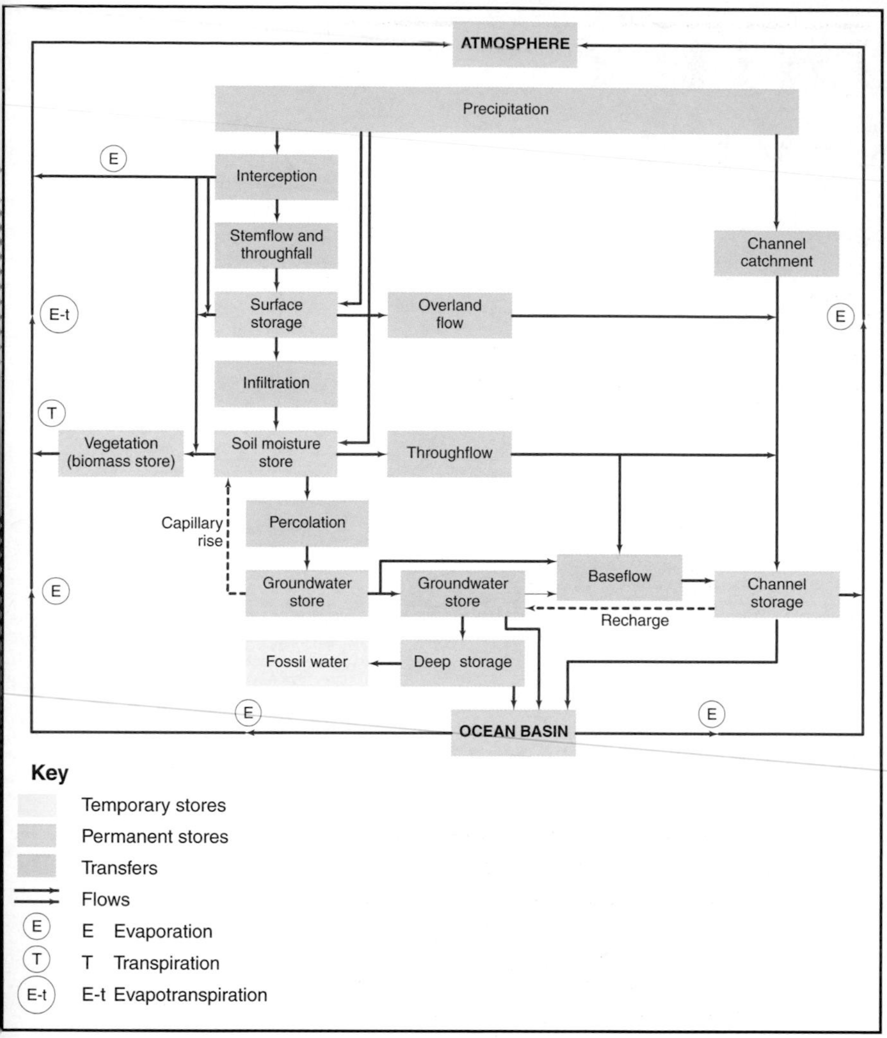

FIGURE 1.2 The hydrological cycle as a closed system

ACTIVITIES

The following activities are based on Figures 1.1 and 1.2.

1 Describe the similarities and explain the differences between the processes of evaporation and transpiration.

2 a) What environmental factors influence evaporation and transpiration rates?
b) Explain briefly how some or all of the factors you mentioned in a) above can affect evapo-transpiration.

3 Explain precisely the nature of each of these precipitation types and the environmental conditions under which they tend to occur: fog, frost, hail, rain, sleet and snow.

4 Precipitation may take place under three different sets of environmental conditions; the resulting forms of precipitation are usually described as convectional, orographic (relief) and frontal (depression).
a) Using annotated sketches only, make it clear that you understand the basic principles involved in the formation of each of these three types of precipitation.
b) What is the common factor which leads to precipitation taking place in all three sets of circumstances?

5 a) Draw a simplified global hydrological cycle in sketch form. Annotate your diagram appropriately and give it a suitable title.
b) Working from memory only, write a short definition of each of the terms which you have included in your sketch.

6 Study Figure 1.1 and refer back to page 2.
a) What do you understand by the term 'system'?
b) Explain what is meant by the concept of a 'closed' system.
c) Compare and contrast the diagram of the hydrological cycle (Figure 1.1) and its representation as a system (Figure 1.2).
d) What are the advantages and disadvantages of these two ways of modelling reality?

7 The hydrological cycle is considered to be a 'closed' system. Such systems may, however, be considered to have inputs and outputs.
a) What is an 'input' to any system?
b) What is an 'output' to any system?
c) List the input(s) and output(s) to the global hydrological cycle?

8 Explain the meanings of the following terms: capillary rise, precipitation, recharge.

9 The systems diagram (Figure 1.2) shows both 'transfers' and 'flows'. With specific reference to this diagram, explain the difference between the two terms.

10 Identify from the systems diagram (Figure 1.2):
a) The various routes that precipitation can follow to reach the sea/ocean basin.
b) The ways in which water may be lost from channel storage.

c) The routes by which water droplets may flow from the soil moisture store.

11 Explain carefully the difference between:
a) Infiltration and percolation.
b) Throughflow and groundwater flow.
c) Groundwater storage and deep storage.
d) Groundwater flow and base flow.

12 Water may be stored at various points throughout the hydrological cycle. Make a copy of the table opposite. For each of the 'stores' listed:
a) Indicate in the column headed 'Time span', whether you consider this to be :
brief (i.e. hours – days)
short term (weeks)
medium term (months – years)
long term (hundreds – thousands of years)
infinite (millennia)
b) On the same copy of the table, provide a definition and field example for each store.

Store	Time span	Definition	Field example
Channel			
Soil moisture			
Surface			
Biomass			
Deep			
Groundwater			
Fossil water			

13 The systems diagram (Figure 1.2) includes a store known as 'fossil water'.
a) What is 'fossil water'?
b) If fossil water is effectively 'lost' to the hydrological cycle, why do you think that we include it within the system, rather than considering it to be an 'output'?
c) Under what conditions (considered over millennia) might fossil water return to the active hydrological cycle?

Drainage basins

At a very simple level, a drainage basin may be modelled as a black box system, i.e. one in which inputs and outputs are represented, but the processes at work within the system are shown only as a black box; such an approach is shown in Figure 1.3. However, whilst such a diagram may show a basic understanding of the system, it fails to facilitate an understanding of the processes which are involved – and understanding processes is what a lot of geography is actually about.

The systems diagram of the global hydrological cycle (Figure 1.2) is a form of **cascading system**; it shows where mass and/or energy moves from one sub-system to another but does not indicate how the transfer occurs. In a similar way, a drainage basin can also be represented as a cascading system.

There are other ways of representing the processes on-going in such a drainage system. One is to consider the drainage basin as a hierarchy (sometimes this is known as a hierarchical system) which focuses upon the components of the drainage basin in terms of scale. Such an approach is illustrated in Figure 1.4. Using a model like this does not show an understanding of process – but can be used to show an understanding that sub-systems 'nestle' within larger sub-systems in much the same way that Russian dolls fit together one inside the other from very tiny to large!

In practice, we can 'focus-in' to study in depth any aspect of a unit, and regard it as being a discrete unit in its own right – provided that we clearly understand what belongs inside the sub-unit and what remains outside it (Figure 1.5). This means that we have to define very clearly the inputs and outputs of our study unit. Within any single drainage basin, therefore, a biogeographer may focus upon soils and vegetation, whilst a geomorphologist may be more interested in the topographical aspects of the same system. With this in mind, Figure 1.6 represents the vegetation and

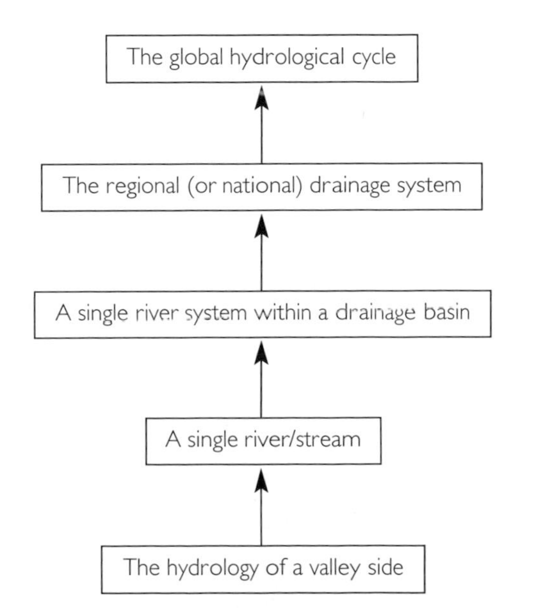

FIGURE 1.4 A drainage basin represented as a hierarchical system

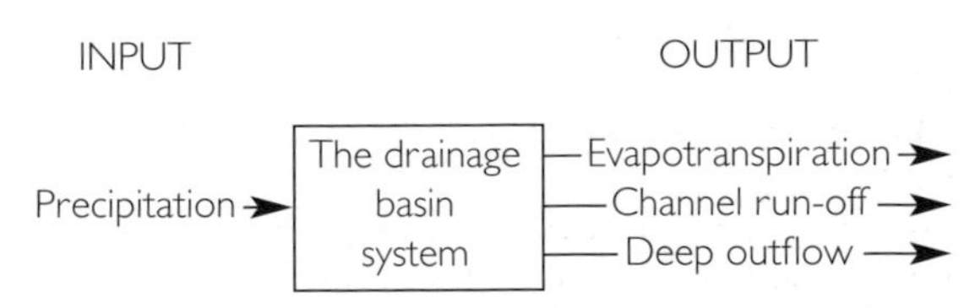

FIGURE 1.3 A drainage basin represented as a black box

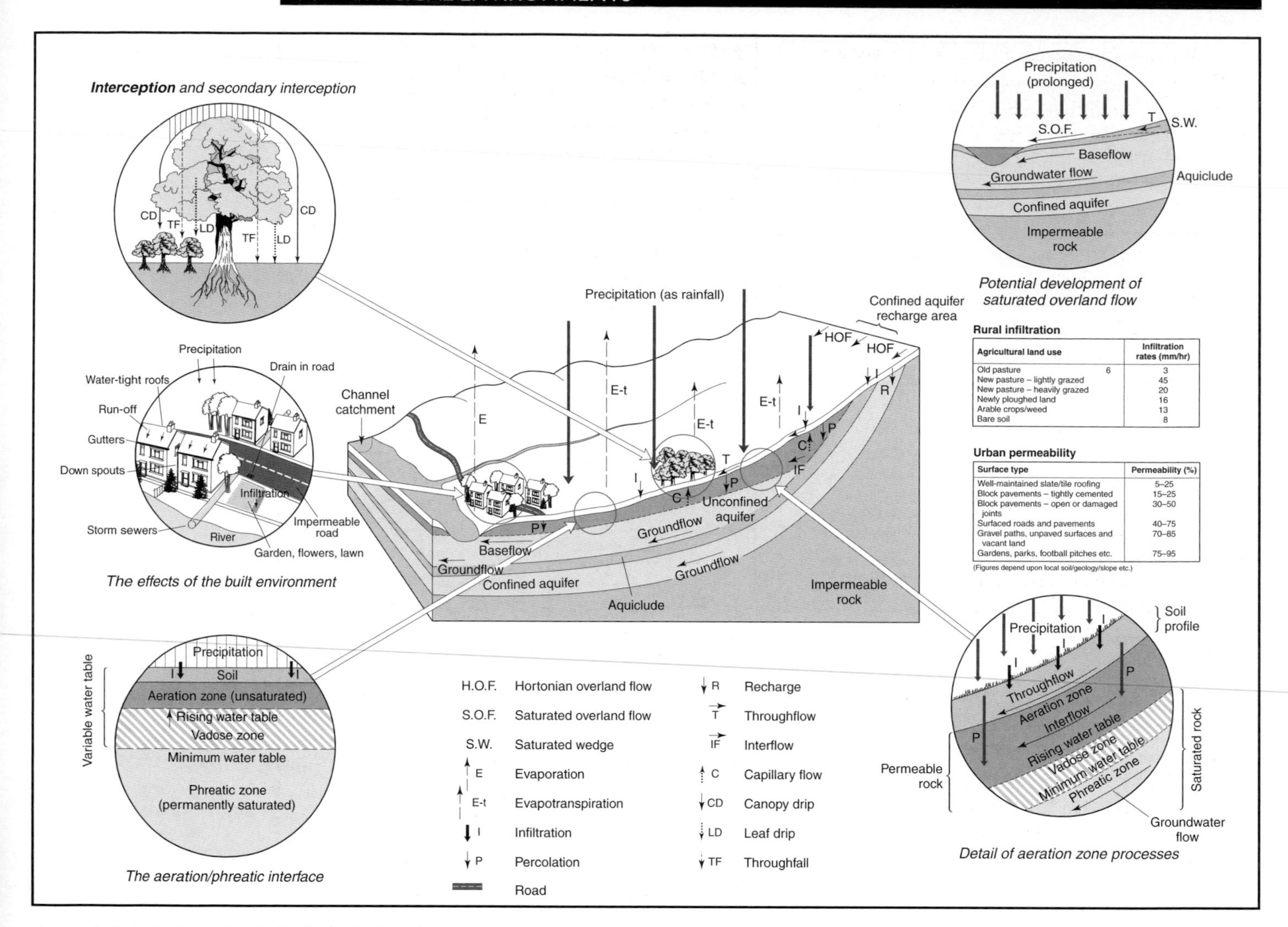

Rural infiltration

Agricultural land use	Infiltration rates (mm/hr)
Old pasture 6	3
New pasture – lightly grazed	45
New pasture – heavily grazed	20
Newly ploughed land	16
Arable crops/weed	13
Bare soil	8

Urban permeability

Surface type	Permeability (%)
Well-maintained slate/tile roofing	5–25
Block pavements – tightly cemented	15–25
Block pavements – open or damaged joints	30–50
Surfaced roads and pavements	40–75
Gravel paths, unpaved surfaces and vacant land	70–85
Gardens, parks, football pitches etc.	75–95

(Figures depend upon local soil/geology/slope etc.)

FIGURE 1.5 A drainage basin hydrological cycle

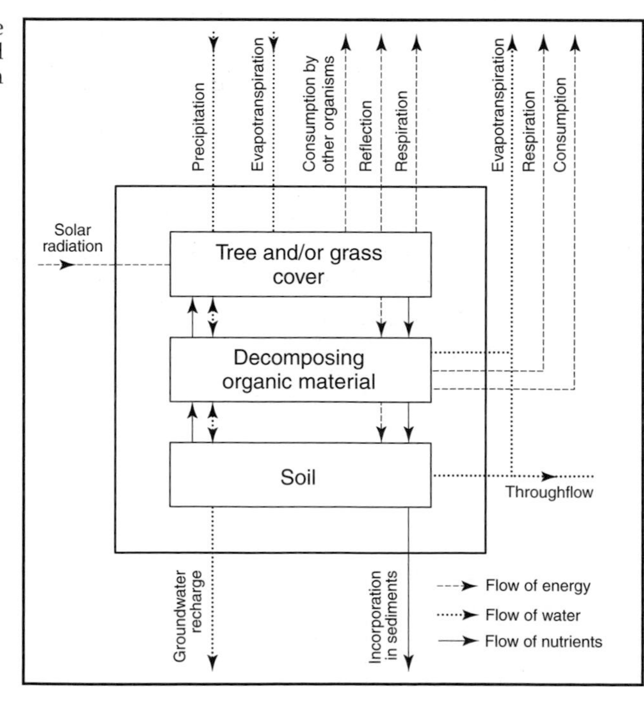

FIGURE 1.6 The vegetation and soil sub-system

soil sub-system of a hypothetical drainage basin. Notice that, at this scale, the diagram represents flows and transfers which have not been included in our consideration of the drainage basin so far. In this way, we can see that inputs and outputs, transfers and stores within any unit are dependent upon the scale and depth of the study.

ACTIVITIES

1 Write brief definitions of each of the following terms, many of which are included in Figure 1.1. Ensure that your work shows clearly that you understand each term.

- infiltration capacity
- infiltration rate
- overland flow
- permeable rock
- phreatic zone
- throughfall
- throughflow
- vadose zone
- water table
- zone of aeration

2 a) Write a short piece of prose which shows that you understand the process of interception and its effects upon the hydrology of a river basin. You should make reference to: canopy drip; leaf drip; throughfall.
b) Now extend this original piece of work with a second paragraph to show that you understand the concept of secondary interception. Ensure that your writing makes clear the source of the water which is being intercepted.
3 Using Figure 1.5 draw a systems diagram to represent the hydrology of a drainage basin. This system is an open system. You will need to identify inputs and outputs as well as the transfers, flows and stores. Include in your model the process of 'interflow' and effects of the urban development. If you wish, you could also incorporate into your diagram the effects of other human interventions such as dams, reservoirs and water abstraction.
4 a) Describe and explain how the development of human settlement and other construction projects (such as roads) interferes with natural hydrological processes.
b) What are the consequences for downstream river discharges of large scale urban developments?
c) Make a list of other human interventions that may disturb the hydrological cycle.
d) Which (if any) of the activities you considered in b) and c) above may be expected to produce negative feedback when considered from the point of view of river discharge?
e) What are the effects on the overall system of:

- Negative feedback?
- Positive feedback?

5 a) Explain the meanings of each of the following terms used in Figure 1.5.

- aquiclude
- aquifer
- confined aquifer
- impermeable rock
- recharge
- recharge area
- unconfined aquifer

b) Aquifers provide important sources of water for drinking, agriculture and industry across the globe. On your own or with a partner, investigate and describe in detail an example of such usage of artesian water. You may like to consider:

- The source of precipitation/water which fuels the aquifer together with details of its recharge capabilities. Does usage exceed recharge?
- Uses of the water.
- Likely future trends in consumption.
- Advantages and disadvantages of using underground water – and of relying on this water source (consider the geo-political considerations if your example is overseas).

Prepare your work with a view to making a class or group presentation of your findings. Prepare appropriate maps and other illustrative data for display, and consider using hand-outs for those who will watch your presentation.
6 Figure 1.7 shows the outline of a model of the hydrology sub-system of a drainage basin. Below are the labels which have been omitted from the diagram:

- deep infiltration
- channel storage
- infiltration storage
- interception storage
- surface infiltration

On your own copy of Figure 1.7, label each store correctly and identify which arrows represent the flows of water and water vapour. Also, add the following labels:

- base-flow
- evapotranspiration
- channel run-off
- precipitation

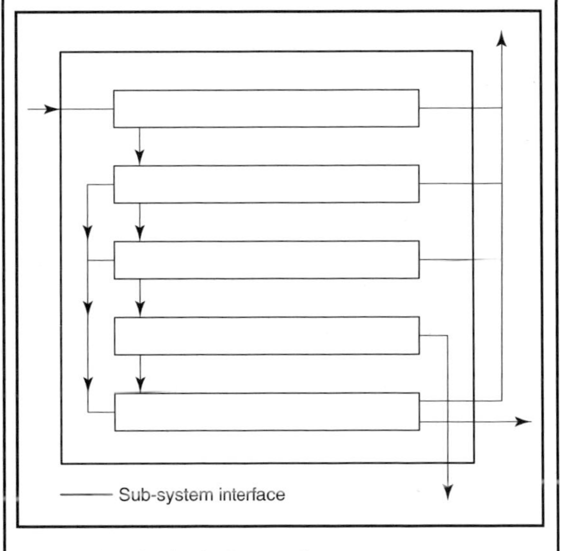

FIGURE 1.7 The hydrology sub-system

Dendritic resembles a tree – with a main 'trunk' and spreading branches. It is found in areas where the underlying geology is homogeneous. This means that the effect of geology upon the surface pattern is neglible.

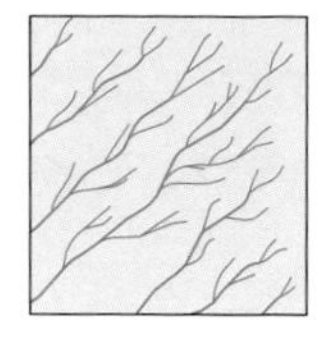

Parallel drainage occurs when a number of trunk rivers (or streams) flow in the same (parallel) direction.

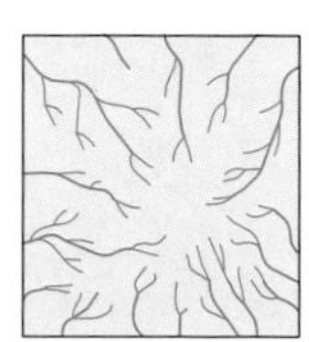

Radial drainage is usually associated with a domed landscape feature – or is the relict feature of such a landscape which remains long after the up-folded landscape has been eroded. Trunk rivers drain outwards from the central region.

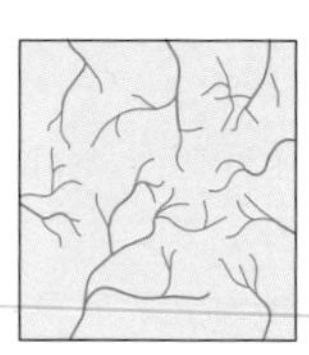

Annular drainage can easily be mistaken for radial drainage. This pattern may also be associated with a domed landscape – by may also occure in a basin. The significant factor is that annular drainage develops where there are bands of weaker rock which occur in a circular pattern. Streams flow *around* the central hub, rather than away from it.

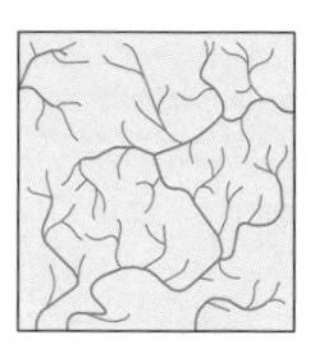

Rectangular patterns are characterised by right-angle bends. it is most usually found where the underlying rocks are jointed or faulted; these features 'guide' the course of surface drainage.

Trellised drainage can, at first glance, be mistaken for a radial pattern. The main difference is that here the major tributaries are very long and are parallel. It develops where sedimentary rocks have been folded to give long, parallel bands of hard and soft rocks.

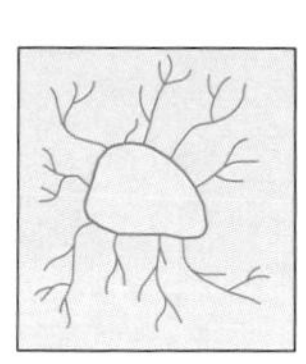

Centripedal patterns are formed when streams flow inwards toward a central depression (basin or caldera) usually creating an inland sea.

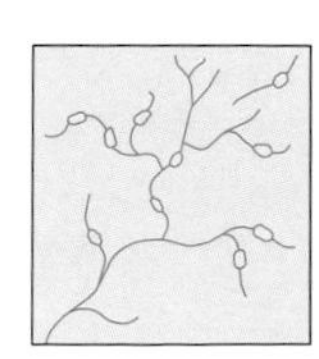

Deranged patterns show no recognisable pattern. They have no linkage to either geology or topography. There are usually found in areas which have recently been glaciated.

FIGURE 1.8 Drainage basin patterns

Drainage patterns

Describing the pattern of drainage within any area or drainage basin has proved to be a very imprecise art. In the field, it is an impossible task to undertake because of the need for an overview of the entire area under study. It is an exercise which needs a detailed map of the area. Until the 1960/70s, geographers described river drainage patterns in words – using a set of standard adjectives such as dendritic (which means 'tree-like') and radial (arranged like rays or spokes in a wheel). Sometimes, in the past, such adjectival descriptors were enhanced with an additional phrase or sentence e.g. 'Annular drainage develops on an up-fold (or dome) where rocks of varying resistance are later differentially eroded.' Some drainage patterns do fit neatly into their 'adjectival category', but some natural patterns need to be viewed very creatively in order to match them with a descriptor. The eight 'traditional' drainage patterns are shown in Figure 1.8.

Descriptors such as these are a very imprecise way of describing what we see – partly because it is very subjective and also because it serves little useful purpose. It does not allow us to compare (or contrast) drainage in different basins, even basins of a similar shape and size. What it can do, however, is draw our attention to other factors such as underlying rock type and structure because these often influence the surface drainage pattern; for example, radial drainage indicates an underlying (or relict) doming of local rocks. Of the eight typical patterns, six can be found within Britain, the remaining two are best exemplified by cases from overseas.

ACTIVITIES

1 Below are eight locations. Using an atlas, match each location to one of the drainage patterns illustrated in Figure 1.8.

- River Lune
- The Weald area of Kent
- The Aral Sea
- The River Thames in the Vale of Oxford (Between the Cotswolds and the Chilterns)
- The English Lake District
- Berriesdale Water, Caithness
- Adirondack Mountains, USA
- Glen Fyne, Argyll

When you have completed the task successfully, you will have a classic example of each archetypal pattern.

The River Lune (see Figure 1.9) exhibits a typical dendritic pattern of drainage although this is truncated to the west by the proximity to the drainage basin of the River Kent.

The drainage basin can also be represented as a **morphological system** showing the components (or sub-systems) of the drainage basins and their linkages (Figure 1.10). Morphological systems are flow diagrams which are simple representations of reality; they do not show processes.

Since the 1960/70s, geographers have moved on to use more quantitive methods which allow comparisons and contrasts to be drawn between drainage patterns – allowing greater understanding of the processes at work. **Drainage density** is the easiest (and most widely used) method of measuring (and then noting similarities and differences within) drainage basin patterns. Drainage density is an expression of the total length of drainage channel in relationship to the total area of land across which it flows. Thus, in the example in Figure 1.11, the total length of all the surface drainage is 74 km whilst the area of the basin is 160 km^2. This gives a drainage density of 74/160 or 0.46 km/km^2.

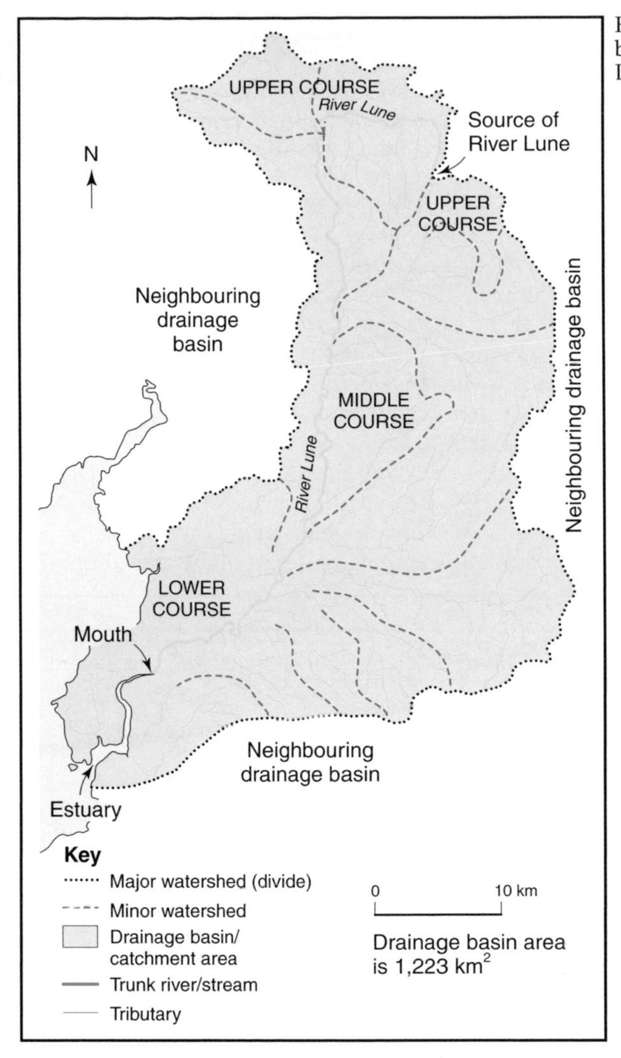

FIGURE 1.9 Drainage basin of the River Lune, Lancashire

ACTIVITIES

2 Study Figure 1.9.

a) Measure the total length of the channel drainage across the entire area of the map (*Hint*: measure each tributary in turn and record this information on a copy of the map. Your final answer needs to be given in kilometres).

b) Calculate the drainage density for the River Lune's catchment area in km/km^2 (*Hint*: the area of the drainage basin is given on the map).

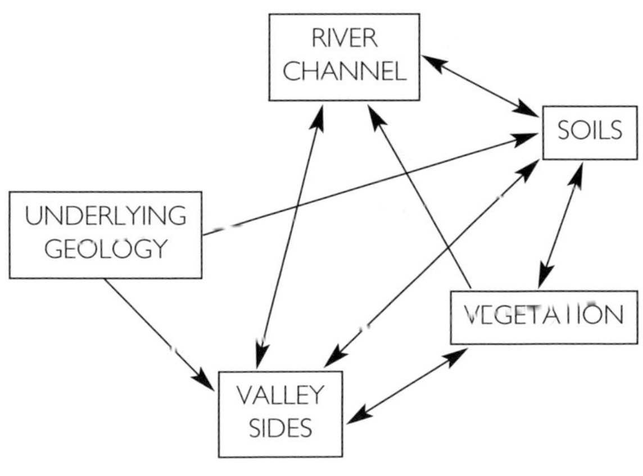

FIGURE 1.10 A drainage basin as a morphological system

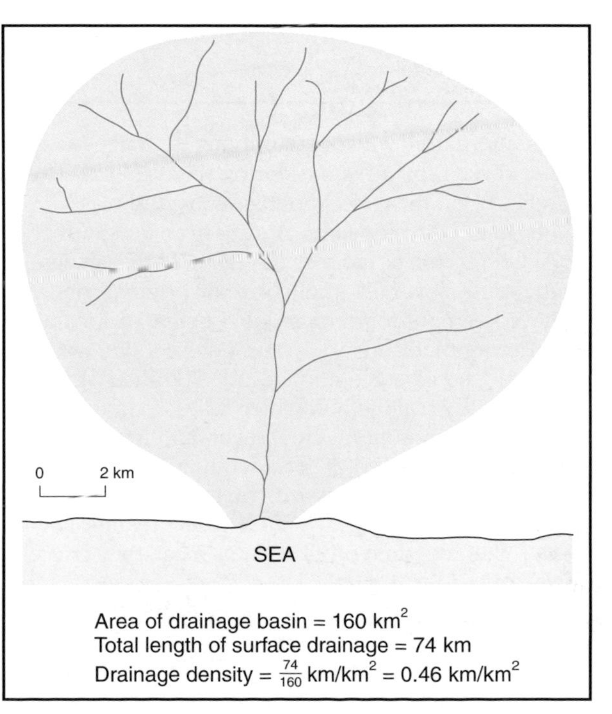

FIGURE 1.11 Calculation of drainage density

In theory, the drainage density of basins of different sizes can be compared. This means that your calculation of the River Lune's drainage density allows you to make comparisons with global river systems such as the River Nile, the Amazon, or your own local river basin. However, in reality, there are some practical difficulties.

ACTIVITIES

3 Either individually or with a partner list ways in which each of the following may affect drainage density calculations:

a) The scale of the map which you use (*Hint:* compare the detail in Figure 1.9 with that available in your atlas).

b) The season during which the map was compiled (*Hint:* consider river discharge in semi-arid and arid areas particularly – but not exclusively).

c) The challenge of calculating drainage density for an area of chalk/limestone. (*Hint:* consider the fact that the drainage basin for the River Lune includes a considerable extent of limestone. Did your original calculation include the underground drainage? What problems might there be in accurately mapping underground water courses?)

FIGURE 1.12 Strahlian stream ordering

Stream Order	Number of 'segments' in this order
1	18
2	9
3	5
4	3

An alternative way of quantifying drainage networks is by **stream ordering** (Figure 1.12). This relies upon the concept of hierarchy and was originally developed by an American geographer called Horton as early as the 1930s. However, since that time, several variations of the original concept have been put forward and the system of stream ordering that is in most common use today was developed by a geographer called Strahler. Again, it is an analytical method which relies upon mapped data – so, once more, the scale and accuracy of mapping is a relevant consideration.

- Headwater streams with no tributaries of their own are considered to be **first order streams**.
- When two first order streams meet, they create a **second order stream**.
- When two second order streams meet, they create a **third order stream**.
- Where third order streams meet other third order streams, **fourth order streams** are created and similarly on up the hierarchy.

ACTIVITIES

4 Now return to Figure 1.9 (which shows the River Lune's catchment area).

a) On an enlarged copy of this map, undertake a Strahlian stream order analysis of this river network. Remember that all headwater streams are first order channels; begin your work by identifying and numbering these on your map. Also note that a first order stream meeting a second order stream does not make a third order one.

b) Copy and complete the dataset below with your results from the River Lune exercise.

Order	Nº of streams in this category
1st	
2nd	
3rd	
4th	
5th	
etc	

c) On a piece of semi-log graph paper, plot this data set, putting 'stream order' along the X-axis and 'number of streams' along the Y-axis. What do you observe about the trend of your graph?

d) What order of stream is the River Lune at its estuary?

e) Researchers working with this methodology suggest that Britain's highest order rivers are only fifth order streams (unless ephemeral streams, rills and gullies are included in the calculations). Study a physical map of Britain and suggest as many reasons as you can why this might be the case.

5 Return to the table you completed in part b) above and the work that you completed for Activity **2**a).

a) Calculate the *total* length of streams in each order across the network and complete the following chart:

Stream order	Total number of streams	Total channel length	Mean length of stream of this order
1st			
2nd			
3rd			
4th			
5th			

b) Graph this dataset using semi-log graph paper. Put 'stream order' along the X-axis and your variable dataset (i.e. mean channel length) along the Y-axis.

c) What do you notice about the trend highlighted by your graph (*Hint*: sketch in the approximate 'best-fit' line)?

Trends such as these which you have, hopefully, identified appear to hold for all rivers – even though their networks seem to develop naturally and sometimes haphazardly. The relationship between the number of streams and stream order (known as Horton's Law of Stream Numbers) which you explored in Activity **5** can be stated as: *'There is a constant ratio between the number of streams of one order and the next highest order'*. This is the **bifurcation ratio**.

ACTIVITIES

6 a) Using the table you completed in Activity **5**b), substitute your data into this equation in order to calculate the bifurcation ratio (BR) of the River Lune:

$$\frac{\text{No of 1st O.S.}}{\text{No of 2nd O.S.}} + \frac{\text{No of 2nd O.S.}}{\text{No of 3rd O.S.}} + \frac{\text{No of 3rd O.S.}}{\text{No of 4th O.S.}} + \frac{\text{No of 4th O.S.}}{\text{No of 5th O.S.}} = \text{B.R.}$$

(Where O.S. represents order of streams and B.R. is the bifurcation ratio)

b) Generally, the bifurcation ratio for most natural networks is a value between 3 and 5. It can be lower, falling to 2.5 for very intricate, dendritic systems, or higher (rising towards a value of 6) for simple networks with a few, large first order streams. Is the bifurcation ratio for the Lune high or low?
c) What does this tell you about the River Lune's network?
d) As a general rule, the lower the bifurcation ratio, the greater the probability that the river will flood relatively often. According to your calculations, is the River Lune liable to flood frequently?
7 a) In the light of all your work in this unit, combined with individual research around the issues involved, discuss with a partner or in a small group why a lower bifurcation ratio indicates the possibility of increased flood frequency.
b) Within this same discussion group, consider why the risk of flooding is greater for a river with a low bifurcation ratio and a generally circular drainage basin.
c) Following from these discussions, consider the flood potential in each of the following scenarios and then sketch an annotated flood hydrograph for each:

- A river with a bifurcation ration of 2.8 and a generally circular drainage basin following prolonged rainfall.
- A river with a bifurcation ratio of 3.9 and an elongated drainage basin following an intense late autumn storm.
- A river with a bifurcation ration of 5.1, and an elongated drainage basin following a storm event which has traversed the basin from the headwaters towards the sea.

d) For each of the scenarios in part c) above, describe how your flood hydrographs might be different if the drainage basins were composed mainly of:

- porous chalk
- impermeable clays

If you are undertaking stream studies as part of your individual investigation, it may be possible for you to develop your work by including drainage density, stream ordering and bifurcation ratio calculations. If your chosen section of stream is liable to flood, you should aim to explain fully why such flooding is a feature of this stretch of the river. It is far better to utilise a range of analytical techniques (rather than relying upon one method) as this gives a more balanced and reliable picture.

Hydrographs

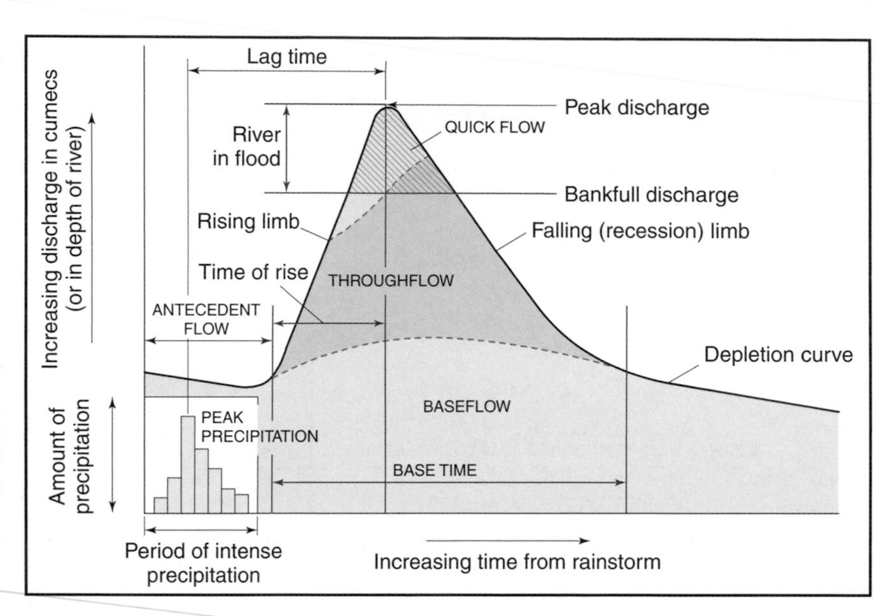

FIGURE 1.13 A model of a stream hydrograph

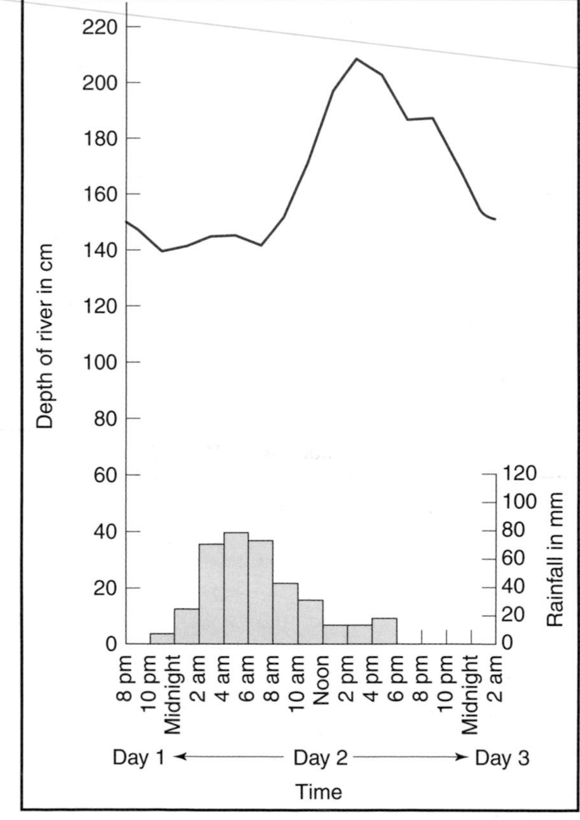

FIGURE 1.14 A storm event, River Swale, North Yorkshire

ACTIVITIES

1 a) Figure 1.13 is an example of a geographical model. What do you understand by the use of the term 'model'?
b) We usually show both precipitation and discharge on a hydrograph. How does the inclusion of both types of information help us to understand the sequence of events during a typical storm event?
c) How does a storm hydrograph aid understanding of the *effects* of a storm event upon stream flow and the surrounding area?

2 Using Figure 1.13 explain the terms:
- antecedent drainage
- bankfull discharge
- base time
- base flow
- flood
- lag time
- peak discharge
- peak precipitation
- quickflow
- recession limb
- rising limb
- throughflow
- time of rise

3 Using Figure 1.14 (River Swale):
a) What (approximately) was the peak depth of the river and at what time did this occur?
b) What was the peak precipitation and at what time did this occur?
c) How long, therefore, was the lag time for this particular storm event?
d) From the graph, estimate:
- the length of the base time
- the depth of the River Swale at baseflow in autumn.

e) Suggest a variety of reasons why the River Swale did not flood on this particular occasion.

4 Using the data provided for the River Swale to help you, draw a sketch graph to show how river depth/discharge might have varied had the same amount of precipitation fallen as snow in January. Annotate your graph appropriately.

5 With the help of an atlas and bearing in mind the direction of the prevailing wind across Britain, explain how each of the following factors is likely to affect the response of the River Swale to an extreme precipitation event:
- drainage basin area
- drainage basin size
- soil permeability
- direction of storm approach
- duration of precipitation
- season
- drainage basin shape
- bedrock permeability
- antecedent conditions
- drainage density
- intensity of precipitation
- vegetation cover

6 Consider the following factors:
- duration of precipitation
- season

- drainage basin shape
- drainage density
- intensity of precipitation
- vegetation cover

For each of the factors explain how it might affect:
a) Ephemeral stream flow in a hot, semi-arid environment following intense precipitation.
b) Stream flow in an Alpine valley downstream of a retreating glacier.
7 Urbanisation dramatically affects storm hydrology – particularly through its effects on run-off, lag time and peak discharge. Where most of the drainage basin is sewered and impervious, the effects are greatest; where some 'natural' surfaces remain, peak discharge is still higher and swifter than normal.
a) Draw separate, labelled sketch graphs to show the likely impact upon storm discharge into a local river:

- Where the majority of the drainage basin is sewered and impervious.
- Where part of an extensive basin has been sewered and built over, but where large areas also remain uncovered.

b) Describe and explain the variations between your two graphs.
c) Describe and account for variations (and any similarities) between your two sketch graphs and Figures 1.13 and 1.14.

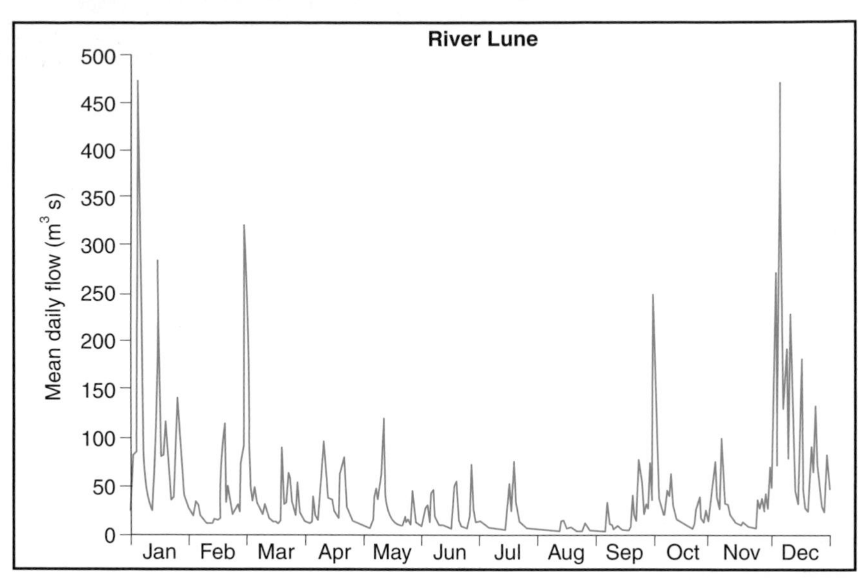

FIGURE 1.15 River Lune hydrograph

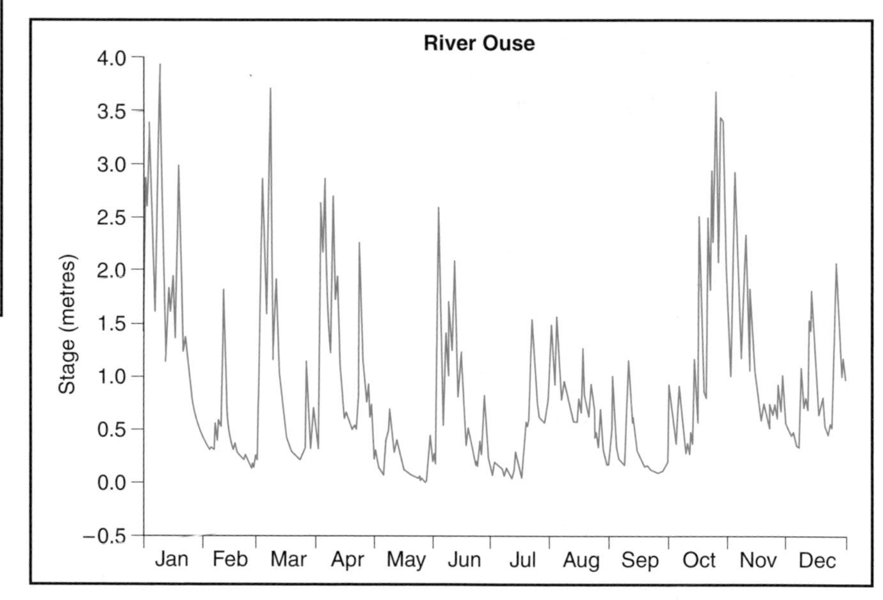

FIGURE 1.16 River Ouse hydrograph

Study 1.1 River basin hydrograph comparison – Rivers Lune, Ouse, Severn and Thames

Figures 1.15–1.18 will increase your awareness of the main environmental and hydrological characteristics for each of the above river basins; they will also help you to make more detailed comparisons and explanations in response to Activity **1**.

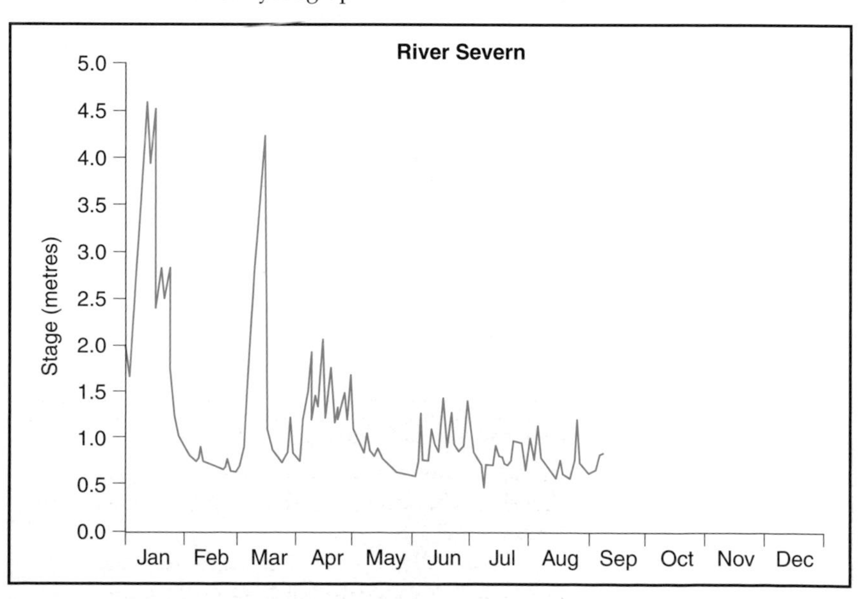

FIGURE 1.17 River Severn hydrograph

ACTIVITIES

1 Use the information provided in this study to compare the patterns displayed by the four hydrographs.
2 Account for the differences which you have identified in Activity **1** – with reference to drainage basin area and relief/precipitation patterns. (*Hint*: you will be able to obtain climate information from an atlas.)

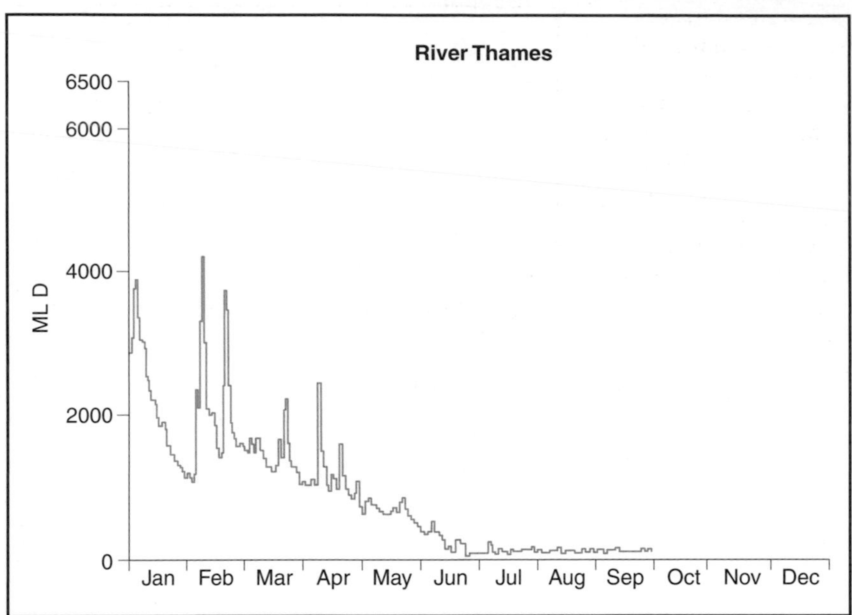

FIGURE 1.18 River Thames hydrograph

Study 1.2 River basin water transfers – Lancashire Conjunctive Scheme

Water is nature's most precious resource. It keeps us alive, nourishes our crops and allows us to lead increasingly healthy and enjoyable lives. Water is equally crucial to the quality of existence of all other forms of life on this planet.

In 1831, when Britain's population was 16.2 million, the average person used 18 litres of water per day. The current equivalent figures for 1981 were 54.8 million and 180 litres per day. The water demands of industry have increased sharply over the same period of time. The steel industry, for example, uses 200 000 litres of water for every tonne of metal it produces and 50 000 000 litres pass through a modern power station every hour.

Britain is very rarely in an overall water deficit situation, the outstanding exception in recent times being the drought of 1976, the subject of Study 3.2. Britain does not experience an annual pattern of seasonal extremes of precipitation or temperature; usually moderate temperatures mean that water losses through evaporation are not high by global standards and much precipitation is retained in underground aquifers (e.g. the London Basin, Figure 1.19) from which evaporation losses cannot take place. Water supply problems within Britain do, however, occur widely on a regional basis due to imbalances between precipitation and population density patterns.

The difficulties faced by some of the ten water authorities in England and Wales established by the Water Act (of Parliament) in 1973 have, in most

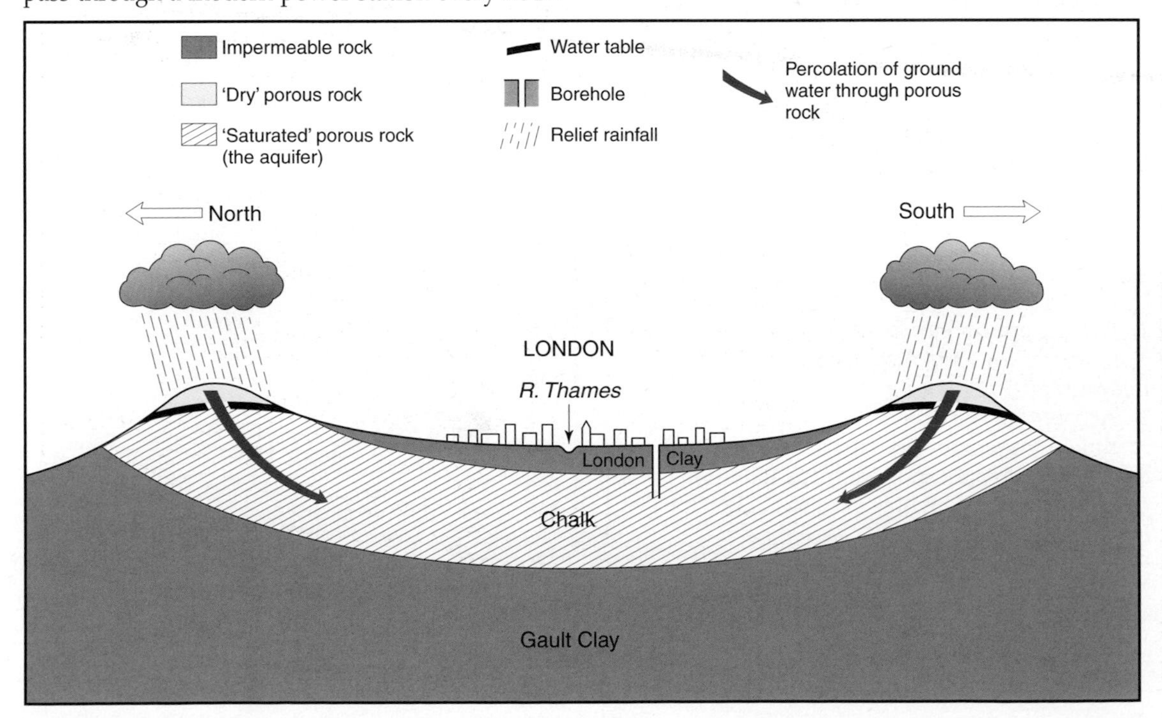

FIGURE 1.19 The London Artesian Basin

cases, been resolved by sensible water transfer agreements; these permit water-deficit authorities to obtain additional supplies from nearby authorities having a surplus. Figure 1.20 shows the most important water storage and transfer systems completed so far, including one of the earliest of these projects which began the transfer of Lake District water to Manchester as long ago as the 1890s. It is quite common for these projects to obtain water from a number of different drainage basins, but there are many small-scale water transfer schemes which only involve the drainage basins of two adjacent rivers. Such a scheme is the Lancashire Conjunctive Scheme, completed in 1980. The word 'conjunctive' was included in its name because this made it clear that different methods of obtaining and moving the water had been adopted for this link between the drainage basins of the Lune and the Wyre (Figure 1.21).

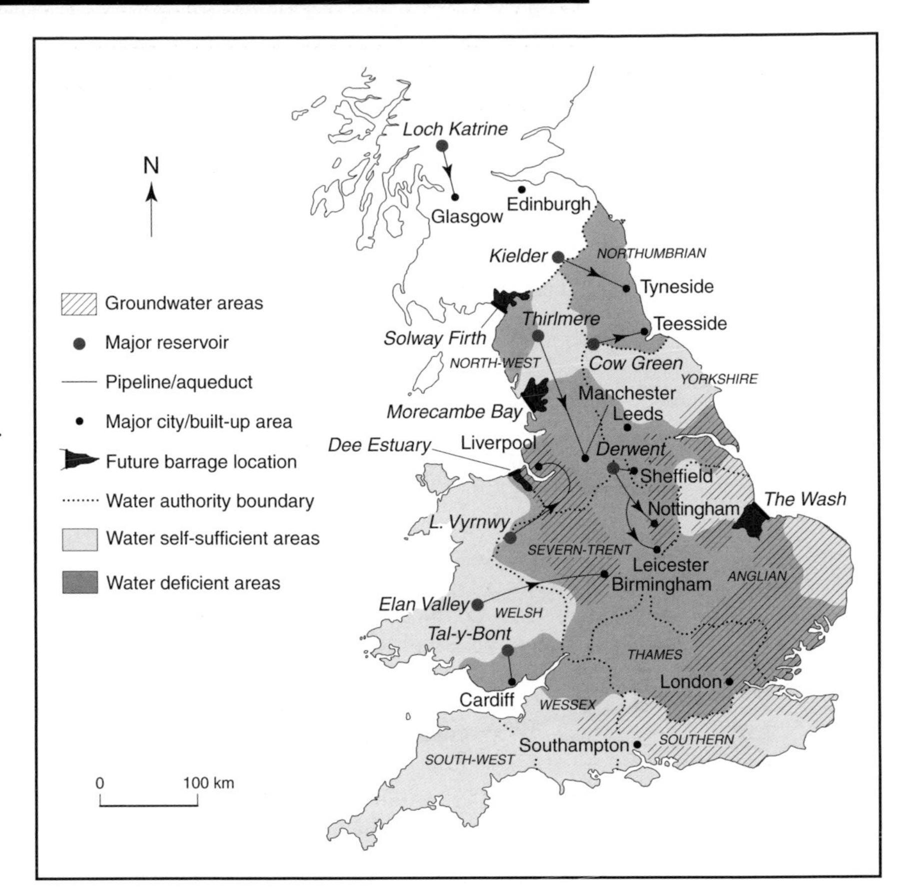

FIGURE 1.20 Water transfers in England and Wales

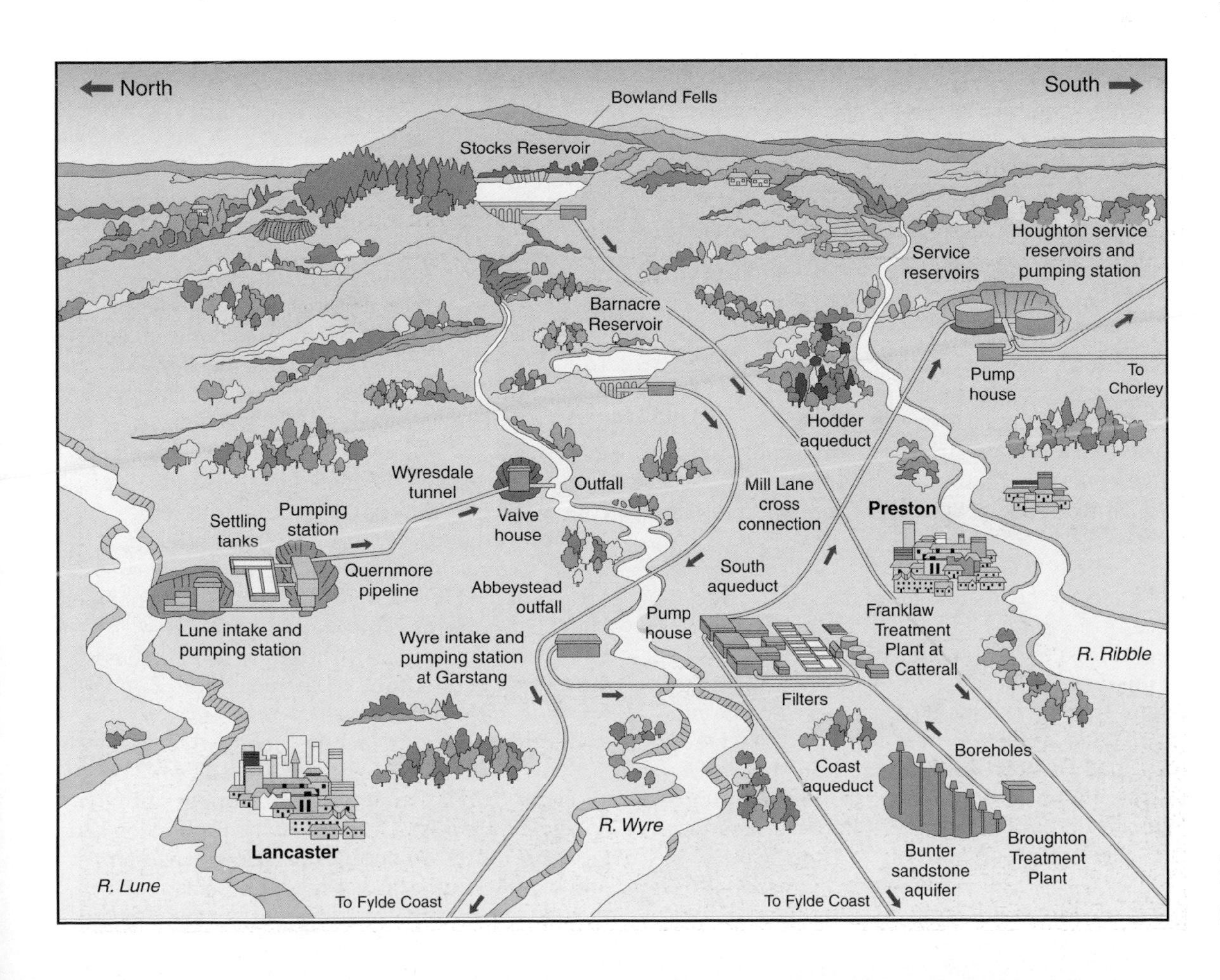

FIGURE 1.21 The Lancashire Conjunctive Scheme

ACTIVITIES

1 Suggest a wide range of reasons for Britain's increasing demands for water.
2 a) Find out what the current figures are for Britain's population and its average consumption of water per person per day.
b) Use simple graphical techniques to display these two pieces of information and the equivalent data given in the text for 1831 and 1981.
c) Comment on any trends which are highlighted by your completed graph(s).
3 Use Figure 1.20 and an atlas map showing Britain's population distribution and density to answer the following two questions:
a) Name two water authorities in England and Wales which are water self-sufficient according to Figure 1.20 and two which are not.
b) How do the two deficient authorities supplement their own natural water supplies in order to meet the demands placed on them?
4 Do there appear to be any general patterns of distribution for the water deficient and self-sufficient authorities? If so, describe them and provide brief explanations for the patterns which you have identified.
5 Use the information in Figure 1.21 to complete an enlarged copy of the outline map in Figure 1.22.
6 In what ways might the Lancashire Conjunctive Scheme be regarded as:
a) environmentally sensitive?
b) cost-effective in the long term?

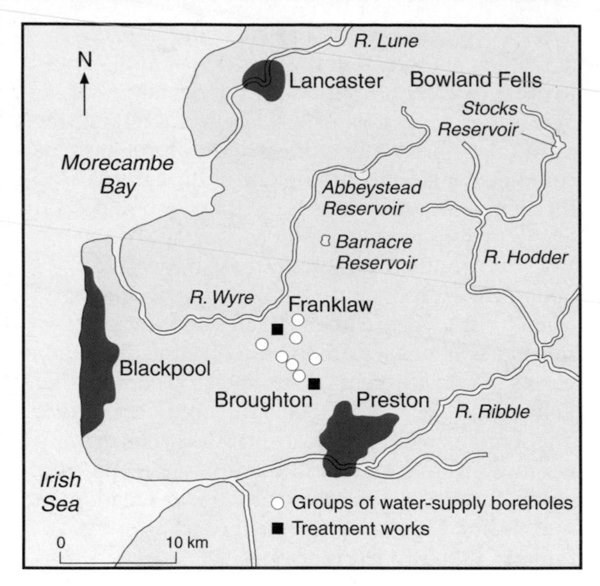

FIGURE 1.22 Lancashire Conjunctive Scheme map

Study 1.3 Aquatic ecosystems – Bosherston Lakes, Pembrokeshire

Bosherston Lakes are owned by the National Trust and managed jointly by the trust and the Countryside Council for Wales as a National Nature Reserve. These lakes support an unusually diverse range of flora and fauna and have been protected by SSSI (**Site of Special Scientific Interest**) status since 1959.

Originally tidal creeks in small valleys eroded by meltwater in the final stages of the last **Ice Age**, the three 'arms' of the lakes at Bosherston (Figure 1.23) were turned into fish ponds between 1790 and 1840 by the Cawdor family – the former owners of the estate. This development was achieved by building a small dam at their most southerly point where, even today, seasonal excess lake water discharges as an intermittent stream leading to the sea during the winter months of higher precipitation and reduced lake-surface evaporation.

The lakes' western and central arms (usually called the Lily Ponds) contain water which is very clear and rich in lime, as their chief source is underground springs drawing on Carboniferous limestone. The total catchment area of the lakes is 18 km^2. Large 'meadows' of the Bosherston White Water Lily *nymphaea alba* cover their surface, whilst their bottoms are carpeted with stonewort – an important food source for the local coot population. The eastern arm is dominated by dense stands of curled and fennel pondweed – the result of **eutrophication** due to its phosphate-enriched feeder streams; these streams form in the northern section of the lakes' **catchment area**, which includes both **Carboniferous limestone** and **Old Red Sandstone** (Figure 1.24). The entire lake system is very stable and uniform in character; for example, one of its major water sources comprises the underground streams, whose emergent temperature is always within one degree of 20°C.

Figure 1.25 illustrates the many important natural and artificial features of the Bosherston lake system. Figure 1.26 outlines the lakes' ecosystem and includes some bird species whose habitats include the surrounding woodlands as well as the

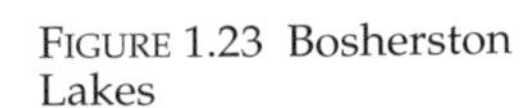

FIGURE 1.23 Bosherston Lakes

aquatic areas; it also includes wildfowl such as coot and pochard which winter on the lakes. A key component of the Bosherston ecosystem is the phosphorous contained in the lake sediments. This is released as part of the lakes' annual natural cycle. Figure 1.27 shows the global carbon and nitrogen cycles of which the Bosherston ecosystem inevitably forms a part.

FIGURE 1.24 Carboniferous limestone and Old Red Sandstone

FIGURE 1.25 Main features of Bosherston Lakes

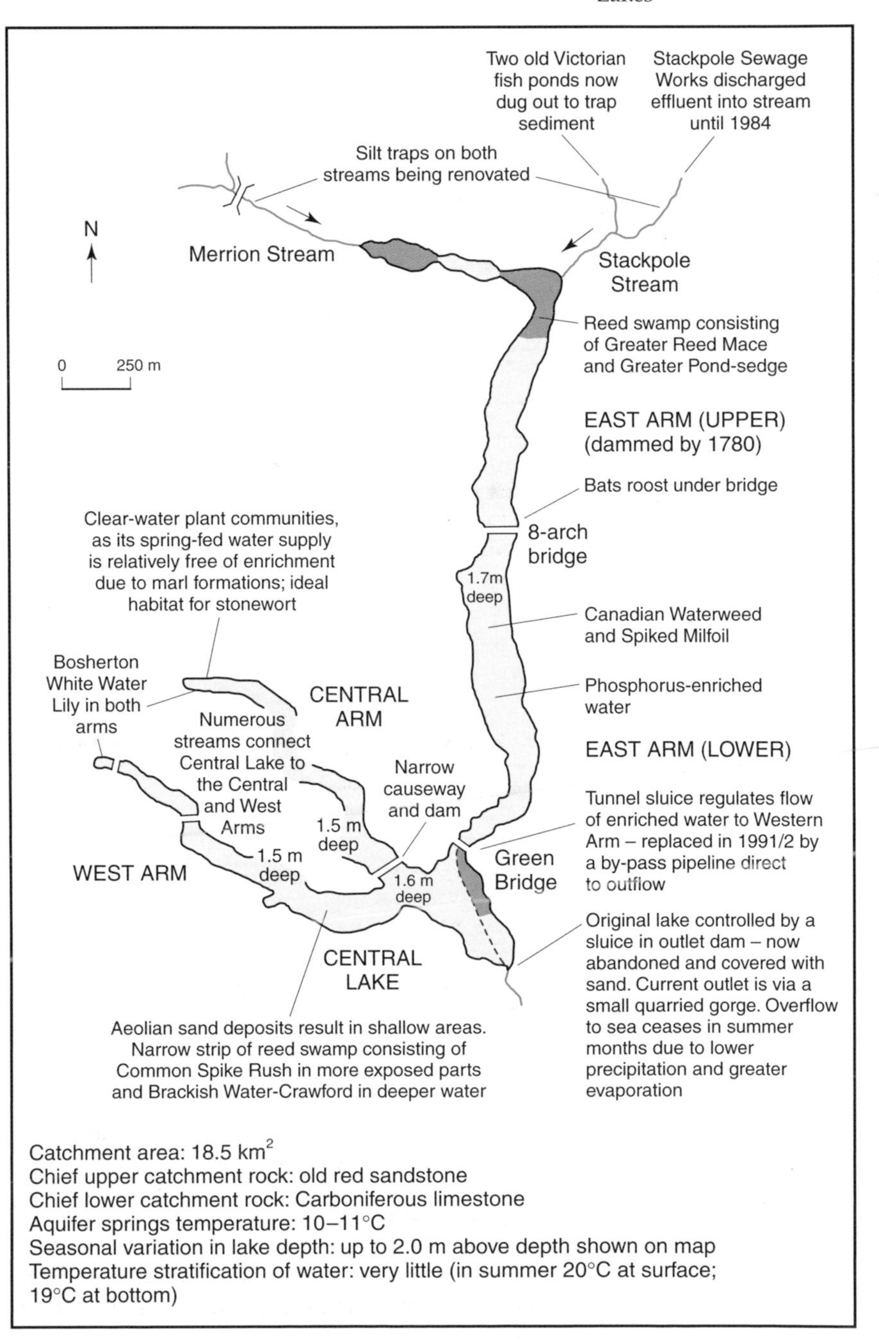

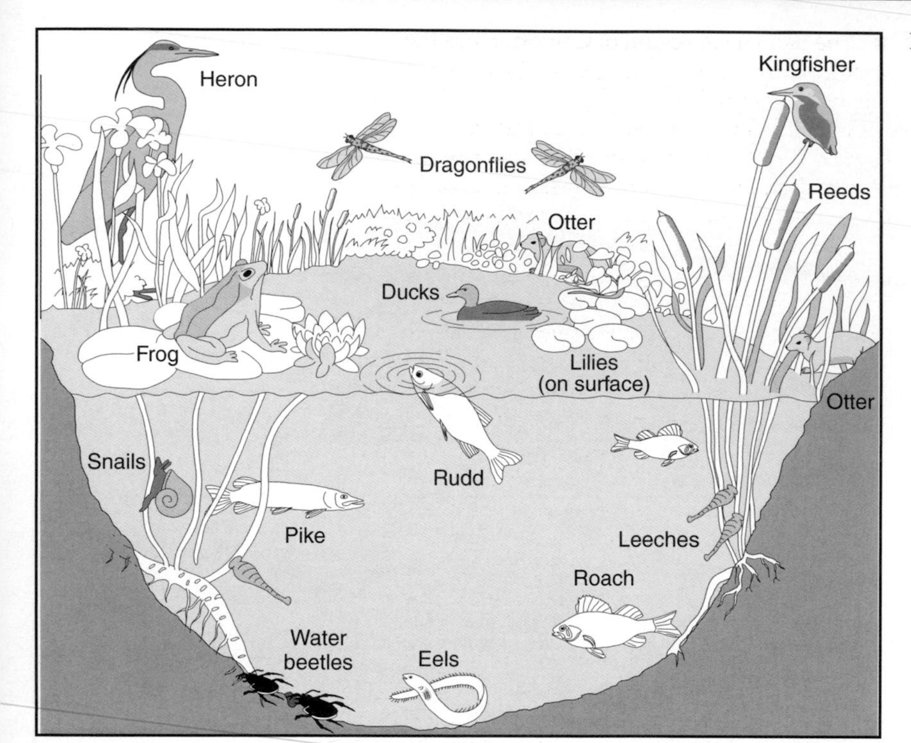

FIGURE 1.26 Bosherston Lakes ecosystem

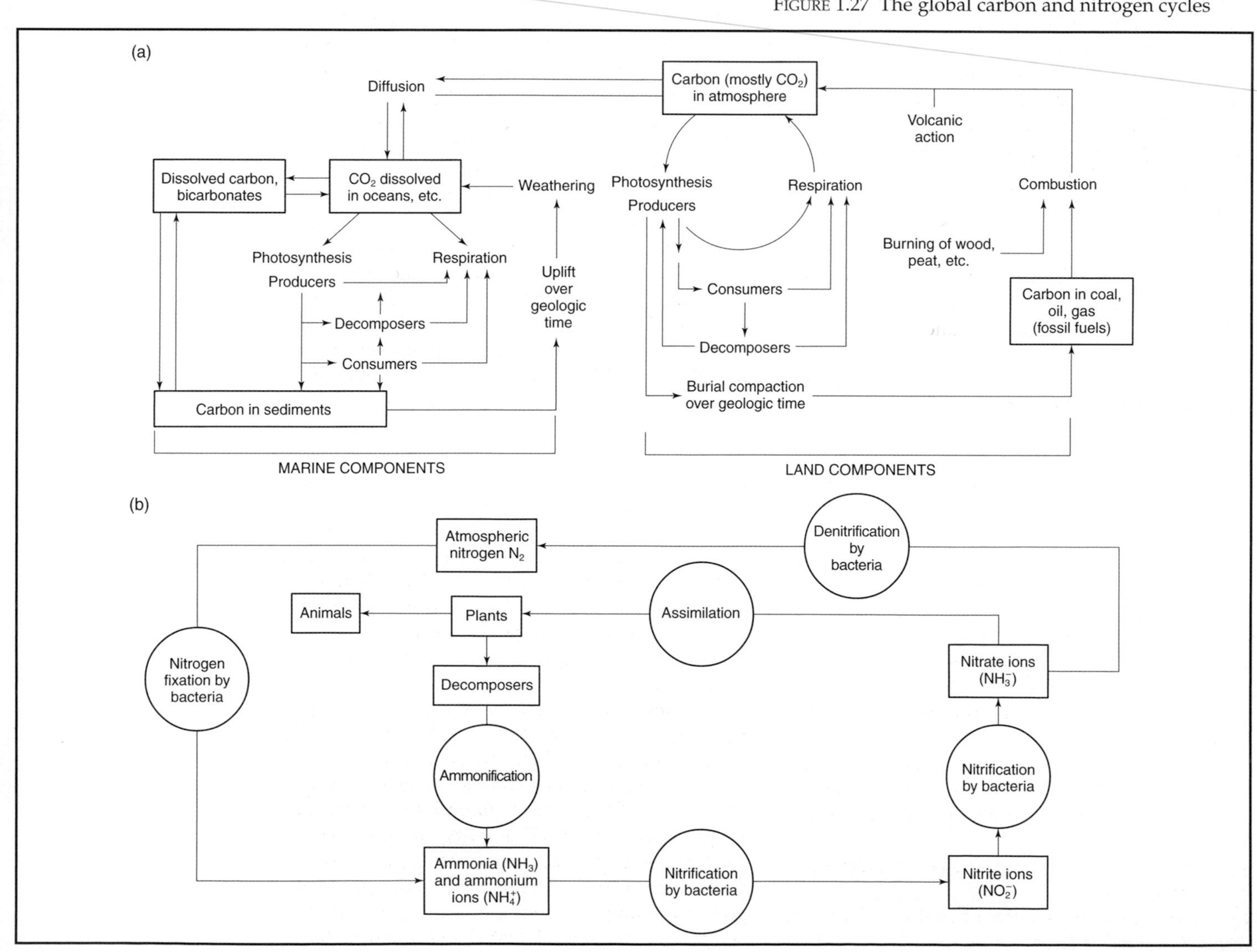

FIGURE 1.27 The global carbon and nitrogen cycles

ACTIVITIES

1 Explain briefly why the Bosherston Lakes were considered so important as to justify the award of SSSI status.
2 Some people regard theses lakes as being artificial. To what extent do you believe this description is appropriate?
3 Use detailed information from the figures and the text in this study to state why these lakes have such a diverse range of wildlife.
4 According to Figure 1.27, what are the chief similarities and differences between the carbon and nitrogen cycles?
5 Compare the Bosherston Lakes ecosystem with that of a pond, march or stream ecosystem which you have chosen from within your local school or home area.

Study 1.4 Wetland conservation – Mitigation banking, California

Wetlands may be defined as 'areas where saturation with water is the dominant factor determining the nature of soil development and the types of plant and animal communities living in the soil and on its surface; they are areas where the water table is at, near or above the land surface for long enough to promote the formation of hydric soils or to support the growth of hydrophytic plants; they include both fresh and saltwater marshes, swamps, mudflats and fens; they may be permanent or seasonal in character'.

There are a number of reasons why wetlands are considered to be worthy of conservation:

- They provide aquatic and terrestrial habitats, which together are able to support a wide range of both animal and plant species.
- They provide food, breeding and rearing habitats and cover for shore birds as well as waterfowl. They also provide food and spawning habitats for freshwater and marine species of fish.
- They stabilize lakeside banks prone to erosion and reduce wave energy on coasts. Their vegetation filters nutrient-rich sediment which aids deposition and the formation of new land.
- They are able to retain excess precipitation and river water and release it in a controlled way over a period of time, thus producing recharge for groundwater and reducing the risk of flooding.
- They provide valuable recreational facilities for many groups of people including anglers, boating enthusiasts, birdwatchers, photographers and students. They are potential contributors to the local economy.

The following statistics demonstrate the decline in California's wetlands up to the present time, as well as the increasing significance of those which remain:

- Only 5 per cent of California's coastal wetlands still remain intact.
- California's wetlands provide habitats for 41 rare and endangered species.
- The San Diego region of California has the greatest number of endangered wildlife species in the whole of the USA.

In 1993, Governor Wilson of California legislated for the Department of Fish and Game to establish wetland mitigation bank sites in its Central Valley region. A wetland mitigation bank can be any wetland area which has been created, restored, enhanced or merely preserved. Mitigation bank areas are wetland sites which are set aside to compensate for future developments in other areas which involve the destruction of long-established wetlands. The value of each bank is determined by assessing the value of the new or improved site. Mitigation banking has evolved in response to laws by both California State and the USA Federal authorities which require that *every individual, company or public authority whose activities destroy or degrade the environment may be required to set aside and/or restore habitat in order to offset the adverse impacts of the proposed activity. For each hectare impacted, a project proponent may be required to set aside or restore at least one hectare.* This legal requirement is in keeping with California's much-publicised aim of achieving 'no loss of wetlands in the short term and an increase in wetlands in the long term'.

Mitigation banking has several advantages:

- Developers are freed from having to create an alternative wetland site in their own area, which may be far from ideal for this purpose.
- Mitigation banking provides some flexibility in achieving environmental conservation priorities. For example, lost wetlands can be 'exchanged' for ancient woodlands in a region experiencing a deficit in such habitats.
- Mitigation wetlands are often much larger than the individual parcels of wetland which they replace. By their very nature, larger habitats have the potential to support much more diverse ecosystems. Also, the owners of larger wetlands are usually more able to obtain the services of highly qualified, up-to-date specialists who can advise on the most appropriate habitat improvement strategies.
- It allows landowners to recoup a much higher value for their land, which might otherwise be constrained due to environmental considerations.
- It takes many years to create new wetlands and enhance those of poor habitat quality. Mitigation banking allows new developments to proceed without the delays which would occur if the developer was obliged to create a replacement wetland first.

Wetland mitigation banking has many potential benefits, as is illustrated by the following example. In 1974, 600 ha of ecologically-rich habitat on the San Vicente Ranch was donated to the Boys and Girls Clubs of East County Foundation, in the San Diego area of California. The organisation spent years trying to put San Vicente to some form of profitable use, including residential development, a golf course and a recreational water park, but the steepness of the land defeated all these projects. It was then decided to preserve it as open space and it remained as such until 1996, when it became one of the first 40 conservation banks in California. The inclusion of the property into the mitigation scheme led to the foundation receiving many millions of dollars – money which will greatly benefit its young members in the foreseeable future.

ACTIVITIES

1 Make a series of brief, easily-revisable notes on California's wetland mitigation banking scheme, based on the information in this study.
2 Comment on the distribution of California's wetland sites, as displayed in Figure 1.28, taking particular notice of the mitigation site distribution.
3 Explain what is meant by the statement: 'larger habitats have the potential to support much more diverse ecosystems'.
4 This study has focused on the potential advantages of mitigation banking. Following group discussion, note down some likely disadvantages of this scheme, starting by considering whether the full range of ecosystem components in a destroyed wetland habitat is likely to be reproduced in a replacement mitigated wetland.
5 a) Prepare a wetlands survey for a chosen area; include current wetland areas as well as some examples of those which no longer exist.
b) Use the information which you have gathered to prepare a wetlands strategy for your chosen area.

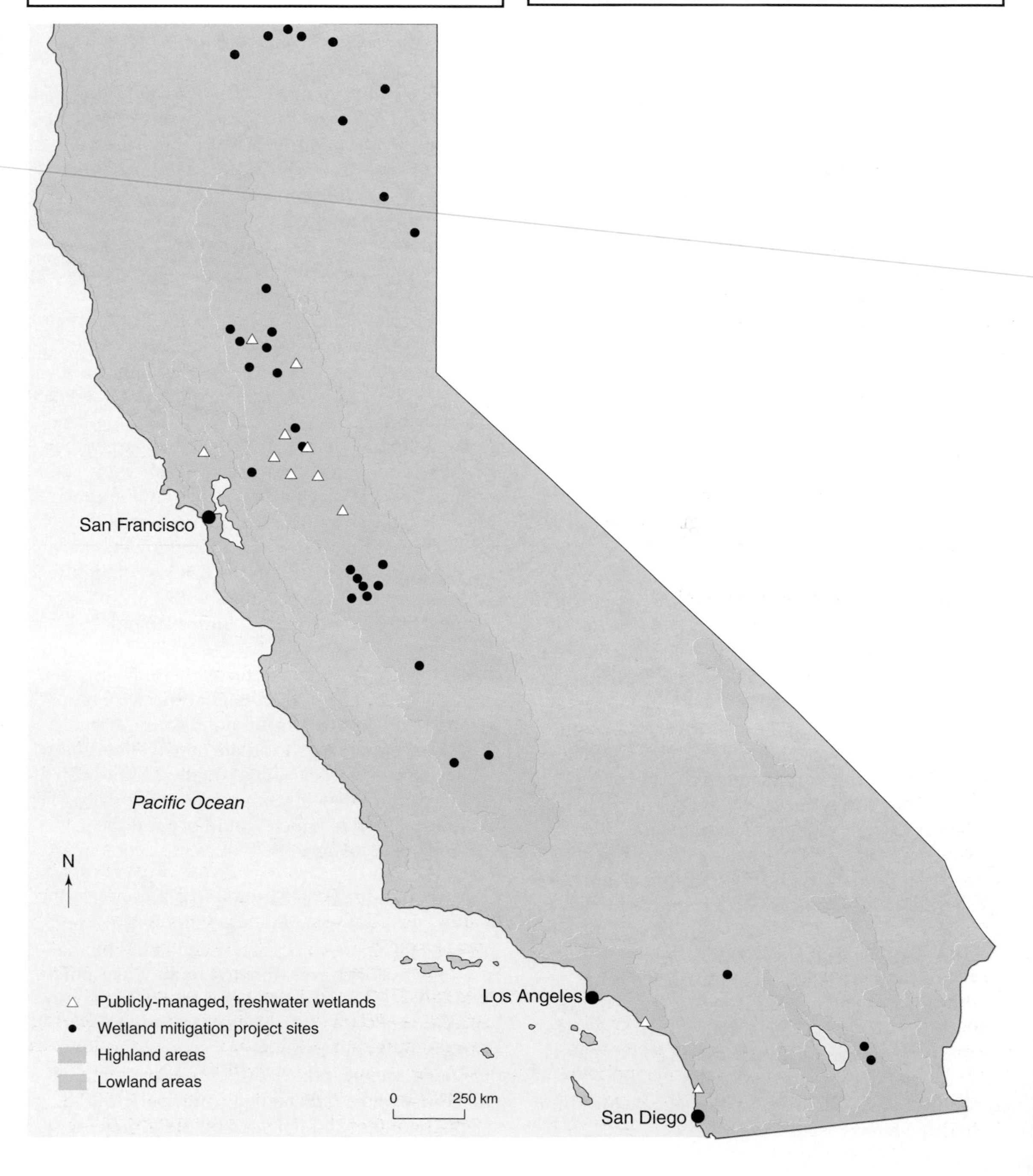

FIGURE 1.28 California wetlands

Study 1.5 River water over-extraction – The Aral Sea, Russian Federation

The hydrology of river basins is constantly changing, but the Aral Sea in central Asia represents an extreme case of reduced surface storage due to human intervention. The name 'Aral' is historically significant, because it is the local Kazakh word meaning 'island' – and islands were once one of the main features of the Aral Sea, when it was the world's fourth largest inland sea. It is now more than ten years since the world first really became aware of the Aral Sea crisis – described by the United Nations as 'the most staggering disaster of the twentieth century'.

The statistics in Figure 1.29 provide evidence of the scale of the changes which have affected the Aral Sea and its basin in recent decades. Figure 1.30 puts the Aral Sea within its environmental context. This map shows the natural components of the sea's catchment area, as well as the human interventions which have led to the massive reduction in its surface area.

The 'disappearance' of much of the Aral Sea has had many serious and wide-ranging consequences. Some of the most important of these consequences are listed below:

- The climate of the Aral Sea region is becoming increasingly more continental, with shorter, hotter, drier summers and much longer and colder, but snow-free winters. The sea used to exert a moderating effect on the local climate. Summer temperatures now reach 40°C and the main volume of the basin's surface water consists of thaw-water from glaciers in the mountainous watershed areas. The crop-growing season has decreased to an average of 90 days per year and the annual frequency of local dust storms has increased to more than 90 days.
- Muynak, which used to be a thriving fishing port on the Aral Sea shoreline, is now little more than a dusty ghost town surrounded by desert-like sand, salt and scrub (Figure 1.31). Nearby 'fields' are so white with salt that they appear to be covered by snow. Its once adjacent shore is now over 80 km away, leaving bare expanses of salt from which 15–75 million tonnes of sand and dust polluted with fertilisers, herbicides and pesticides are eroded every year from its surface and transported up to 250 km by the increasingly strong winds. Many of Muynak's buildings are either decaying or already derelict and many of its families now live in squalid shacks. Long-term unemployed adults wander aimlessly about the streets in search of something to occupy their time. Employment prospects for the young adults are bleak and many have turned to alcohol as means of escapism.
- The old coastline is now littered with stranded and decaying fishing trawlers, as all commercial fishing ceased in 1982. Some of the derelict trawlers have been dismantled for scrap metal, whilst others act as rusting playgrounds for local children (Figure 1.32).
- The Aral Sea first split into two separate lakes and one of these new lakes is itself currently in the process of further sub-division (Figure 1.33). Local ecologists predict that the entire former sea area could disappear by 2015, unless radical counter-measures are taken. One member of the World Bank Uzbeck mission (Werner Roider) summarised the seriousness of the situation in these words: 'For the Aral Sea to return, people would have to stop all irrigation for ten years. People would die upstream just to refill the sea! What is the value in that?'
- Illness and infant mortality rates are increasing rapidly and both of these statistics for the Aral Sea area are amongst the highest in the world – due mainly to inadequate diet, lack of safe drinking water, and rapidly deteriorating quality of life.

	Unit of measurement	1970s	1990s	Percentage change
Lake surface area	km^2	66 000	36 300	−45
Lake surface level	m. above global sea level	53	36	−75
Lake volume	km^3	N/A	N/A	
Lake water salinity	grammes/litre	10	40	
Native fish species	number	24	4	
Fishing catch	tonnes/year	40 000	5000	
Crop growing season in surrounding area	days/year	170	90	

FIGURE 1.29 Size reduction of the Aral Sea

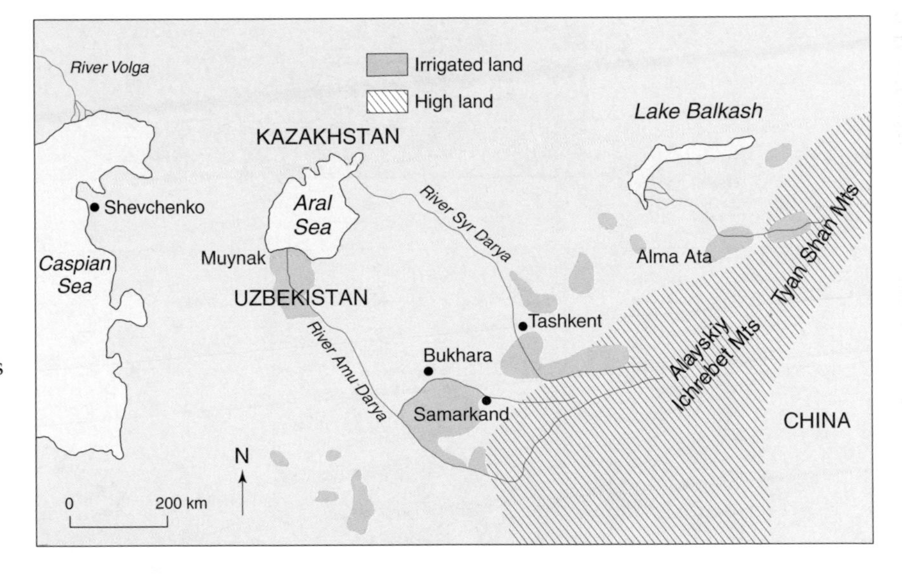

FIGURE 1.30 The Aral Sea drainage basin

FIGURE 1.31 Muynak ghost town

FIGURE 1.32 Stranded trawler in the Aral Sea

Drinking water is at least three times more saline than is considered healthy; it also contains unacceptably high levels of metals such as strontium, zinc and manganese which trigger anaemia and other serious long term medical disorders. Doctors at Muynak's hospital battle constantly against cancer, tuberculosis and hepatitis. The last 15 years have witnessed 3000 per cent increases in chronic bronchitis and diseases of the kidneys and livers; arthritic disorders have increased by 6000 per cent! 97 per cent of pregnant women are anaemic, 30 per cent of all births result in serious medical problems and 10 per cent of births are stillborn. Tuberculosis is now endemic; 40 per cent of the region's prison population suffer acutely from this condition, which can thrive only within very poor and malnourished communities.

■ Approximately half of the region's indigenous bird and animal species are either already extinct or on the verge of extinction – killed off by the fall in the level of the water and its rising salt content. The damage to aquatic fauna has also been very extensive, due to the disappearance of the fertile deltas which existed where the major rivers used to enter the sea. The biologically rich lagoons, marshes and wetlands totalling 555 000 ha in area have also largely disappeared. The mineral content of the water has increased four-fold to 40 grams/litre, resulting in the total loss of at least 20 of the sea's 24 fish species.

FIGURE 1.33 The break up of the Aral Sea (1961–2000)

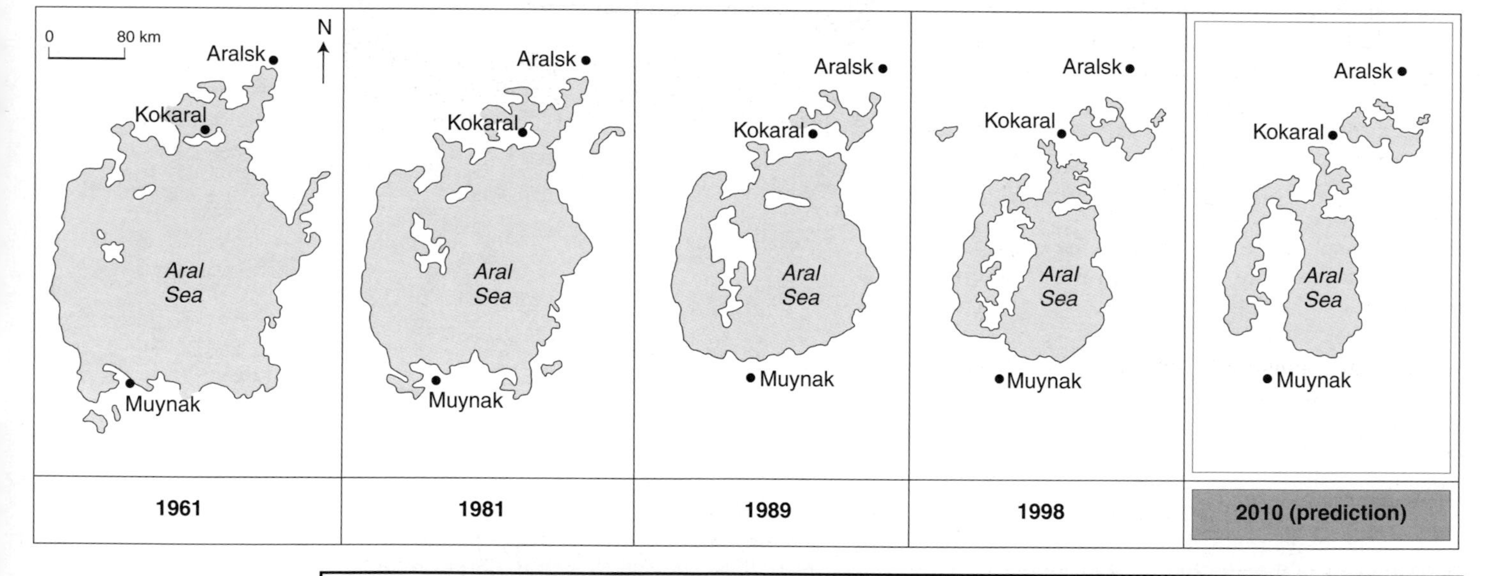

ACTIVITIES

1 a) Complete a copy of the table in Figure 1.29. Use the formula below to calculate the information for the last column. The entries for the first two lines has already been made – so that you can check the accuracy of your calculations.

$$\text{per cent change} = \frac{\text{Earlier statistic} - \text{Later statistic}}{\text{Earlier statistic}} \times 100$$

b) According to your completed table, which two statistics indicate the greatest percentage change?
c) After group discussion, write down possible reasons why these two statistics should indicate such great changes.

2 Quote information from this unit which is likely to support the following statements:
a) That the Aral Sea catchment area has an especially delicate regional water balance.
b) That the human interventions within its catchment area would lead inevitably to a great reduction in the size of the Aral Sea.

c) That it is now impractical to attempt to return the Aral Sea to its former size.
d) That every effort should now be made to arrest the sea's reduction in size as much as possible. (It is a good idea to list your items of information for this last part of the activity under a series of separate headings such as human, economic and environmental considerations.)
3 Figure 1.34 displays a range of initiatives which might be helpful in reducing some of the problems currently faced by the Aral Sea region.
a) Discuss their likely viability, then select a combination of initiatives which you believe likely to be both effective and feasible.
b) Give reasons for your rejection of those initiates which you opted to discard.

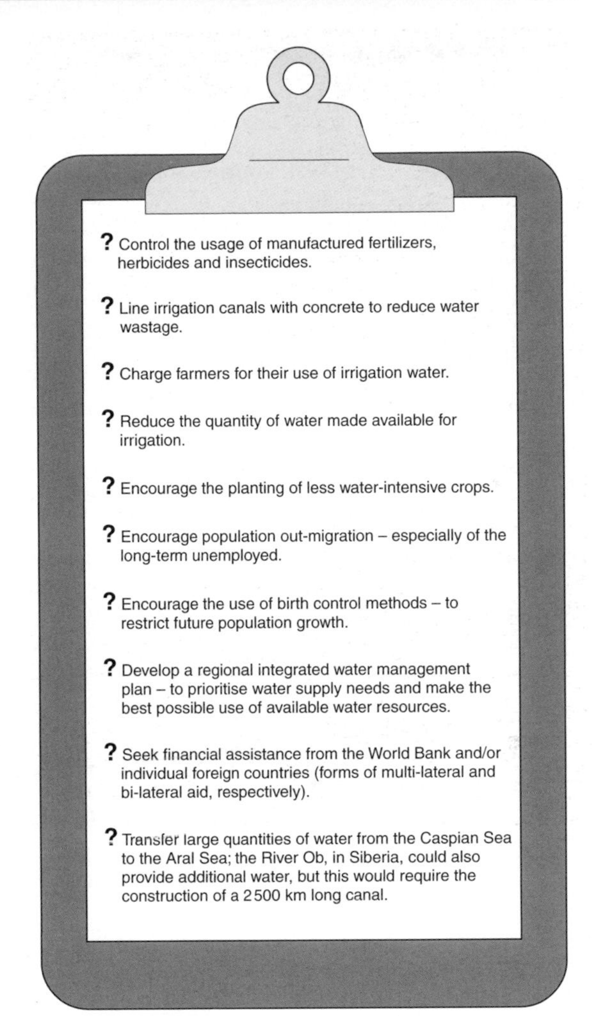

FIGURE 1.34 Reducing the problems in the Aral Sea area

Study 1.6 Water supply related issues – The Middle East

Water is crucial to life and a lack of it may trigger instability within communities of any scale. This study examines a range of difficulties and potential conflicts due to water supply related issues currently faced by one of the world's most politically volatile regions, the Middle East (Figure 1.35).

Figure 1.36 provides a useful introductory summary of the present-day regional tensions within the Middle East; it also identifies those areas which appear to be at the greatest risk of increasing instability due to the issue of water supply. Figure 1.37 is an account of water pollution in Israel and emphasises the seriousness of that country's lack of ability to provide an adequate and safe water supply of its own. The third account (Figure 1.38) concerns Turkey's determination to dam the River Tigris at Ilisu – in spite of considerable internal and international opposition. Recently revised official maps of the affected region actually show the dam already in existence! – clear evidence that the Turkish Government will continue to be ruthless in achieving its goal of building this $2billion joint water supply and hydro-electric power generating project.

ACTIVITIES

1 Annotate an enlarged copy of the map in Figure 1.35 so as to highlight the potential water supply-related conflict 'flash-points'.
2 Outline the main reasons for Israel's current water supply difficulties, and suggest appropriate strategies for their resolution.
3 a) According to Figure 1.38, what appear to be the main internal and international concerns regarding the building of the Ilisu Dam in Turkey?
b) Report on the progress of the Ilisu Dam debate and its possible future construction with the help of other newspaper articles and appropriate websites.

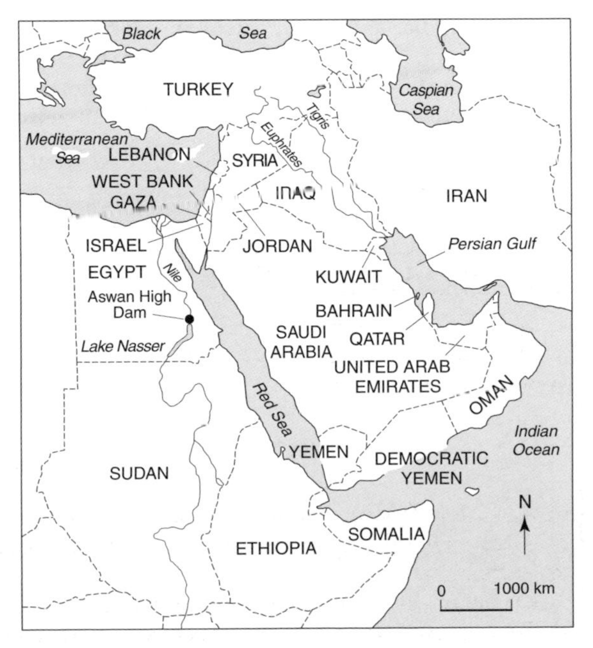

FIGURE 1.35 The Middle East

Water Wars in the Middle East

Because of differences in climate, some regions of the world have an abundance of water, whereas other parts have a shortage. As populations, agriculture, and industrialization increase, there is increasing competition for water – especially in dry regions such as the Middle East.

The next war in the Middle East may well be fought over water – not oil. Most water in this region comes from three shared river basins: the Jordan, the Tigris–Euphrates, and the Nile. Water in much of this arid region is already in short supply, and the human population in this region is predicted to double within only 25 years.

Disputes among water-short Ethiopia, Sudan, and Egypt over access to the water from the Nile River basin are increasing rapidly. Ethiopia controls the headwaters that feed 86 per cent of the Nile's flow; thus it has control over the downstream flow and water quality as the river flows north, through Sudan and along the whole length of Egypt, before emptying into the Mediterranean Sea. Ethiopia plans to divert more of this water; so does Sudan. This could greatly reduce the amount of water available to desperately water-short Egypt, whose terrain is desert except for the thin strip of irrigated cropland along the Nile and its delta.

Between 1995 and 2025, Egypt's population (which is growing by 1 million every nine months) is expected to increase from 60 million to 98 million, greatly increasing the demand for already scarce water. Egypt's limited options appear to be to go to war with Sudan and Ethiopia to obtain more water, slash population growth, improve irrigation efficiency, or suffer the harsh human and economic consequences of not taking any of these measures.

There is also fierce competition for water among Jordan, Syria, and Israel, which get most of their water from the Jordan River basin. In the mid-1960s, Israeli air strikes destroyed facilities built by Syria to divert the headwaters of the Jordan away from Israel. The 1967 Arab–Israeli Six Day War, won by Israel, was fought in part over access to water from the Jordan (through the Golan Heights) and to water from the West Bank aquifer.

Israel uses water more efficiently than any other country in the world. Yet, within the next few years its supply is projected to fall as much as 30 per cent short of demand because of increased immigration.

Jordan, which gets about 75 per cent of its water from the Jordan River system, must double its supply over the next 20 years just to keep up with projected population growth. In 1990, King Hussein declared that water was the only issue that could cause him to go to war against Israel.

Syria, which has long predicted water shortages by the year 2000, plans to build a series of dams and to withdraw more water from the Jordan River to supplement what it gets from the Euphrates River. This will decrease the downstream water supply for Jordan and Israel. Israel warns that if the largest proposed dam is built, it will consider destroying it.

Some 88 million people currently live in the water-short basins of the Tigris and Euphrates rivers, and by the year 2020 the population there is projected to almost double to 170 million. Turkey – located at the headwaters of these two rivers – has abundant water, and it plans to build 22 dams along the upper Tigris and Euphrates to generate huge quantities of electricity and irrigate a large area of land. These dams will reduce the flow of water to downstream Syria and Iraq.

Clearly, distribution of water resources will be a key issue in any future peace talks in this region. Resolving these problems will require a combination of regional cooperation in allocating water supplies, slowed population growth, and improved efficiency in the use of water resources.

Living in the Environment by G. Tyler Miller Jnr, 1998

FIGURE 1.36 Water problems in the Middle East

Valley of death

It took the Chosen People 2,000 years to end their exile and return to the Promised Land. It has taken them only 52 years to turn the land of milk and honey into a country of foaming rivers, carcinogenic water and dying fish.

Surrounded by enemies, Israel has become aggressively self-sufficient, greening its deserts and increasing industrialisation.

But in doing so, it has poisoned its own land so badly that, if Christ were born today, his first miracle would be to survive his own baptism. In fact the holy River Jordan, where Jesus was baptised, is officially an "effluent channel" – or stinking ditch – by the time it reaches the Jordan Valley.

Most of the rivers in Israel are now so badly polluted tha fish can live in them for minutes only, and the fishermen's flesh is rotting. Already, admits Dalia Itzik, the country's Environmental Minister, 40 per cent of water piped to Israeli and Palestinian homes is "undrinkable". Some scientists have already warned that carcinogens are turning up in tap water.

It seems extraordinary that the Israelis, who have fought so long and so passionately for their land, could have polluted it so completely in such a short time. But it was their determination to turn this desert into a habitable land that is killing their life source.

The fact that Israel not only has no water conservation policy, but no environmental policy to speak of has been a significant contributor to its downfall.

"The system is collapsing. Years of neglect and misunderstanding of the situation has left the system rotten," says Ofer Ben-Dov, Israel's Greenpeace representative.

There is no doubt that Israel is on the edge of an environmental catastrophe that will not only destroy the livelihoods of thousands of its people, but also threaten the long-term chances of peace between Israelis and Palestinians.

Yet despite the evidence on their doorsteps, the poisoning of the Promised Land could have been continued for years had it not been for the case of Lieutenant-Colonel Yuval Tamir and 30 of his fellow naval commandos.

Tamir, 42, served with the elite SEALE commandos – the equivalent of Britain's Special Boat Squadron – for 22 years. Now, instead of fighting their country's enemies, Tamir and his comrades are battling a more invidious foe – cancer, which they are convinced was caused by them being forced to train in the heavily polluted Kishon River.

According to Greenpeace and the University of Exeter, the Kishon is a poisonous brew of heavy metals and other carcinogens. One environmental campaigner recently joked that there were no biological dangers from human waste, such as dysentery, typhoid or cholera, because "not even bacteria can live in Kishon".

Yet for years no one thought to question the wisdom of sending Israel's commandos to train in water so thick with waste that commandos say it "felt like a solid" and took four hours to wash off. . . .

Israeli fishermen in Kishon harbour have also been dying at an alarming rate. Lawyers who are representing them in what are expected to be big damages claims against the Israeli Government, Haifa municipality and several chemical factories and refineries that dump their raw toxic waste into the Kishon say that at least half a dozen have died from cancer. Another 200 are being screened.

The Kishon is now seen as a national disgrace. Its acidity level is so high that, in the years when it has flooded, it has rotted away a concrete bridge and girders. Fishermen are quick to roll up their trouser legs to show what this toxicity does to their shins – it rots them too. . . .

The Kishon is probably the most polluted river in Israel. But a fall into the Yarkon, which runs through Tel Aviv, could be fatal.

Three years ago four Australian athletes died when a bridge collapsed into the toxic soup that runs through what is known as "Israel's Central Park". Two died from their injuries, while two more perished after swallowing and inhaling the green poison that masquerades as a local amenity. Scores of other people were hospitalised. The Alexander River is as bad as the Kishon, and the holy River Jordan is not much cleaner.

Environmentalists are calling on the Israeli Government to implement as soon as possible plans to start desalinisation schemes to make the waters of the Mediterranean drinkable. Failure to do so, they claim, will undermine the peace that Barak hopes to sign with Yassir Arafat, the Palestinian leader, before the end of this year. . . .

Israel and the Occupied Territories are suffering from a drought. Pumping from the Sea of Galilee, which supplies a third of the water for both communities, has been stopped. The coastal aquifer, an underground lake, is dropping by 70cm a year and is in imminent danger of sucking in salt from the Mediterranean. It is already badly polluted.

The level of the mountain aquifer . . . is also dropping fast. Israeli officials last week rushed to Turkey to begin negotiations to import water. But they admit that this would take at least ten months and huge investment, and that the problem is in the here and now. . . .

The Times, 4 July 2000

FIGURE 1.37 Water pollution in Israel

'Oh yes, we're going to build the dam!'

... The Ilisu dam project will provide electricity for Turkey but will drown ... dozens of villages. In London, Tony Blair is keen on the Ilisu and his government has said it is "minded" to provide $220 million in export credit for it. But then again, the Prime Minister has never been to this place and, therefore, does not know what it is like. If he came, he might change his mind. Freedom of movement and freedom of expression do not seem to exist here...

Back in London everyone says the story of the Ilisu is a complicated one, and it is true that the decisions about whether it will be built involve huge sums of money and power politics. These judgements will be made by people in London, Washington, Berne and Ankara. ...

The Turks regard the Ilisu dam as something that already exists. They talk of it in this way and the reservoir actually appears on Turkish tourist maps. To them, it is clear that it is just reality that is lagging behind. And maybe it is. Certainly, international politics bode well for the Ilisu. Turkey is a country much in demand, with its huge army and strategic location. Britain wants Turkey to be part of the European Union, and this project could help to pave the way. America is also keen to please: it has used the country as a base from which to bomb Iraq for years. ...

But these are not normal times for Turkey. For the past 15 years Ankara has fought a war against the Kurdish rebels of the PKK. It has been a bloody and brutal affair on both sides – in total at least 30,000 have died since 1984 – and it has been fought in the villages and hills and valleys of the South East. The Ilisu would flood 300 square kilometres of this land, destroy 80 hamlets and villages and displace up to 36,000 people. Many Kurds see the Ilisu project as part of the strategy to destroy their culture and way of life.

Plots aside, however, no one denies the war has been expensive and in the 1990s Turkey found that it simply could not afford to build the Ilisu by itself. ... Turkey had to find its $2 billion elsewhere, and these days international money for a dam does not come without strings attached. The Swiss-led consortium that stepped forward in 1998 pulled in companies from America, Britain, Italy and Germany and they, in turn, have all sought export credit from their governments. ...

And so, in at least five countries, hearings have been held and reports organised. There have been many questions. Why did Turkey want to dam the Tigris just 65 km from the Iraqi border? Why would anyone want to build a dam in a war zone anyway? What did the people on the ground think? The companies in question, including Balfour Beatty in the UK, turned to Turkey for the answers. But some questions, and particularly the last one, could not be answered. For, it seems, the people on the ground had never been asked what they thought. The Ilisu has been on the drawing board since 1954 but no one in the area had ever received so much as a postcard about it. Altinbilek says this is the way it works in Turkey: first you sign the dam contract and then figure out what to do with the people.

Not so for the rest of the world. International standards are clear that people must be consulted as soon as possible. The fact that these particular people are Kurds makes it worse. The Swiss moved first, refusing to approve the money until an international expert and a panel on resettlement was appointed. Other countries expressed worries, too, and on December 21 the UK announced that it was "minded" to grant the export credit but that a resettlement plan must be in place first. ...

The South East is a dry area – only 7 per cent is irrigated – but the land along the river is the exception and the stone houses are flanked by green fields and fruit trees. The source of the Tigris is north of Diyarbakir; it flows for 450 km in Turkey and then into Iraq. The Ilisu would be built 65 km north of the border and would create a long, thin reservoir for 215 km to the north....

The town of Hasankeyf appears ... as something out of another time. The cliffs and hills around the town are riddled with 5,000 caves, some of which have doors and numbers. ... It has been inhabited since Assyrian and Urartu times and ... was annexed by the Ottoman Empire in 1516. The population is now 5,500 and it is by far the largest single town that would be flooded by the Ilisu....

The mayor of Hasankeyf says that 100 per cent of the people on the town are against the dam....

The Turkish authorities have emptied dozens of villages in this area as part of their war against the Kurdish rebels. This is a big complication for any Ilisu resettlement plan. No one knows how many people lived in the emptied villages, though the Government now puts the figure at 12,000. All should still receive compensation of some sort for the Ilisu, though there seems to be no way of tracing them....

There is great difficulty in finding out even basic facts about the Ilisu. A few months ago it was accepted that 36,000 people were involved (16,000 in villages, 20,000 who have left). That figure has been continually revised. Semor now says the real figure is about 25,000 (12,739 in villages and towns, 12,000 who have left). The real answer? No one knows....

The Times, 17 April 2000

FIGURE 1.38 Ilisu dam, Turkey

Study 1.7 Local storm events – Lynmouth, Devon and Todmorden, Lancashire

Many lowland areas flood quite regularly, usually on a seasonal basis, due to their nearness to a river whose water level rises significantly as a result of winter rainfall and spring snow-thaw. The River Severn is a classic example of this hydrological pattern, as any of the older residents of Shrewsbury will be quick to remind you.

A less common type of river flooding takes place much more suddenly, and does not follow any seasonal or inter-year patterns. Floods of this type are the result of 'local storm events' – brief but torrential downpours, which may or may not have been preceded by a period of unusually heavy rainfall. One example of a local storm event took place in the small Pennine town of Todmorden, Lancashire in early June 2000. One month's average rainfall (Figure 1.39) saturated the surrounding area in only a few hours and raised the level of the local river so high that it burst its banks and breached a protective river wall (Figure 1.40). A Flash Flood Warning was issued by the Environment Agency, but only after local residents had managed to contact the agency and advise it of their rapidly deteriorating situation. Although no one was killed, the immediate damage to both domestic and commercial properties in the narrow, linear-shaped valley town totalled over £12 million.

A much more distressing **flash flood** event took place in 1953, in the small North Devon holiday resort, retirement centre and fishing village of Lynmouth. Lynmouth lies at the confluence of two minor rivers – the West Lyn and the East Lyn – which share a combined catchment area of approximately 100 km^2 within the Exmoor Hills (Figure 1.41). These hills are composed of impermeable, erosion-resistant Old Red Sandstone rocks of the Devonian period and rise to 500 m above sea level within a distance of only 10 km from the coast. The twin rivers occupy steep-sided, gorge-shaped valleys in which the dominant process is erosion, and interlocking spurs are one of the most common landscape features. Eroded material is quickly transported to the coast where accumulated deposits have formed a small arcuate-shape deltaic fan. Ancient stone walls offered some protection against flooding to houses and roads in Lynmouth village adjacent to the river, but many stretches of these walls were supported by buttresses which stuck out and so could easily be carried away by the force of the deluge. The small size of the catchment area, the nature of its rocks and the relatively short lengths of the Lyn rivers were all major factors in the account which follows.

During the first two weeks of August 1952, it had rained very heavily for 12 days, causing Exmoor to become increasingly saturated. Eventually, no more rain could soak into the thin surface soils covering the catchment area's rocks and the excess water transferred quickly to the rivers as run-off. On 15 August 1952 an intense downpour led to a further

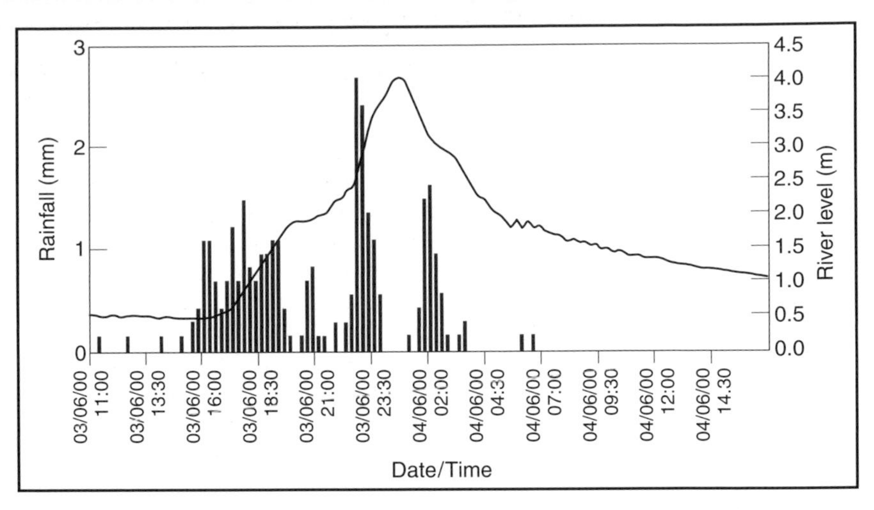

FIGURE 1.39 Storm hydrograph for Todmorden, June 2000

FIGURE 1.40 Todmorden in flood

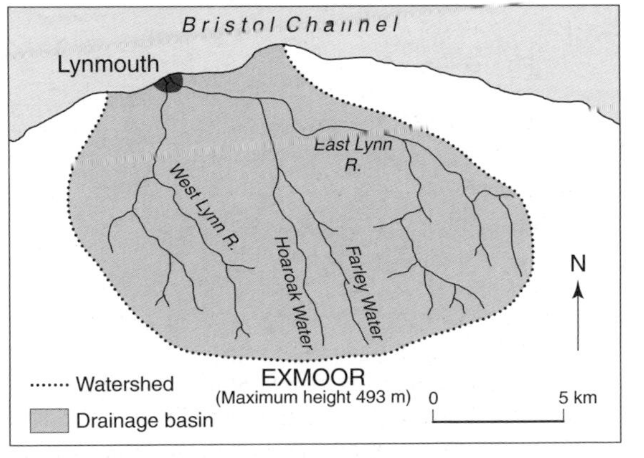

FIGURE 1.41 Catchment area of the East Lyn and West Lyn rivers

230 mm of rainfall – almost ten times the amount of normal rainfall and representing a weight of over a quarter of a million tonnes of water per square kilometre (higher rainfall figures had been recorded in Britain on only four previous occasions). The heaviest rainfall during that critical period was

closest to the sources of the 14 tributaries and the main channels of the two Lyn rivers. The dams controlling their upper course flows burst, allowing violent pulses of water to surge down-valley. At one stage, the River Lyn rose 4 m in only five minutes and it was not until 3 am the following morning that the water level finally began to recede. The main features of the events which followed are listed below:

- A low bridge across the West Lyn with a 9 m arch acted as a dam as it became blocked first by trees and then boulders. The water upstream of the bridge eventually burst its banks and changed course before joining the East Lyn. When the bridge finally broke, a huge surge of water 15 m high and moving at 20 km/hr swept through the village.
- Boulders weighing up to 15 tonnes each were dislodged from the sides of the gorge and transported to the beach. The main hotel in the village was destroyed by a single 7.5 tonne boulder.
- Road surfaces were totally removed and any softer surface materials were eroded to form gulleys, 7 m deep in parts, as far down as solid rock.
- Hundreds of uprooted trees, telegraph poles and motor vehicles (including a 36-seater bus) were carried by the river and greatly increased the damage potential of the rushing water. Almost a quarter of a million tonnes of debris was left by the water within the village area itself. Drowned animals from Exmoor farms were among the flood debris, which was piled 8 m high in places.
- 34 of the 200 people in the village at the time of the flood were killed, 22 of them local people. The fatalities included three boy scouts, whose camp was swamped by a river of mud, and a postman who was on his delivery round at the time.
- 90 houses were either destroyed or so badly damaged that they had to be demolished; others were flooded and their contents ruined. The High Street was virtually destroyed by the combined force of the floodwater and its transported material. Harold MacMillan, the British Housing Minister at the time, described the street as looking like 'the road to Ypres' (Figure 1.42). Damage to property totalled over £200 million.
- 130 cars were damaged beyond repair – 100 of them lost out to sea. 19 fishing and pleasure boats were also destroyed.
- The electricity, water supply and sewage disposal infrastructures were completely wrecked.
- The underground fuel tanks at a garage were uncovered, severed from their foundations and swept away without trace.
- The village was sealed off to the public for almost three weeks whilst the worst of the damage was removed.
- The Army alone provided over 1000 volunteers for this clearance work.

FIGURE 1.42 Lynmouth – destruction due to the 1952 flood

Several measures were taken to prevent a recurrence. These included:

- A new bridge with higher arches was built so that a much greater volume of water could flow freely under it.
- More trees were planted so as to reduce the rate of soil erosion on the valley sides.
- Sections of river channel were widened, deepened, straightened and even re-routed.
- Escape channels were built to allow floodwater to disperse more quickly.
- Sensitive river monitoring equipment was installed upstream to warn villagers of rising water levels.
- Centre of the village was re-shaped with new, replacement buildings much further away from the stream banks – to allow greater volumes of water to flow through it unhindered and so reduce the likelihood of further damage to property.

The violence of the flood led people to undertake the task of reconstruction without undue regard to cost – the immediate priority being to greatly reduce the risk of a recurrence on such a scale. The validity of this **risk assessment** for the Lynmouth area has, however, not yet been put to the test. The local people hope that it never will.

ACTIVITIES

1 Make brief case study notes of the Lynmouth flood. These notes should be in an easily-revisable format and be sub-divided under appropriate headings of your own choice (e.g. 'Catchment area' and 'Hydrological processes involved').

2 a) You are going to produce a 'Local Storm Event Survey' of a village or small town which you know well; this settlement can either be in your local area or a more distant place which you have visited often. Having made your choice of

settlement, obtain atlas and O.S. maps which will provide essential information about the locality.
b) Your 'Local Storm Event Survey' should cover many aspects of the town, its surroundings and the potential risk from flash floods such as those which occurred at Lynmouth and Todmorden. It should conclude with a list of preventative measures which, you believe, would minimise your chosen town's risk of future flooding.
c) Display your completed survey in such a way that staff and fellow students will be able to suggest refinements which might increase its usefulness.

Study 1.8 River flooding in LEDCs – Bangladesh

Every area liable to flooding (and indeed every flood event) is unique in terms of its potential for causing damage. Some causes of flooding (e.g. relief) are permanent and their impact is predictable. Other factors, such as land use and natural vegetation cover, may involve change over time and so tend to have a much less predictable impact on the environment. A third group of factors (e.g. exceptionally heavy localised rainfall) are of short duration and are unpredictable. Figure 1.43 illustrates many of these factors.

The damage potential of a flood is made even more complicated by a number of additional factors. These concern the area actually under water and some of the more important of these factors are:

- Location (e.g. whether densely or sparsely populated and whether mainly urban or rural).
- Size of area affected by flooding.
- Depth of flooding.
- Duration of flooding.
- Seasonal timing (especially important in rural areas, where crop planting and harvesting are critical phases in the agricultural year).

On 29 August 1998 Bangladesh experienced its worst river flooding in living memory – due to the simultaneous rising of the water levels in all three of its major river systems – the Ganges, the Brahmaputra and the Meghna. A substantial rise in the levels of these rivers is an annual, predictable feature, but the abnormal flooding of late summer 1998 was a direct result of monsoon rains (Figure 1.44) well in excess of the norm, and six crucial monitoring points recorded all three rivers to be at their highest levels for over 100 years.

Whilst monsoon-induced river flooding can cause great disruption and may lead to serious loss of life, its effects can be almost insignificant when compared with those of other floods which are caused by cyclonic rainfall. The following facts indicate the enormous scale of the 1998 flood disaster in Bangladesh which was chiefly **cyclone**-induced:

- At its peak, almost 70 per cent of the country was covered by floodwater.
- Two-thirds of the capital city, Dhaka was flooded, with most of its suburban areas being at least knee-deep in water; raw sewage was observed floating on the surface in many of Dhaka's residential zones. The retreating floodwater left behind pools of stagnant, contaminated water which represented a very serious health risk to the surviving urban population.
- 30–40 million people were rendered homeless and the homes of a further 5 million (45 per cent of them children) were seriously damaged.

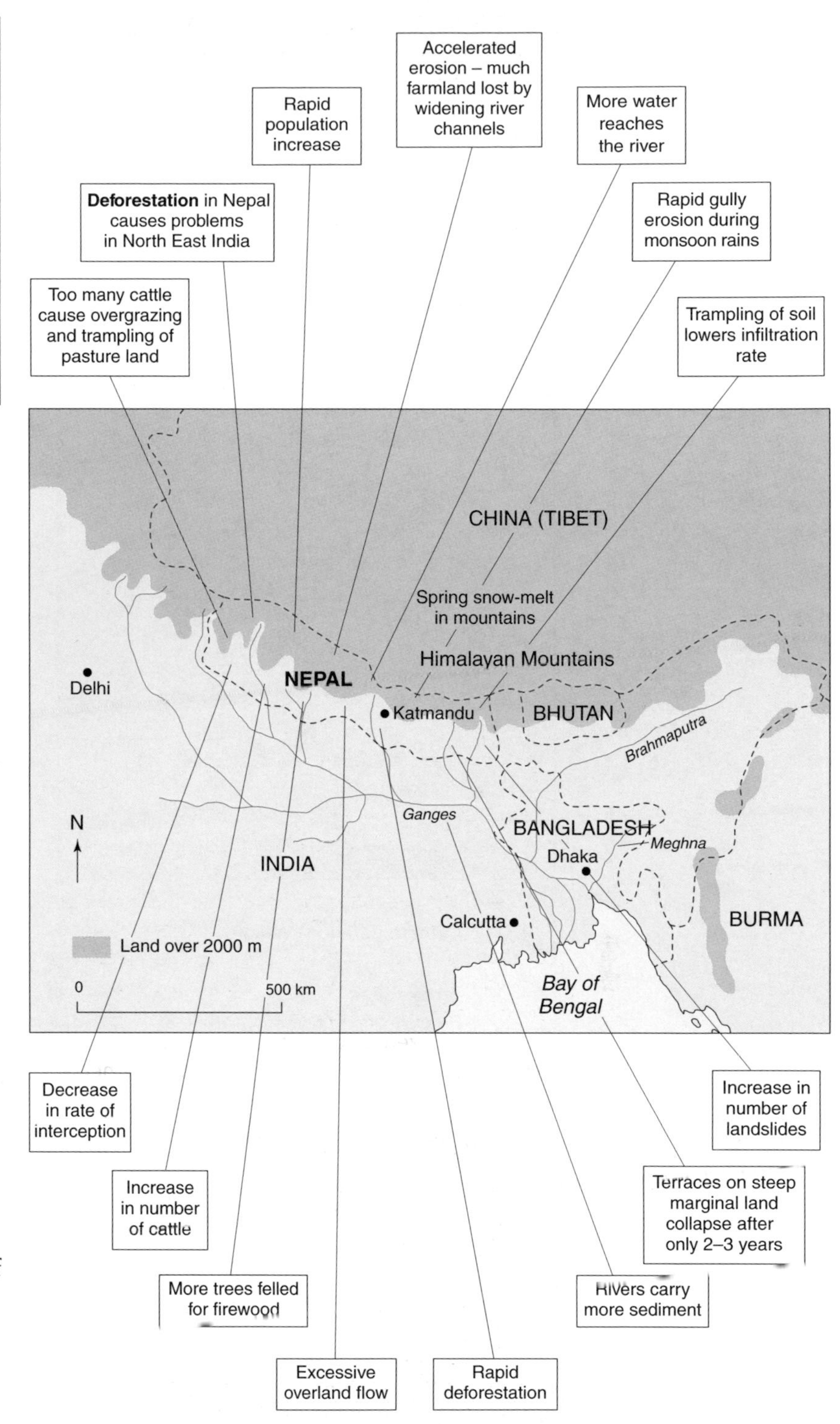

FIGURE 1.43 Flood factors in Bangladesh

FIGURE 1.44 Monsoon rain patterns in south-east Asia

January

1 The Himalayas separate the Asian High Pressure Zone from the Punjab High Pressure Zone.
2 Winds blow from the Asian High Pressure Zone across the Equatorial Low Pressure Zone to the more intense Australian Low Pressure Zone.
3 Winds blow from the Punjab High Pressure Zone to the Equatorial Low Pressure Zone.
4 Indian sub-continent has dry season – due to offshore winds.

July

1 The Himalayas separate the Asian Low Pressure Zone from the Punjab Low Pressure Zone.
2 Winds blow from the Australian High Pressure Zone across the Equatorial Low Pressure Zone to the more intense Asian Low Pressure Zone.
3 Winds blow from the Horse Latitude High Pressure Zone across the Equatorial Low Pressure Zone to the more intense Punjab Low Pressure Zone.
4 Moisture-laden onshore winds bring heavy monsoon rains to Indian sub-continent.

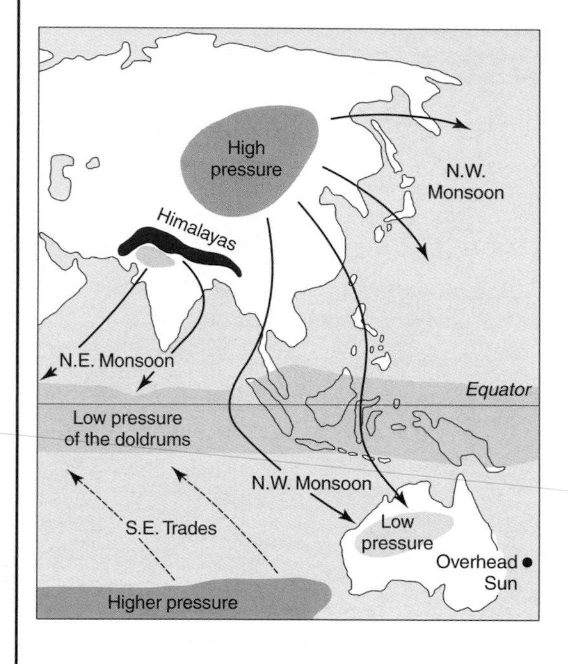

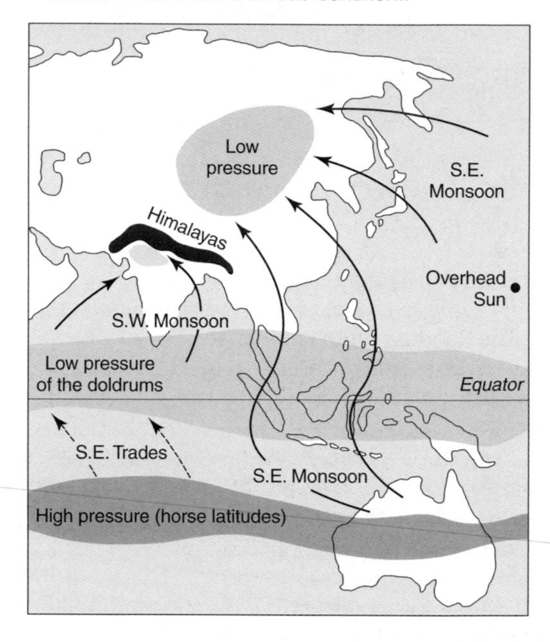

Climate data for Dhaka

	J	F	M	A	M	J	J	A	S	O	N	D
Temperature °C	20	22	27	30	30	30	29	29	29	28	23	19
Precipitation mm	18	31	58	103	194	321	437	305	254	169	28	2

■ The number of deaths which resulted from the flood exceeded 1000.

■ The World Heath Organisation (WHO) confirmed that 227 deaths were due to diarrhoea and that more than 250 000 confirmed cases of the condition had been reported. Its chief causes were polluted water and rotting food; children aged 12–35 months proved to be at greatest risk from diarrhoea.

■ Outbreaks of skin disease, jaundice and measles were especially common where people had clustered closely together for long periods of time (especially in flood shelters, which were often occupied continuously for up to three months).

■ Leg sores frequently developed and cases of snakebite increased sharply when people had been forced to wade in water for long periods of time.

■ In the worst flood-affected areas, victims often changed their eating habits from three to two meals per day. Also, the amount of food they took at each meal usually became less than half of their normal, pre-flood intake.

■ 1 million tonnes of wheat and 3 million tonnes of rice were destroyed by the floodwater.

■ 90 per cent of the nation's stock of sweet potato seeds was destroyed.

■ 42 500 drinking water 'tube wells' had to be repaired to restore them to normal use.

■ Some of the worst affected rural people were habitual borrowers of 'microcredit' (small-scale loans) in order to survive the rest of their subsistence farming year. Farmers whose harvests were devastated by the floods could rarely afford to repay their loans and the small banks and wealthy private individuals who had provided this essential financial support also suffered severe hardship.

■ The incidence of 'distress-selling' of land, animals and other property increased – especially of domestic animals.

■ The Bangladeshi Government had little option but to request $880 million of immediate foreign aid – aid which would lead to an even greater long-term financial burden which the country could ill afford.

Floods pose complex challenges to engineers, administrators and planners. Effective flood management must take into account not only technical and economic but also social, administrative, environmental, political and legal factors.

Structural measures for changing flooding patterns were known as 'flood control projects'. However, it is now realised that absolute control of flooding is rarely possible, either in practical or economic terms, and there is a growing understanding that it is simply not possible to achieve zero flood tolerance. Put another way, 'safe' can no longer be interpreted as meaning 'risk-free'! Structural measures are, therefore, now described as 'flood damage mitigation measures' or 'flood management measures'. Even if total control were achievable, many experts are still arguing about what is really causing the increased flooding in major drainage basins such as the Ganges and its neighbouring rivers. In particular, there is an on-going debate about whether upstream flood-prevention measures (e.g. **re-afforestation**) are more effective than protective measures, such as heightening river embankments further downstream.

The flooding which took place in 1987 and 1988 was so devastating that it provoked renewed international interest in flood control issues within Bangladesh. By 1989, a five-year Flood Action Plan had been funded jointly by the World Bank and a number of developed countries, and its measures started to be implemented during the following year. These measures included:

- Strengthening the western embankment of the Brahamaputra in the north of the country; raising and strengthening 260 km of embankments along the Jamuna River.
- Raising embankments and replanting mangrove swamps on the coast.
- Improving flood control and riverbank protection for Dhaka and other large urban areas.
- Allowing more controlled flooding, by dividing land up into separate compartments then surrounding them with protective embankments through which floodwater can pass by opening sluice gates built into them.
- Improving flood forecasting and warning systems – including extending the national network of radar stations designed to monitor weather changes and transmit flood and cyclone warnings.
- Undertaking cyclone protection studies of the most vulnerable coastal areas.
- Developing 'education in disaster preparedness' programmes in schools and adult communities.
- Improving disaster preparation facilities by building escape centres and roads on higher ground; also providing school buildings with some degree of flood protection.
- Increasing the use of 'watershed management techniques' such as building dams to control riverflow, dredging river channels to reduce their flood potential, extending afforestation/re-afforestation schemes and encouraging more valley-side farmers to build terraces and use contour ploughing.

Without doubt, the most popular method of flood management in Bangladesh has been the heightening of natural embankments (also known as **levées**). This is cheap to do and is reasonably effective in providing extra protection against moderate flooding. But preventing floodwater from spilling over onto the surrounding floodplain means it remains confined it to the river channel – causing the water level to rise even higher than before and more silt to be deposited on the river bottom instead of the flood plain. It is not unusual for Bangladeshi farmers to remove the upper sections of heightened embankments so as to restore the normal flow of fertile silt-laden river water to their fields. Embankment raising can increase meander migration and the maintenance needed to repair erosion damage is a constant burden on both local and regional communities.

ACTIVITIES

1 a) Identify the many reasons which contribute to an annual pattern of river flooding in Bangladesh. This is probably best done by listing reasons under appropriate headings such as relief, climatic and human factors.

b) In your own judgement, which of the factors that you have just listed can best be described as:

- Steady-state – leading to little, if any, short or long-term changes in the annual flooding pattern of Bangladesh.
- Capable of leading to significant short-term changes in the flooding pattern.
- Likely to lead to significant but much longer-term changes.

2 The map in Figure 1.45 locates those districts in Bangladesh with highest percentages of total area affected by the 1998 flood as well as those with the lowest such percentages.

a) Describe and compare the distributions of both sets of locations.

b) Suggest reasons for each of the distribution patterns.

3 A sample of the first 748 deaths caused by the 1998 flood produced the following results:

Cause(s) of death	Number of fatalities
Snakebite	54
Drowning/boat capsizing	244
Trapped under mud or walls	12
Electrocuted	34
Diarrhoea	127
Jaundice	20
Colds/fevers	28
Others	229

a) Suggest some likely causes of death in the 'Others' category in this table.

b) Display the information in the table graphically, using an appropriate method of your choice.

c) Comment on the nature and possible reasons for any patterns which you are able to identify with the help of your completed graph.

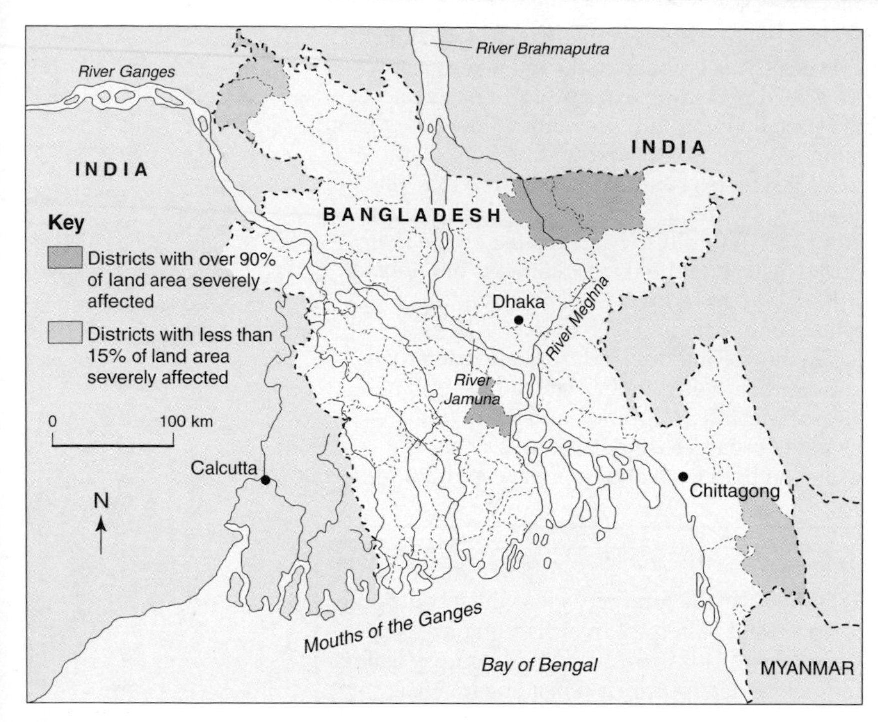

FIGURE 1.45 The effects of the 1998 flood on Bangladesh

Study 1.9 Cyclone-induced flooding – Mozambique and Madagascar

Early in 2000, a very severe and unseasonal drought was followed by a series of cyclones and **tropical storms** which produced devastating flooding in south-east Africa and the offshore island of Madagascar (Figure 1.46). Brief details of their impact on Madagascar are given below, followed by summaries for Mozambique.

■ The rains which normally begin in late October were well below the average for that time of year and drought conditions caused difficulties for farmers throughout the region until January 2000 – by which time the land surface was much too dry and hard to absorb the deluge which was to follow.

■ Towards the end of January, a deep low air pressure depression created a severe wind storm followed by heavy rains in the western and southern parts of Madagascar as well as Mozambique and its mainland neighbouring countries. In the Madagascan west coast town of Morombe, a whole year's average precipitation fell in only 36 hours.

■ Cyclone Eline (which was referred to as Leon before moving westwards into the south-west Indian Ocean area) passed over the east coast of Madagascar just south of Vatomandary on 17 February. It had a diameter of 450 km and produced winds of up to 200 km/hr as well as a further 131 mm of rainfall. 100 villages were flooded and 560 000 people suffered injury and/or property damage as a direct result of Eline/Leon.

■ Tropical storm Gloria passed down the centre of Madagascar from 2 to 5 March, following an unpredictable, Z-pattern course, bringing 100 km/hr winds and 263 mm of rain. Gloria finally dissipated 300 km inland from the island's southern tip.

■ Cyclone Hudah struck northern Madagascar on 2 April, before moving westwards across the Mozambique Channel the following day. Winds up to 300 km/hr were recorded and widespread damage was caused to crop plantations of vanilla, coffee and cloves; over 300 000 people suffered damage to farms and other properties. Shipping was disrupted over a very wide area. Figure 1.47 shows the structure of a typical cyclone and highlights those parts of it which are especially destructive at gound level.

Mozambique suffered much greater devastation

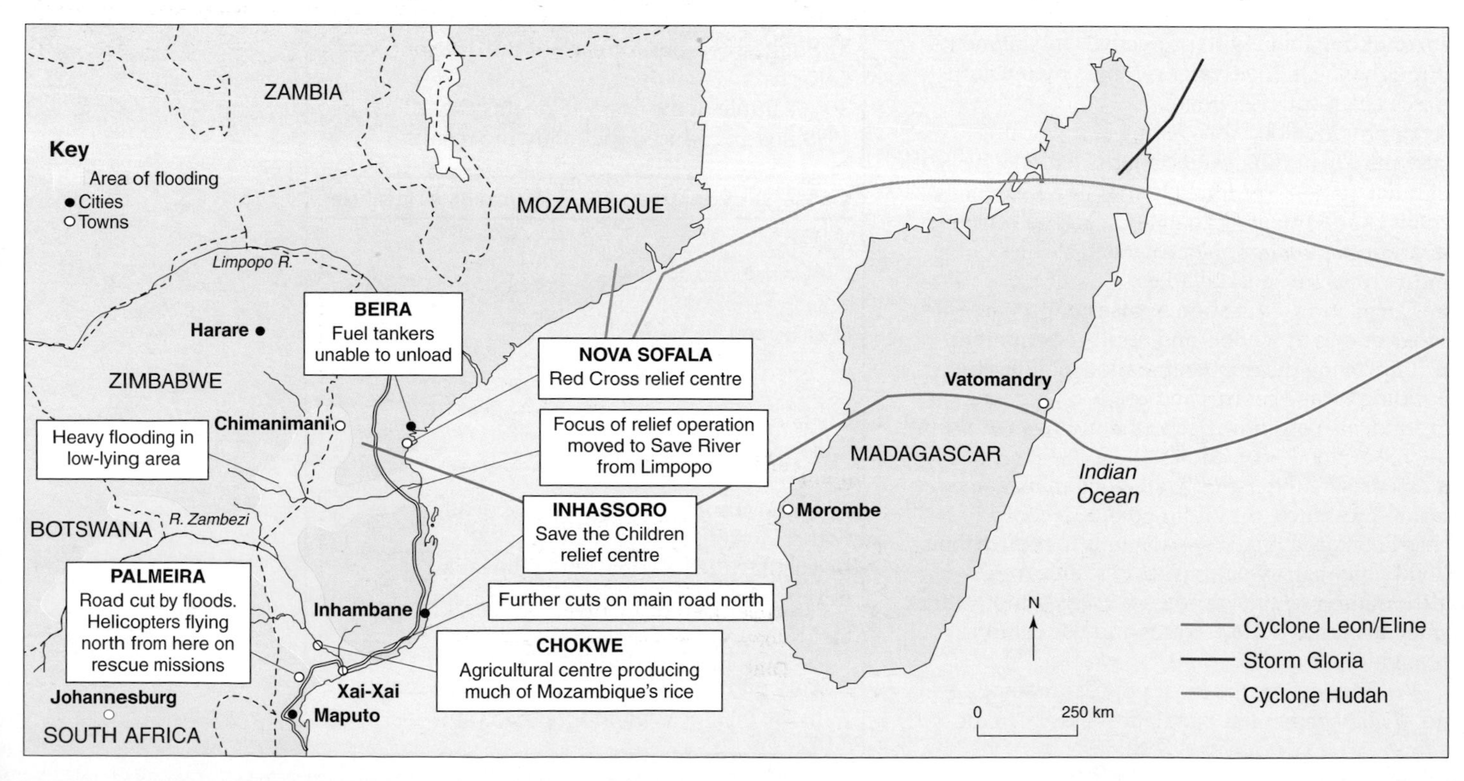

FIGURE 1.46 Cyclones and tropical storms in south-east Africa

than Madagascar, due to the more prolonged rainfall which preceded the cyclones and the vulnerability of its coastal and valley lowlands to floodwaters from neighbouring countries, all of which had also been subjected to torrential rain. In the worst affected areas, 500 mm of rain fell in just 24 hours – the region's highest rainfall for 50 years. Cyclone Eline and the storms which followed it pounded coastal provinces with further rain and high winds.

The Zambezi and Limpopo rivers have tributary networks extending over ten other states on the African mainland, including Zambia, which opened the spillway gates of the bulging Kariba Dam. Although this was a necessary emergency measure, it created high pulses of water along the Zambezi's entire downstream network. At the peak of the flooding, the Limpopo became 16 km wide instead of its usual 300 m and estimates put the volume of its moving water at 11 000 cubic metres per second (which has the equivalent weight of the same number of small family cars). In the city of Chokwe, floodwater levels in the Limpopo rose by 5 m in less than one hour and, in just one of many similar incidents, 33 passengers died when their bus was swept off a bridge. A suspension bridge over the Limpopo became an emergency 'island' for stranded vehicles and pedestrians as the surrounding water level rose (Figure 1.48). The total regional death toll exceeded 650. One Mozambique woman became instantly world-famous after giving birth whilst clinging to a tree branch only a short distance above the swirling floodwaters below her. An additional hazard for rescue teams and local people were land-mines floating erratically in the swirling water after being uncovered by surface erosion. It is estimated that 1.9 million of these devices had been planted in the rural areas on the fertile and hence more densely-populated Mozambique lowlands.

The damage to Mozambique was particularly ill-timed because the country was just beginning to become economically viable – in spite of having the lowest per capita **Gross Domestic Product (GDP)** of any country in the world. In 1999, it had produced its first national food surplus in 25 years and local industries were showing signs of becoming profitable. The Kruger National Park attracts many foreign visitors (and their much-needed foreign currency) every year, but most of its transport and accommodation infrastructures suffered flood damage which took over £7 million to return them to full use. An unlikely bonus for the Kruger game wardens was the way the floodwaters had flushed out many of the River Sabie's reed beds and rotting trees which had been blocking its main channels. These had previously created silting which required expensive dredging to clear and kept the wildlife so far from the river banks that visiting tourists could hardly see them.

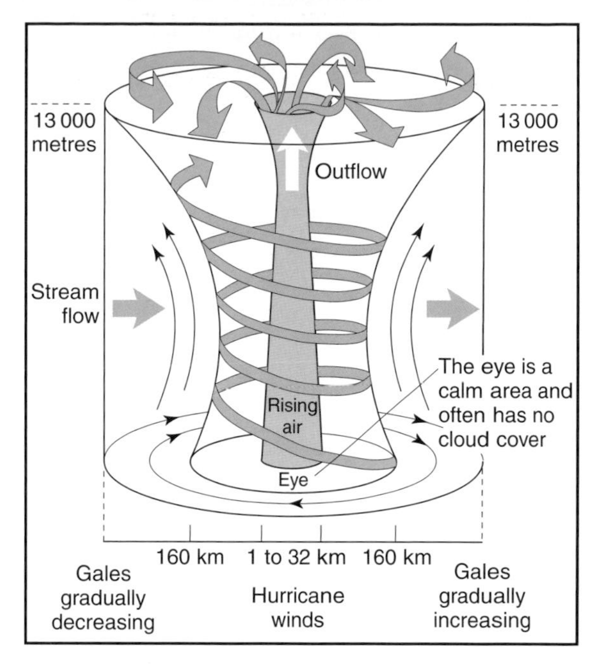

FIGURE 1.47 Structure of a cyclone

FIGURE 1.48 Flooding in Mozambique

ACTIVITIES

1 According to Figure 1.47, what are the most destructive parts of a tropical cyclone?

2 a) Discover for yourself detailed information about the populations and economies of Madagascar and Mozambique.

b) Discuss the extent to which Madagascar and Mozambique were likely to be able to rectify the damage they both suffered during the first four months of 2000.

3 Governments and relief organisations were heavily criticised for their slowness to respond to the region's appeals for flood relief help. What measures might they be expected to take to make them much more effective when responding to similar appeals in the future?

Study 1.10 River basin management in LEDCs – River Nile, north-east Africa

The Nile (6693 km) is the longest river in the world and drains about 10 per cent of Africa's total land surface. The area of the Nile Basin itself (3 031 000 km^2) is shared amongst nine states; of these, Sudan has sovereignty over 62.7 per cent and Egypt 9.9 per cent. However, in a typical 12-month period, it is

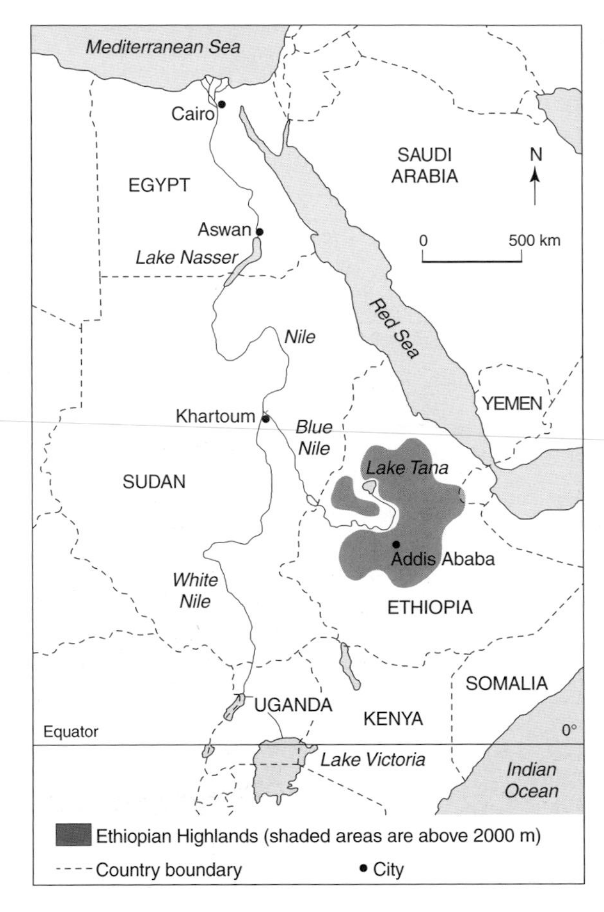

FIGURE 1.49 The drainage basin of the River Nile

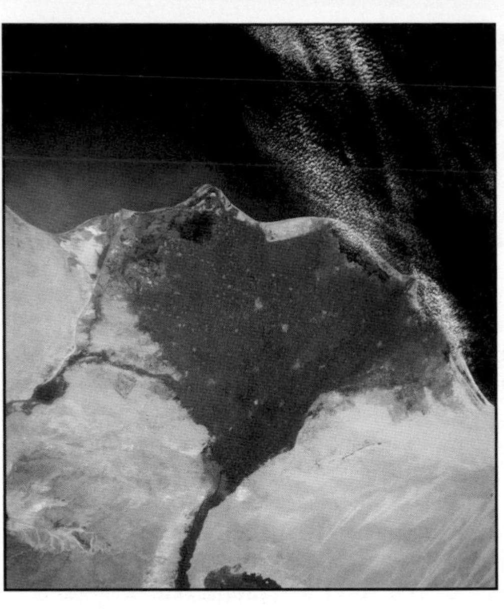

FIGURE 1.50 The Nile Delta

Ethiopia which provides 86 per cent of the Nile's river water – a contribution which rises to 95 per cent during the flood season which follows the late summer rains in the Ethiopian Highlands. At flood peak, the river's discharge may be as much as 16 times greater than its base flow. There is, however, considerable annual variation in discharge, two notable extremes being 155 billion cumecs in 1878 and only 12 billion cumecs in the much leaner years of 1913–14. There is a similarly dramatic contrast between the discharge patters of the two major tributaries of the Nile – the White Nile and Blue Nile, whose confluence point is at Khartoum (Figure 1.49).

The fertile, silt-laden floodwaters of the Nile have created agricultural wealth for Egypt and it was the envy of early Mediterranean civilisations. It is hardly surprising that the Greek occupation of Egypt in 332 BC was centred on the largest and most fertile area of all – the Nile Delta. Satellite images show the Nile as a narrow, dark green ribbon meandering through a seemingly unending yellow desert waste, until broadening into its classic, fan-shaped delta on entering the Mediterranean Sea (Figure 1.50).

The Aswan High Dam was built 600 km south of Cairo, in a **gorge** through which the Nile flows (Figure 1.51). The new dam's chosen location was 6 km south of the original Aswan Dam and actually within its reservoir area. This earlier dam was built by the British, in an attempt to increase food production for Egypt's already rapidly expanding population. This dam was later heightened in 1907–12, and again in 1929–34. Such a concentration of dam building within a very short stretch of the river's course is explained by the presence of granite in both its bed and its banks. The riverbed granite's hardness provided an ideal foundation for the High Dam. Similarly, the granite cliff banks formed natural buttresses able to withstand the enormous combined lateral pressures of the dam and the reservoir behind it.

FIGURE 1.51 The Aswan High Dam and Lake Nasser

Construction of the High Dam was only possible due to considerable foreign investment. In 1956, however, the USA withdrew its financial support after discovering that Egypt had started to buy arms from the Soviet Union. The Soviets were later persuaded to underwrite one-third of the project's costs; they also trained Egyptian workers in the skills of dam-building and provided their own specialist engineers and equipment to speed up construction. It was appropriate, therefore, that the dam's formal opening ceremony in January 1971 was hosted jointly by President Sadat of Egypt and President Podgorny of the Soviet Union. The following facts give some idea of the scale of the project:

■ It created a reservoir (later named Lake Nasser, after a famous Egyptian president) which is 16 times as long as Lake Windermere; the reservoir is 560 km long and has a surface area of 3 031 000 km^2 (Figure 1.53). It extends southwards beyond Egypt's border with Sudan. Although capable of holding vast quantities of water, it is estimated that one-quarter of its total capacity is lost annually through surface evaporation. In fact, only one-third of all water entering Lake Nasser actually flows northwards out of it. It would have been technically possible for Egypt to extract water from the sea – but processing it in sufficient quantities for existing agricultural, domestic and industrial uses would have been far too costly for an LEDC such as Egypt to consider. An additional factor was the prediction of a national water deficit of 16 per cent for the year 2000, which would have required a colossal investment in desalination equipment. Lake Nasser drowned a number of the waterfalls (called cataracts) which had previously hindered navigation along the Nile and, in doing so, greatly restricted the tourist potential of the river.

■ The Aswan High Dam traps 98 per cent of all sediment transported by the Nile. So much sediment is now being deposited in the lake that a new delta is forming at its southern end – at the expense of the Mediterranean coast delta, which is now denied the volume of silt needed to sustain its present area. In a typical year before the Aswan High Dam was built, the Mediterranean **delta** received 120 million tonnes of fertile river-born silt. Coastal erosion removed 350 ha of deltaic land between 1984 and 1993 whereas only 125 ha of river sediment was deposited during the same period of time; this net rate of loss has led to the delta retreating by about 25 m annually. Figure 1.52 compares the Lower Nile's sediment budget before and after the building of the Aswan High Dam.

■ The purchase of manufactured fertiliser to compensate for the loss of fertile silt is estimated to have cost the Egyptian farmers on the Nile floodplain at least £50 million each year. Silt was the local traditional raw material for brick-making and the output of Nile Valley brick works dropped by one-third during the 1970s due to the High Dam's impact on the river's sediment regime.

■ The silt which used to reached the Mediterranean not only built up the delta but also carried nutrients which provided food for large offshore shoals of sardines. Catches of this fish are now 95 per cent lower than before the High Dam was built and over 3000 jobs in Egyptian sardine processing plants have been lost as a result. Increasingly bigger (and more expensive) boats have to be bought in order to fish further away from the nutrient-depleted and increasingly polluted

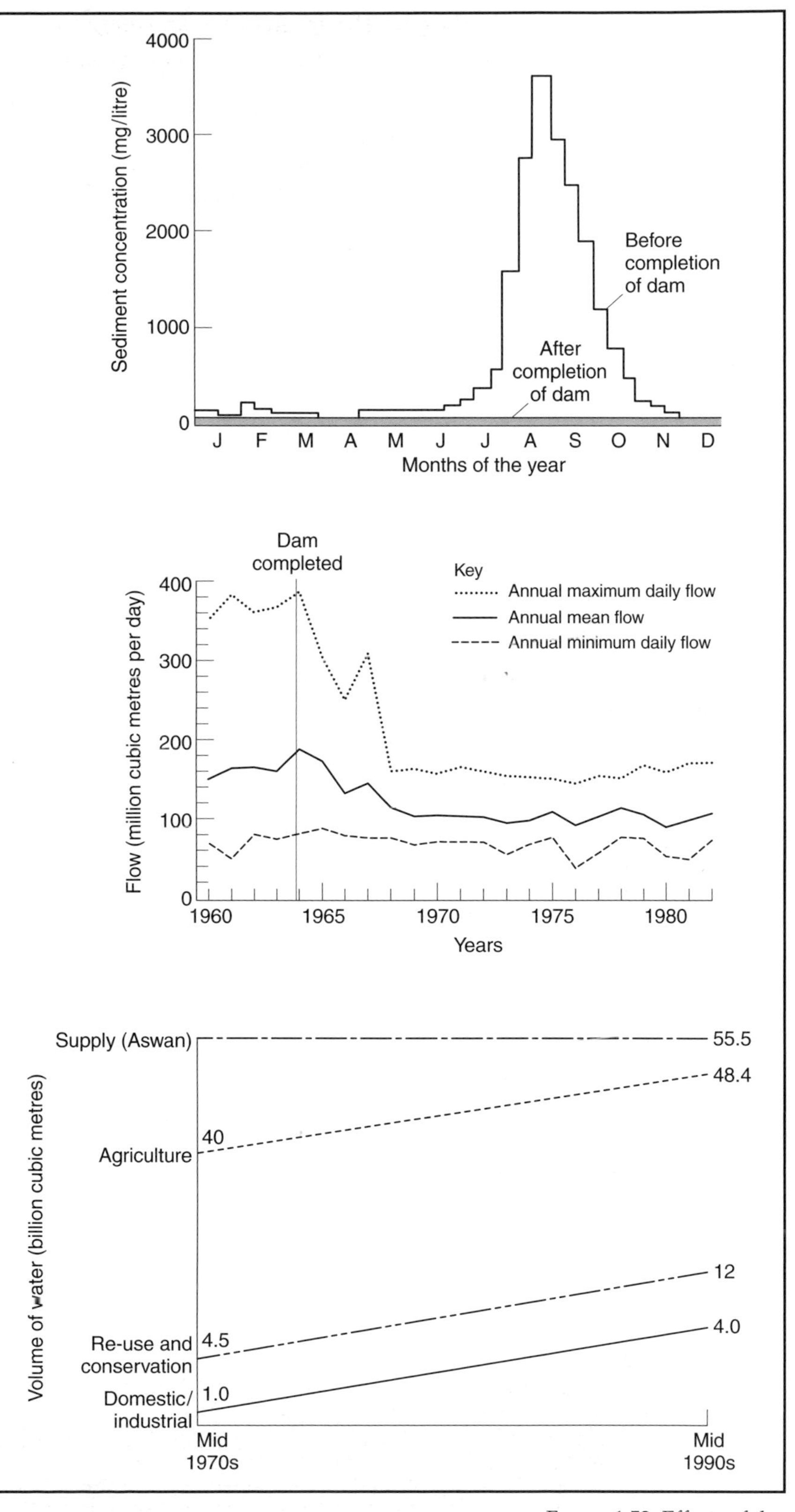

FIGURE 1.52 Effects of the Aswan High Dam on the River Nile

FIGURE 1.53 Lake Nasser

coastal waters. Fishing on Lake Nasser has, however, become a lucrative growth industry, together with its job-creating tourist and recreational activities.

- For at least three months every year, no fresh river water flows from the Nile into the Mediterranean Sea. Because of this, long barrages have been built across the mouths of the delta's distributaries to reduce the risk of seawater flowing upstream into the river channels. In spite of these measures, salt water has been able to seep at least 13 km inland to replace lost river/groundwater.
- 60 per cent of Lake Nasser's water is used for irrigation. This has allowed 800 000 ha of desert land to be put to agricultural use to offset, to some extent, the 600 000 ha of much more fertile delta land lost due to urban sprawl. As only 3 per cent of Egypt's surface area is productive farmland and the country has one of the highest population growth rates in North Africa, it is clearly vital that the best possible use be made of every tract of potentially productive land. Water availability throughout the year, instead of only during the brief flood season, has allowed perennial irrigation to take place and more than one crop to be grown each year on about 350 000 ha of long-established farmland. It has also created a 30 per cent increase in the cultivatable land in Egypt, raised the Saharan water table from 15 m to 3 m below the surface, and made vitally-important desert oasis artesian wells in Algeria much more productive.
- Abu Simbel was unearthed in 1817. This impressive sandstone structure consists of four seated statues of Ramses – each almost 21 m high. To save the monument from being completely submerged by the rising waters of Lake Nasser, UNESCO sponsored an internationally-funded enterprise to cut the statues into 30-tonne blocks and reassemble them 61 m higher up the cliff face from where they were originally carved 3200 years ago. Twenty other monuments from the Egyptian part of Nubia and four monuments from the Sudan also formed part of this $34 million site relocation scheme.
- Lake Nasser flooded a large area of sparsely inhabited land previously occupied by 100 000 people. Egypt agreed to compensate the north Sudanese for their lost farmland and riverside settlements, and allow them to extract enough reservoir water to meet their own domestic and irrigation needs.
- A number of geologists believe that the huge weight of Lake Nasser has destabilised the region and significantly increased the risk of earthquakes; the region's earthquake of November 1981 is believed by many to have been initiated, at least in part, by a temporary imbalance within the Earth's crust 'prompted' by the creation of such a large body of water.
- Lake Nasser has been invaded by the water hyacinth *Eichomia crasspipes*. Such weeds grow profusely and are very difficult to combat, even with potentially highly polluting herbicides. In certain areas, these create a dense, matted bed of vegetation 5 m thick and strong enough to support an adult elephant's weight! They form barriers to navigation, block irrigation and drainage channels and increase the loss of lake water through their transpiration process.

In common with many other regions in which major water initiatives have heightened political sensitivity, the damming of the River Nile during the twentieth century has created serious tensions between member states of the area. The potential for unrest was recognised quickly by the British (under which Egypt existed as a protectorate). The British Government entered into a number of treaties whose chief objective was to ensure that none of the upper basin states could assume a right to interfere with the downstream flow of Nile water; partners in such treaties include Italy (1891) and Ethiopia (1902). Major water-related concerns show little sign of ceasing and a number of similar initiatives have taken place in more recent times, most noteworthy being the 'Agreement for the Full Utilisation of the Nile Waters' in 1959 and current attempts to form an 'International Nile Basin Association'.

ACTIVITIES

1 Make a full analysis of the statistics displayed in each of the graphs in Figure 1.52.

2 Justify the statement: 'The River Nile has a highly complex hydrological regime which is subject to much variation'.

3 List the many varied reasons why dam building along the River Nile has been a crucial aspect of Egyptian domestic and foreign policies for over 100 years.

4 Argue the case for replacing the Aswan High Dam with another dam at a location much further upstream within the River Nile system, having first decided the most suitable alternative location for such an ambitious development.

Study 1.11 River valley features – River Lune, north-west England

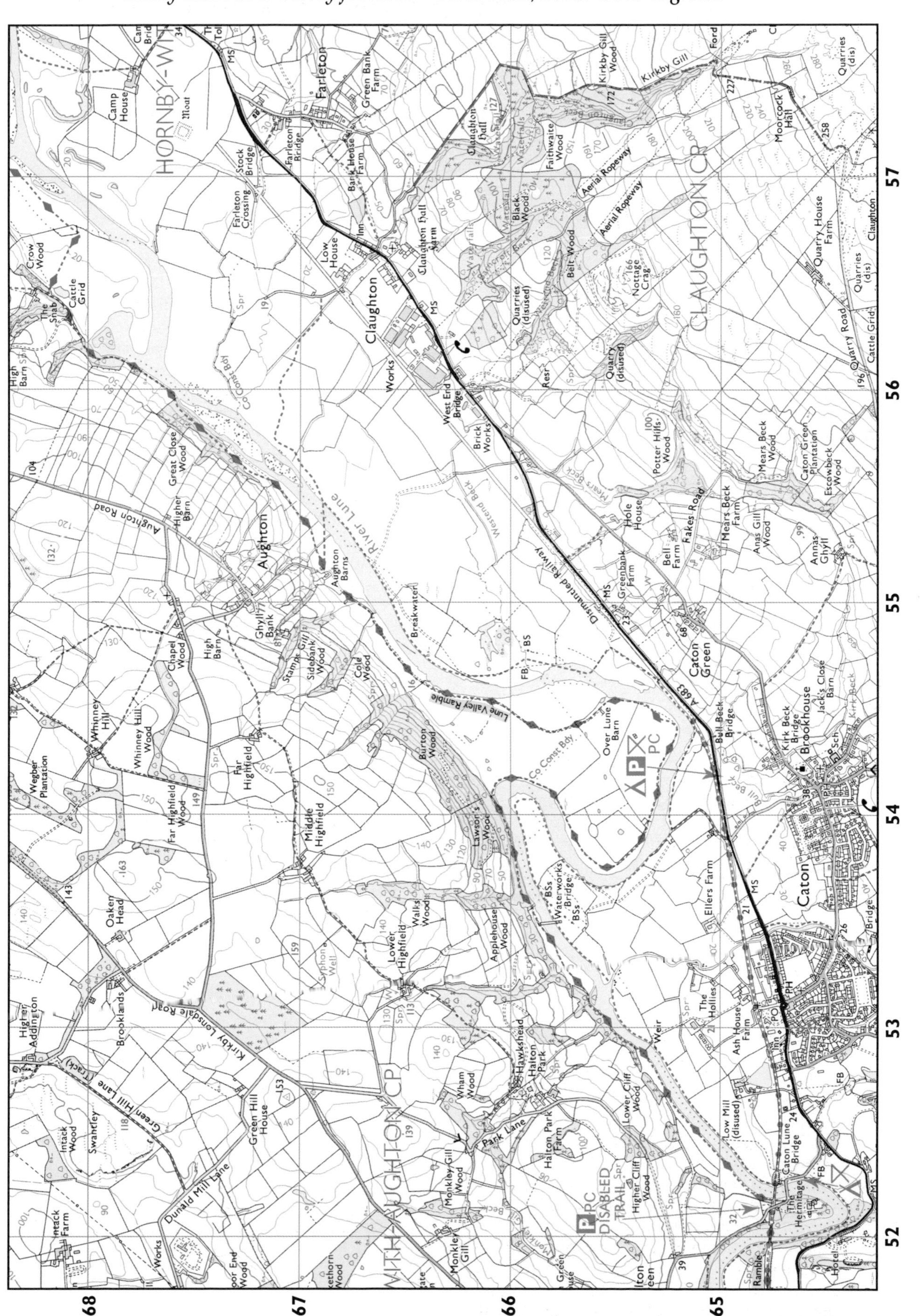

FIGURE 1.54 OS map extract 1:25 000 Outdoor Leisure sheet 41

FIGURE 1.55 The Kingsdale Valley

The river's upper course

One of the Lune's most easterly tributaries is Kingsdale Beck. It is a modest river, originating in many small streams within a long, symmetrical-shaped valley which shares the same name (Figure 1.55). Each of these streams emerges at the meeting point of two very different types of rock. The upper layers of rock in this area are Carboniferous limestone which, being permeable, allows rainwater to percolate down through its joints and along its horizontal lines of structural weakness (its bedding planes) – until reaching the layers of impermeable slate below. Unable to penetrate these lower layers, the groundwater follows the line of unconformity between the limestone and the slate until emerging as a series of springs at very similar heights along both sides of the valley.

FIGURE 1.56 Ingleton Fall Gorge

Kingsdale Beck flows southwards (its name changes to the River Twiss on the way) and into the River Greta which, in turn, joins the main channel of the River Lune at a confluence point 4 km north of Hornby village. Just before flowing into the Greta, the Twiss occupies a very narrow gorge, which is popular with local walkers and ecologists because of its sheltered location, and has created an ideal habitat for ancient oak woodland (Figure 1.56). This 2 km long gorge owes its existence to Thornton Force, a 12 m high waterfall further upstream (Figure 1.57). This waterfall originated a short distance to the west of Ingleton village, at what is now the southern end of the gorge. This is the point where the river crossed the North Craven Fault, where the flowing water was able to quickly erode the softer rocks which lie to the south of the fault line (a process known as **differential erosion**). The inevitable result was a waterfall, which has since migrated northwards, at a very slow and uneven rate, over the intervening 285 million years, leaving in its former path an ever-lengthening gorge.

Figure 1.58 shows the present appearance of Thornton Force, viewed from the northern end of the gorge. Its uppermost layer, including the lip

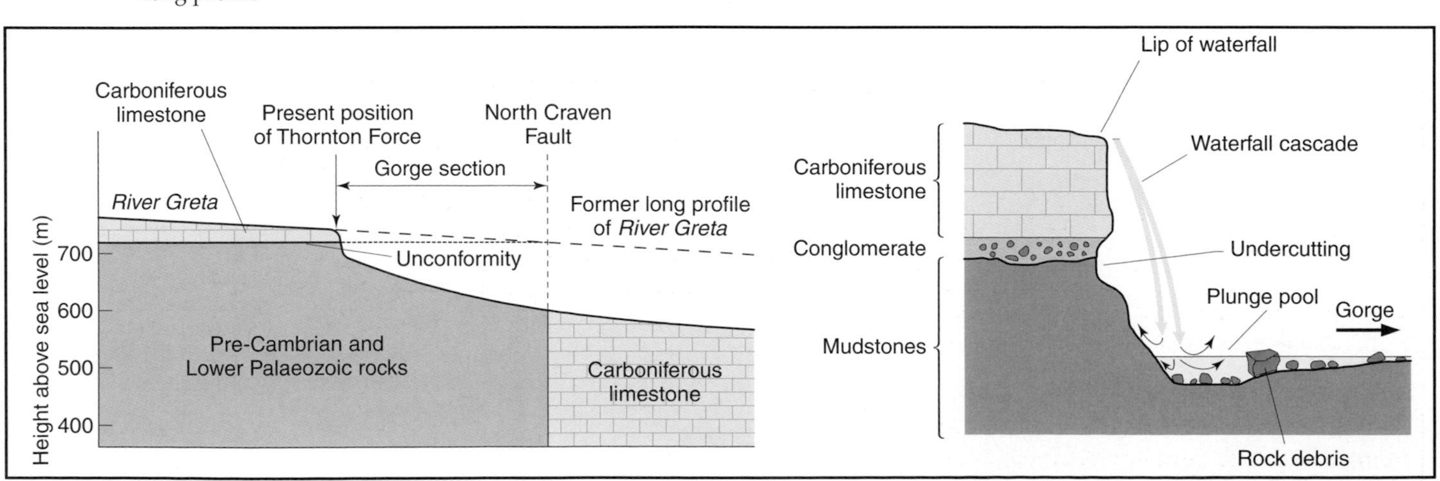

FIGURE 1.57 Thornton Force – cross section and long profile

over which the river cascades, is a continuation of the Carboniferous limestone outcrops of Kingsdale Beck. Its lowest rock layer (which forms its base and contains an almost circular **plunge pool**) is the same impermeable slate which occurs in the bottom of Kingsdale valley (Figure 1.57). Between these two layers is a much thinner band of **conglomerate** – a softer mixture of pebbles, which the falling water has been able to erode relatively quickly and undercut the rock face so deeply that people can walk along its ledge to get behind the curtain of falling water. The plunge pool is deepest where the falling water cascades into it and is surrounded by debris from recent waterfall retreats.

FIGURE 1.58 Thornton Force

ACTIVITIES

1 Explain briefly the relationships between the following features and processes:
a) The existence of the North Craven Fault and the creation of what is now Thornton Force waterfall.
b) The migration of the waterfall and the creation of the gorge.
c) The vertical outline of the waterfall's rock face and the relative hardness of its three component layers.
d) The size of waterfall rockfall debris and its location within the river system.
2 Draw a fully annotated outline sketch to show the key geographical features of the scene in Figure 1.58.

The river's lower course

Many of the Lune's lower course features are clearly identifiable from the Ordnance Survey map extract in Figure 1.54 on page 45. The following activities are based on this extract.

ACTIVITIES

3 Pair-up these river features with their correct grid reference positions on the map extract:

artificial drainage channel	522646
bluff line	533653
braiding	539660
confluence	553664
(part of) the flood plain	557673
meander	572683
ox-bow lake	563670
tributary	565662
waterfall	573680

4 Suggest reasons for the route chosen for the sections of the A683 main road to the east of GR 540650.
5 a) Calculate the width of the Lune's floodplain in a south-easterly direction across the valley from each of these three grid reference points: 523650, 545665 and 562680.
b) Compare your three measurements and suggest reasons for the presence (or absence) of any changes in the width of the flood plain which you have discovered.
6 Account for the numerous waterfalls and the straightness of the stream courses south of Claughton village.
7 a) Measure the amplitude and the wavelength for each of these two meanders:
- The meander in the south-west corner of the map extract (known locally as the 'Crook of Lune').
- The larger meander to the north of Caton village.

b) Calculate the amplitude:wavelength ratio for each meander.
c) Discuss any possible relationship between your two ratios, then record any conclusions agreed by your discussion group.

Study 1.12 Meander migration management – River Mississippi, USA

All rivers tend to **meander** (follow an increasingly curved course) as they progress towards their inevitable union with a sea or a major lake. Most meanders start to form where stream currents are forced to flow around obstructions – resulting in a spiralling movement of the water called helicoidal flow, and varying rates of water speed which are closely linked to their location within the river channel (Figure 1.59).

A mature river's function is to transport the volume of water within its main channel. Meanders slow down that process by increasing the distance over which the transportation has to take place; they therefore make the river less efficient in removing excess water which, under certain conditions, can overflow its banks and cause flooding.

Meander migration causes difficulties because changes in course interfere with current land use patterns, often in intensively developed but vulnerable **flood plain** areas. The technology of river basin management has become highly sophisticated and other studies in this book provide some global examples of both success and failure.

Many whole-basin strategies have relied heavily on meander modification and channel bank stabilisation strategies, some of the most common of these being illustrated in Figure 1.60. Figure 1.61 provides one outstanding example of river basin management, which has utilised meander/levée modification so extensively as to justify an entry in the Guinness Book of Records!

FIGURE 1.59 Features of a typical meander

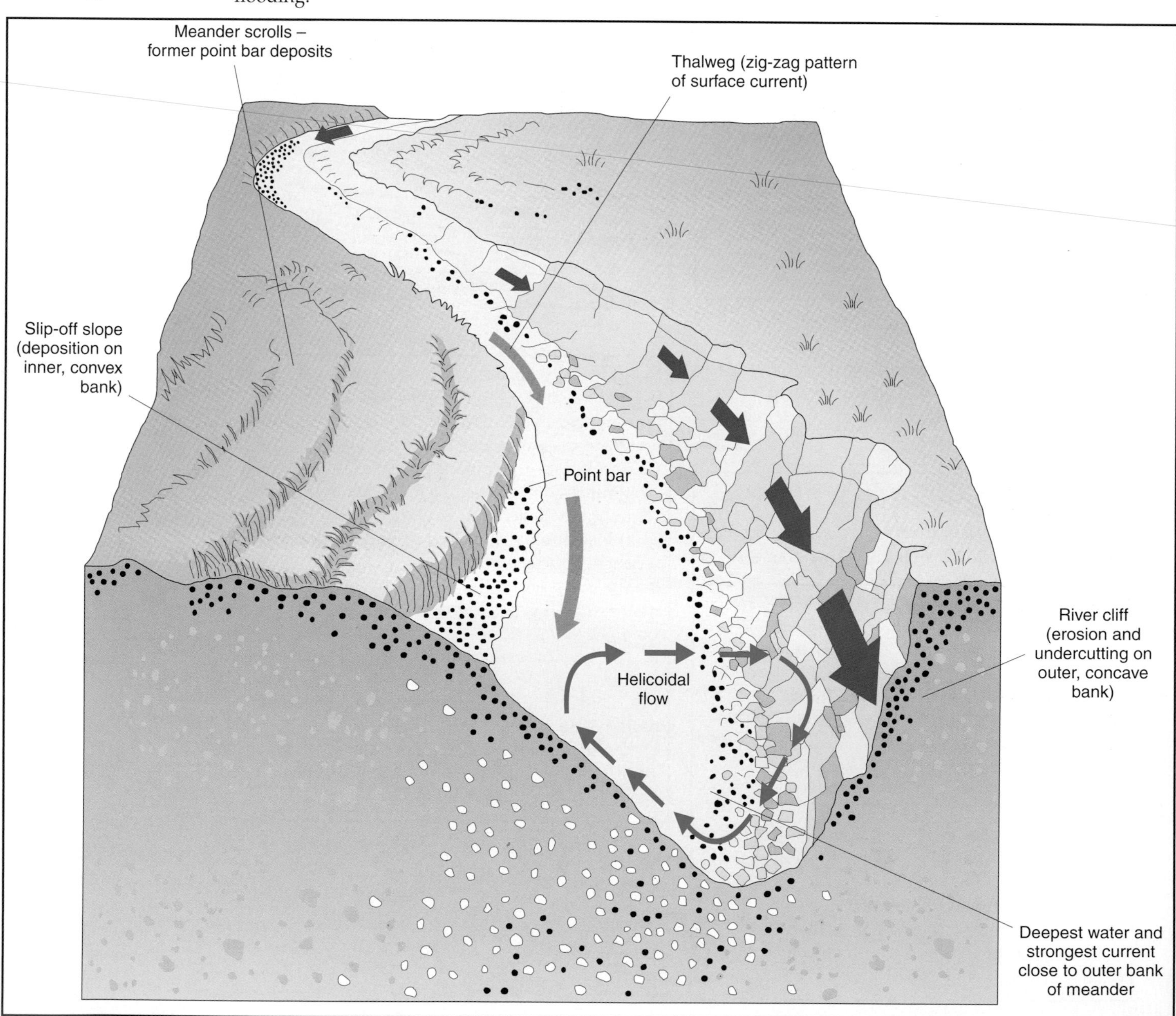

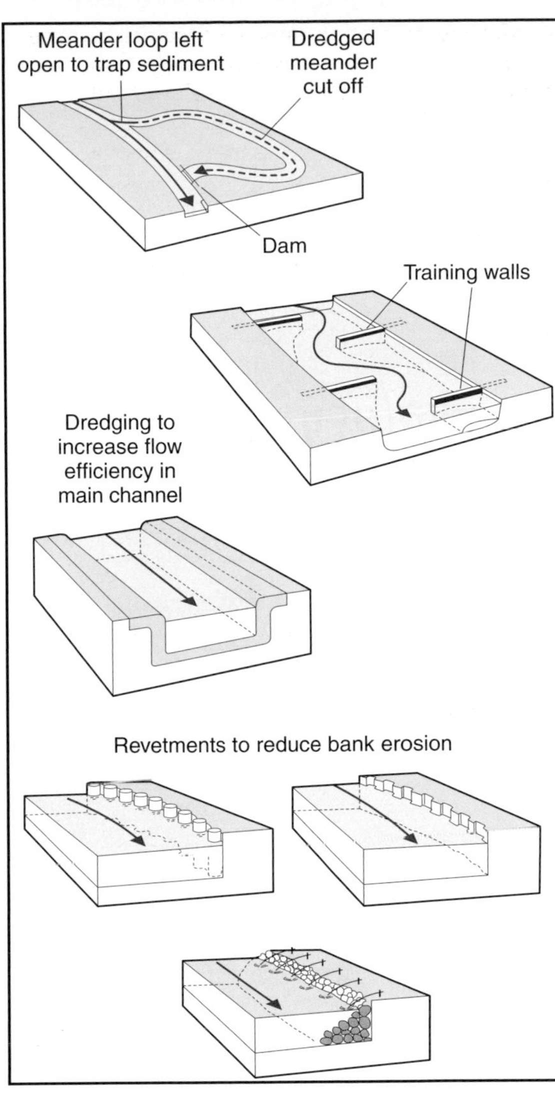

FIGURE 1.60 Meander and channel bank modification techniques

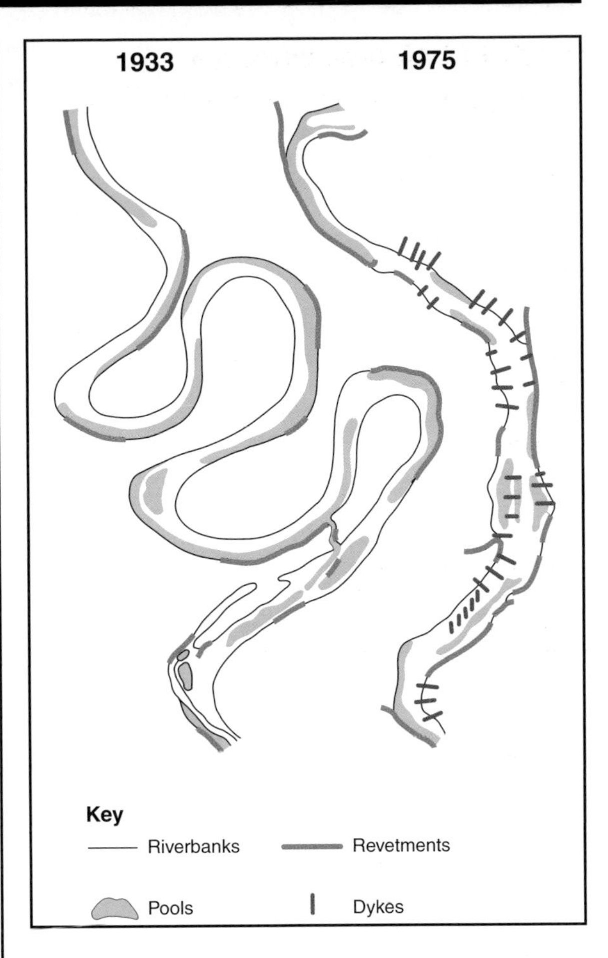

FIGURE 1.62 Meander modifications on the Mississippi

'Largest levées: The most massive ever built are the Mississippi levées begun in 1717 but vastly augmented by the US Federal Government after the disastrous floods of 1927. These extend for 2787 km along the main river from Cape Girardeau, Missouri, USA to the Gulf of Mexico and comprise more than 765 million cubic metres of earthworks. Levees on the tributaries comprise an additional 3200 km. The Pine Bluff, Arkansas to Venice, Louisiana segment of 1046 km is continuous.'

FIGURE 1.61 The largest levées

ACTIVITIES

1 Discuss the purpose and advantages/disadvantages of each of the river management strategies illustrated in Figure 1.60.

2 Use Figure 1.62 as your resource and provide a summary report of the modifications made to the Greenville Reach of the Mississippi between 1933 and 1975.

3 Use information from Internet sites to:

a) Provide information about the major flooding which took place on the Mississippi in 1993, as well as any more recent flood events in the Mississippi–Missouri basin.

b) Obtain additional information about flood-defence measures taken in that river basin.

Unit 2
Coasts

Introduction to coastline morphology

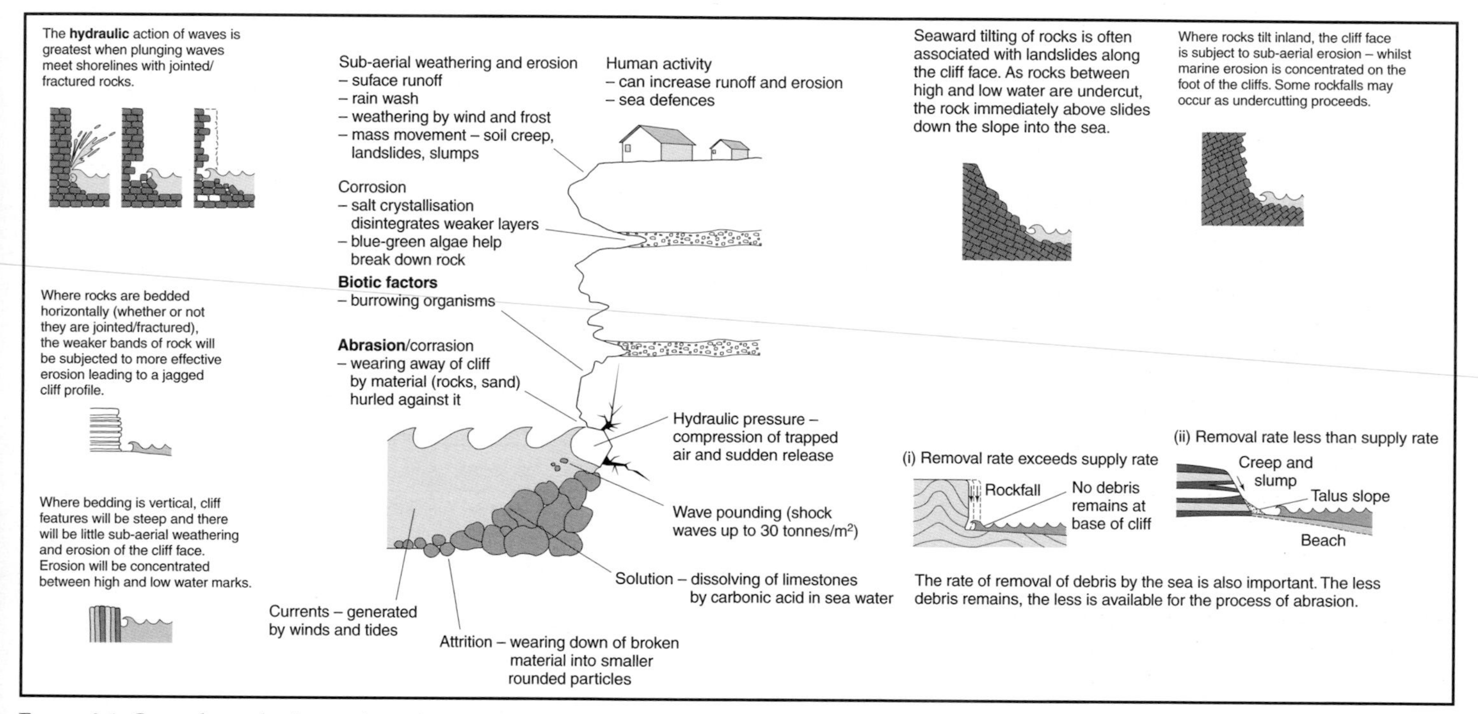

FIGURE 2.1 Coastal weathering and erosion processes

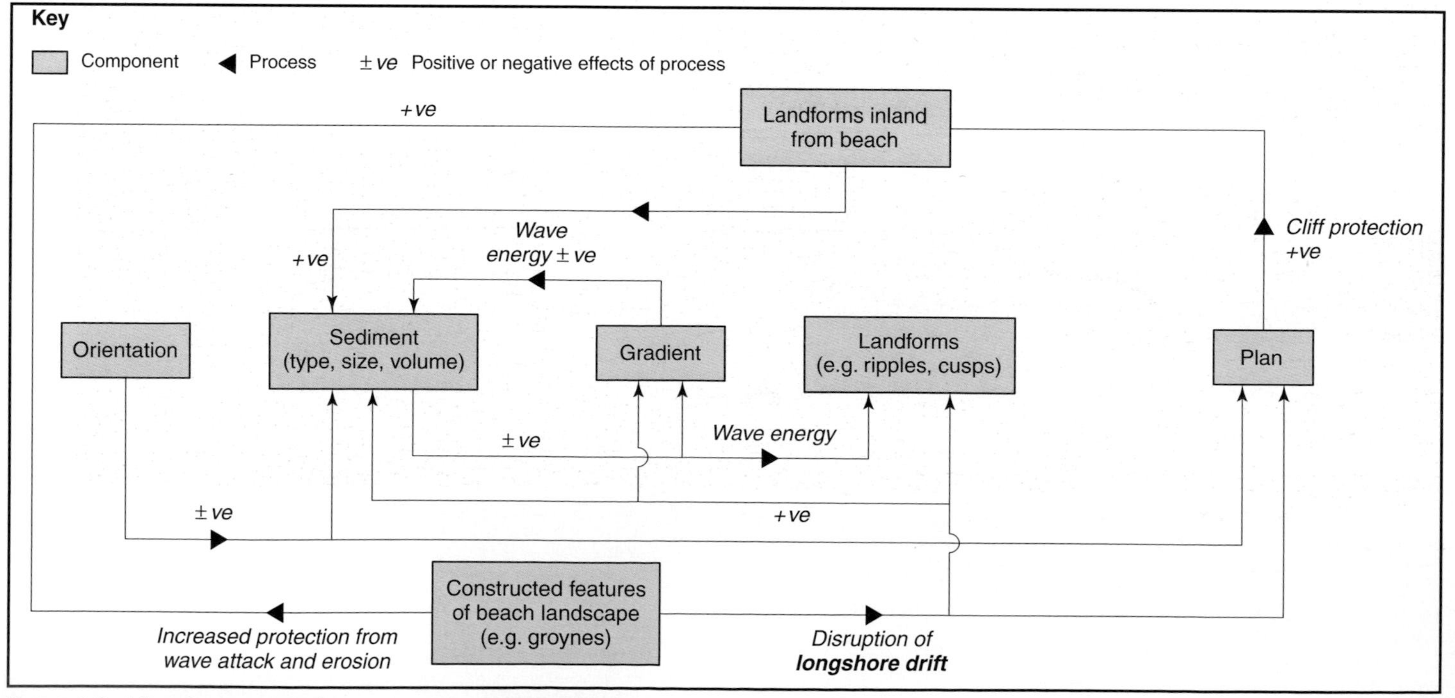

FIGURE 2.2 A beach represented as a process response system

Study 2.1 High energy coastlines – Pembrokeshire

The Pembrokeshire Coast is unique in Britain. It was the only area to be granted national park status (in 1952) because of the outstanding quality of its coastal landscapes. Nowhere in this narrowest and most linear of our national parks is more than 16 km from the sea (Figure 2.3). The outer stretches of the natural harbour of Milford Haven (Figure 2.4) were, however, excluded from the national park area because of their industrial and commercial activities.

Pembrokeshire's deeply indented coastline is partly the result of prolonged attacks by Atlantic gales generated by **prevailing** south-westerly **winds** over its unusually large **fetch**. It is also due to the varying age and hardness of its rocks, some of which are so resistant to erosion that they have resulted in some of Britain's most dramatic stretches of cliff-scenery.

Being remote and sparsely populated, Pembrokeshire is an especially attractive holiday location for families wishing to relax on uncongested beaches and explore its sheltered bays, harbours and coves. Walkers are drawn to the long-distance footpath which gives direct access to 270 km of Pembrokeshire's 370 km coastline. Some outstanding examples of coastal erosion and deposition features occur within the southern section of the Pembrokeshire Coast National Park and are illustrated by the series of photographs in Figure 2.5. The grid reference locations refer to the Ordnance Survey map extract in Figure 2.6. The Pembrokeshire Coast is equally rich in terms of its flora and fauna and one particularly fine example of a pond ecosystem is at Bosherston, whose lily ponds are the subject of Study 1.3.

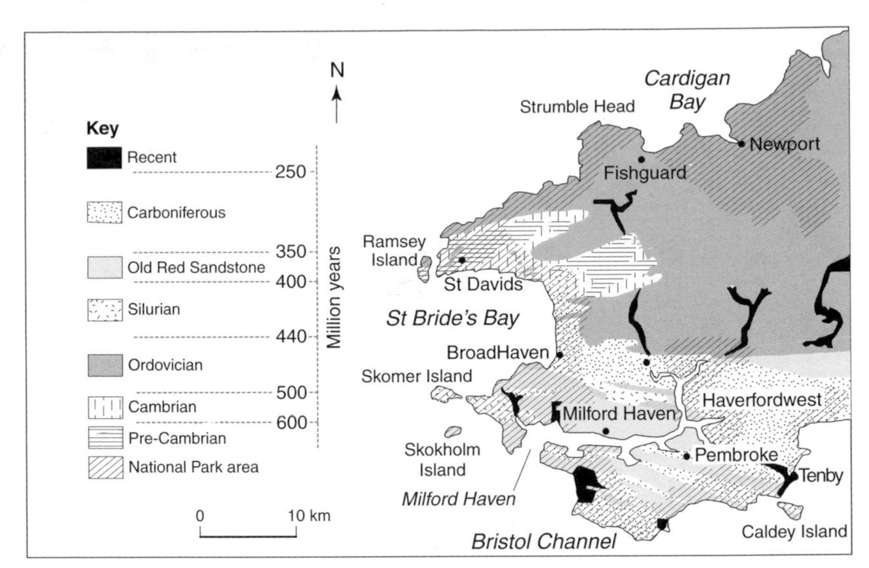

FIGURE 2.3 Pembrokeshire coast National Park

FIGURE 2.4 Milford Haven

FIGURE 2.5 Landforms of coastal erosion/deposition. a) **Arch**; b) **Wave-cut platform**; c) **Stacks**; d) Miniature arch; e) **Wave-cut notch**; f) **Sand dunes**

ACTIVITIES

1 Give specific reasons why certain parts of Pembrokeshire were:
a) excluded from
b) included in
the area designated as a national park.

2 a) With the help of an atlas, calculate the distance of the fetch between Pembrokeshire and the nearest point to it on the South American coast.
b) Make similar calculations for the fetches which are between the following points on the British coast and the nearest land in the directions given:

- Due east from Holderness on England's east coast.
- Due south from Weymouth on England's south coast.

3 With the help of Figure 2.3, describe some of the links between rock type and the shape of the coastline within the Pembrokeshire Coast National Park.

4 Complete two tables with the column headings shown below – one table for all the erosion features illustrated in Figure 2.5, the other for the deposition features shown in Figure 2.5:

Type of natural feature	Named location of feature	6-figure grid reference of feature's location

5 a) Describe the precise locations of any two of the features included in your erosion table, and suggest reasons why they are likely to have developed at these particular places along the stretch of coast shown in the Ordnance Survey map extract on the following page (Figure 2.6).
b) Repeat part a for two features selected from your deposition table.

6 With reference to specific land uses and their Ordnance Survey map locations, suggest a variety of reasons for potential conflict of interests within the map extract area.

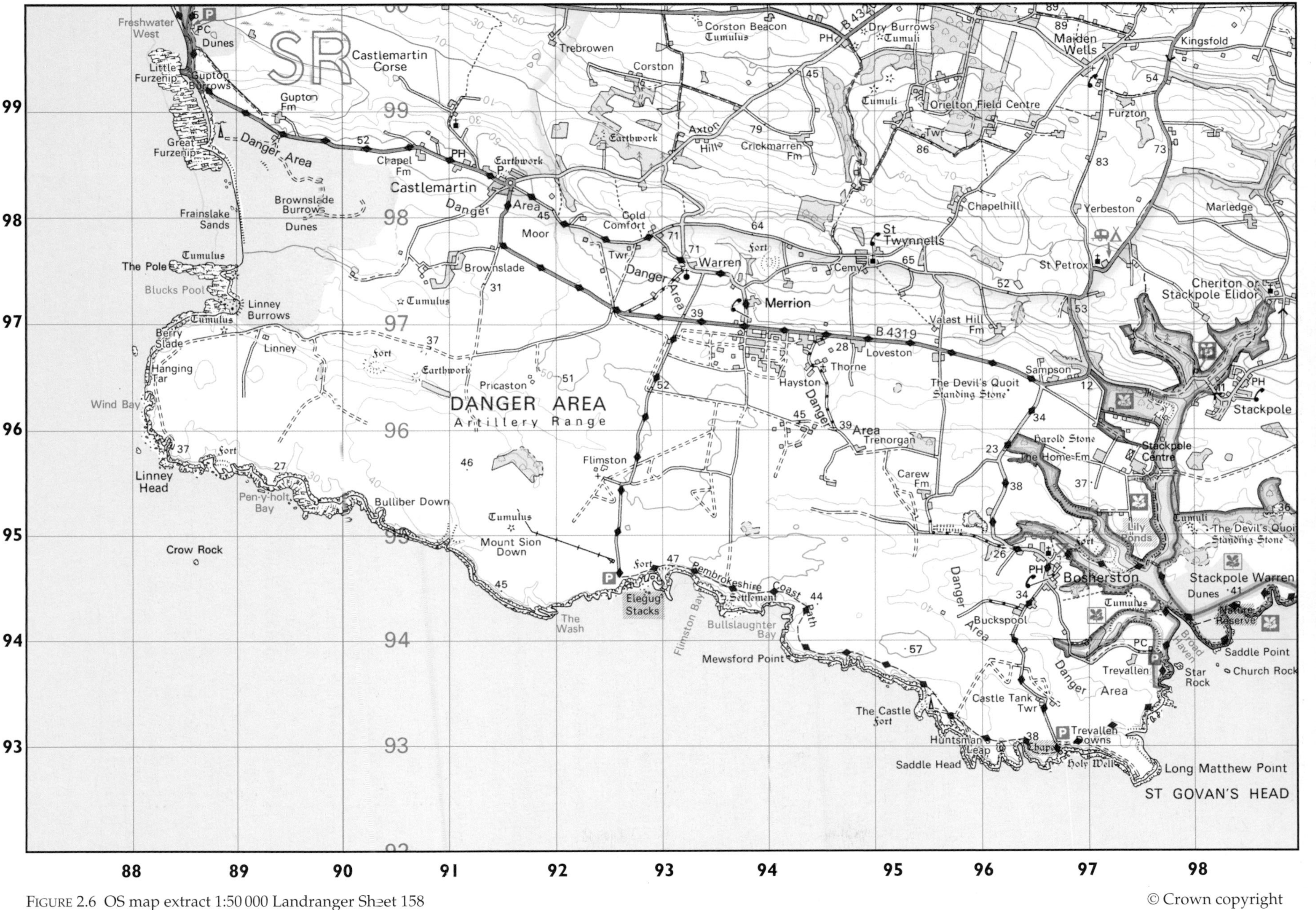

FIGURE 2.6 OS map extract 1:50 000 Landranger Sheet 158

Study 2.2 Localised coastal erosion – Birling Gap, West Sussex

Many stretches of the British coastline are subjected to the natural forces which impact upon them with no protection. This makes coastal properties in such areas so vulnerable that it may be impossible for their owners to sell or even insure them. One community affected in this way is Birling Gap, on the western side of Beachy Head. The newspaper article in Figure 2.7 focuses on the occupants of a short terrace of houses. The occupant of Number 1 – an artist called Jean Cooke – was made homeless

The houses that are falling into the sea

Michael Durham reports

Jean Fawbert's front room feels pleasantly old-fashioned. Like the parlour of a retired mariner, it has ships in bottles, sea shells, watercolours of waves breaking. We are close enough to the sea to hear the surf. Too close, in fact: in fewer than 20 years this house could be washed away.

The sea is an enemy at Birling Gap, a tiny hamlet nestling in a fold of the white cliffs at Eastbourne, East Sussex. For local people, however, there is a second enemy: the National Trust. Normally associated with the upkeep of old buildings and a vanished way of life, the Trust is being accused of standing by while a community disappears.

This enclave of 12 houses and some 30 people, which has 250,000 visitors a year, is being swept away as the Channel advances by about three quarters of a metre annually. The next dwelling to go over the cliff will be Betty Lazarino's cottage, which Wealden Council has declared unsafe and wants to demolish. Geoff Nash's bed-and-breakfast bungalow is also on the brink.

But this week the Save The Gap campaign, a well-connected pressure group, took on the National Trust, the officers of Wealden Council and English Nature as a planning inquiry opened to consider the Trust's refusal to install sea defences. The group includes two members of the House of Lords.

Mrs Fawbert has lived at No 4, The Coastguard Cottages, Birling Gap, for ten years and has known it for as long as she can remember. "It has been in the family for 30 years, since my mother bought it. It's a lovely place to live, I just can't believe it might go," she says. She fears the sea could be gnawing at the foundations of her house within 20 years.

Like other residents, Mrs Fawbert is reluctant to talk about the people whose way of life and homes are under threat. But the nature of the community is part of the fascination of the Birling Gap saga. Two generations ago, the short row of cottages closest to the sea, originally built as a row of eighty by the Admiralty in 1878, became the homes of a remarkable clifftop community, which included many artists.

The six remaining cottages – two have already been claimed by the sea – became a substantial and mutually supportive community, clinging to the chalk downs.

The National Trust owns three of the cottages and most of the land at Birling Gap, a geologically sensitive area and a Site of Specific Scientific Interest that is internationally famous for its coastal erosion. Unfortunately for the residents, the Trust has chosen to put geological erosion before the houses, natural processes before people.

The Birling Gap Hotel, popular with walkers and day-trippers, is also doomed. Further up the lane, Don Holloway has taken to the barricades on behalf of the community, even though it will be another century or two before his own retirement bungalow falls into the Channel. He, too, is very angry with the National Trust.

So what has the Trust done – or not done – to invite this invective? It has adopted the official policy of "managed retreat" from the sea which, according to locals, amounts to surrender. As the sea advances, properties will be demolished and the occupants dispossessed.

Local people have proposed burying huge boulders of French limestone in shingle to stop the sea's advance. Wealden Council, backed by the Trust, refused permission for a retaining wall the length of the beach, but subsequently agreed to a much shorter one. At that point the Department of the Environment called in the papers for ministerial study.

The National Trust is sticking to its guns and says it has carefully balanced the rights of householders against the needs of conservation. It has, it says, "reluctantly" concluded that the Gap's special landscape, character and geology must take precedence.

Paul Pontone, the Trust's spokesman, says a rock barrier could be unsightly and speed erosion further along the coast. It might, in many cases, not work. "It is a personal tragedy for property owners, but houses built on cliffs constantly falling into the sea are clearly very vulnerable," he adds.

In this the Trust is supported by leading geologists and English Nature. To geologists, Birling Gap is an unspoilt example of a glaciated dry river bed, carved through chalk by torrents of water from melting ice 100,000 years ago. The meltwater left deposits of softer, brown pebbles that are eroding faster than the surrounding chalk. The geologists are horrified at the prospect of huge boulders of French limestone being buried at the foot of the cliffs!

"What makes this site so special is that you can study the erosion as it happens. Climate change and rising sea levels are the biggest thing in the environment; you can read the history of climate change here," says Dr Andy King, senior geologist at English Nature. "It's a site of international importance and it would be a tragedy if it were destroyed.

The Times, 21 July 2000

FIGURE 2.7 Effects of coastal erosion

FIGURE 2.8 Birling Gap a) in 1900 and b) in 2000

when the National Trust had the house demolished. The most recent occupant, another artist called Jane Pattison, began living in Number 6 in 1993. The views in Figure 2.8 show how Birling Gap changed during the twentieth century.

ACTIVITIES

1 a) Outline the arguments which the members of the 'Save The Gap' campaign are likely to use to further their cause; you should visit the campaign's website before answering this question.
b) How are the National Trust and its supporters likely to counter the campaign's arguments?
c) In your opinion, which 'side' appears to have the stronger arguments?
d) Give full reasons for the decision you have just taken in answer to c) above.

Study 2.3 Littoral cell coastlines – Holderness, Yorkshire

In August 1998 the British Parliament's Agriculture Select Committee concluded that the millions of pounds spent annually on flood prevention and coastal defences was 'an unsustainable and deluded waste of money'. This committee's formal report to Parliament stated: 'It is time to declare an end to the centuries-old war with the sea and seek a peaceful accommodation with our former enemy'. Peter Luff, MP (the committee's chairperson) added: 'We must now begin to work with nature and not against it; we need to be a little more humble about our relationship with it'.

Luff's committee had focused much of its attention on East Anglia and the adjoining stretches of coastline. Its investigations had concentrated on south-east England, in response to an earlier report (1996) produced by the Climate Change Impacts Group (CCIG) for the Department of the Environment. This report had predicted a 50 cm rise in sea level in the south-east compared to a rise of only 25 cm in the north of Scotland – the difference in levels being due to a higher rate of upwards land movement in the north following the retreat of its

FIGURE 2.9 Spurn Head Spit

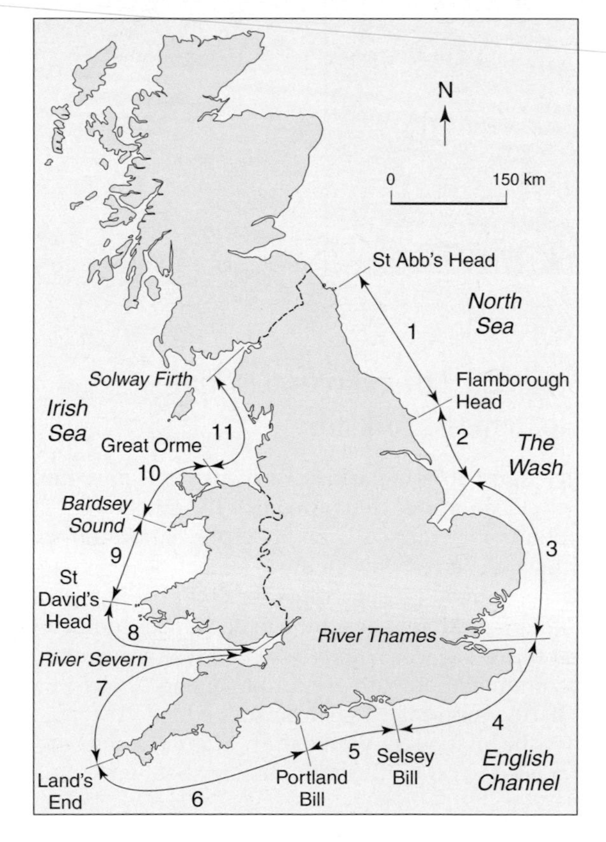

FIGURE 2.10 Littoral cells on the coast of England and Wales

heavy burden of ice at the end of the last Ice Age. It was, the Select Committee argued, now time to give up the fight against the sea along much of East Anglia and the south-east coast and to plan for the phased withdrawal of all but the largest riverside settlements to safer positions behind protective walls and 'washlands'.

Whilst extra protection might be needed to save Britain's cities from the sea's encroachment in the long term, the committee's MPs felt able to support the abandonment of more sparsely-populated coastlines and recommended that the farmers and other land owners who lost property or income due to such measures should be compensated in a realistic way. Farmers' and landowners' associations attacked the report immediately. Even more important than this prediction of a rise in sea level was an increased likelihood of more frequent and more severe **storm surges**. Storm surges are caused by gale force winds blowing over the sea and the CCIG had predicted that these would increase in Britain by up to 30 per cent by 2050. On occasions, surges could coincide with increasing sea levels to flood coastal lowlands with devastating effects.

As a result of such reports, all coastal local authorities are now required to produce and act upon Shoreline Management Plans (SMPs). The chief purpose of SMPs is to encourage a more co-ordinated approach so that authorities will, in future, manage their own stretches of coastline in such a way as to minimise any adverse effects on the coastlines of neighbouring authorities.

The Holderness Peninsula is one of the most striking examples of coastal erosion in Britain. The Holderness Plain occupies the triangle of land bordered by the North Sea coastline, the Humber Estuary and the upland ridge of the Yorkshire Wolds. The most outstanding coastal feature of Holderness is its **spit**, whose southerly tip is at Spurn Head (Figure 2.9).

Holderness forms the major part of one of 11 **littoral cells** into which the coastlines of England and Wales have been sub-divided for management purposes (Figure 2.10). A littoral (or sediment) cell is defined as 'a length of coastline which is relatively self-contained as far as the movement of sand and shingle is concerned and where interruption of such movement should not have a significant effect on adjacent cells'. The SMPs referred to earlier are based on such stretches of coastline. The Holderness coast forms approximately half of Littoral Cell No. 2. This extends down as far as Gibraltar Point, at the most northerly point of the Wash (because it is well-known that much sediment from Holderness is transported southwards as far south as East Anglia). The Humber Estuary forms a littoral cell in its own right due to its size, economic importance and the volume and mobility of its sedimentary load.

About one million years ago, the Holderness coastline formed a line of cliffs almost 32 km west of the present coastline. During the last Ice Age, extensive deposits of soft **boulder clay** accumulated against these cliffs and extended its coastline further eastwards. It is these, more easily eroded clays, which are now threatened by the North Sea's constant attempts to force the coastline into its pre-glacial alignment. Longshore drift now transports material eroded at Flamborough Head southwards along the Holderness shore as far as its first natural break, the Humber Estuary, where it is constantly

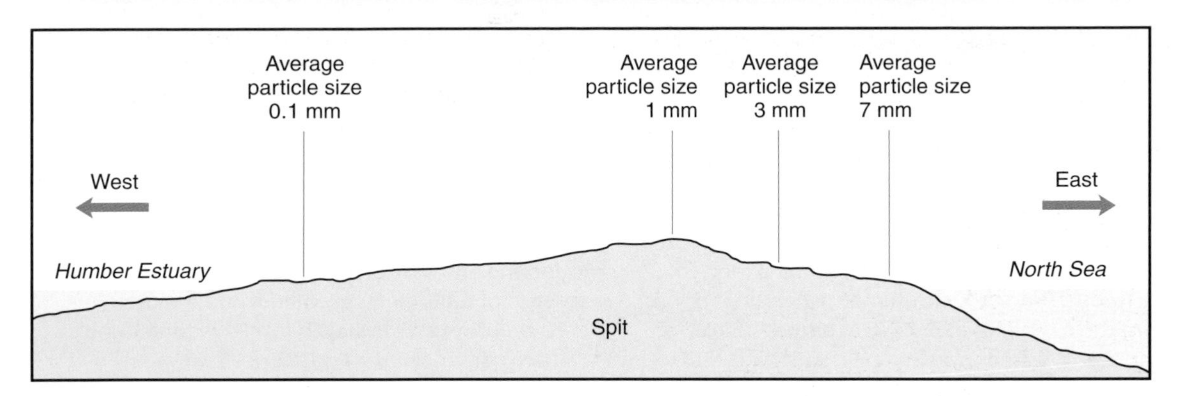

FIGURE 2.11 Cross-section of Spurn Head Spit

deposited on and around Spurn Head Spit (usually referred to by local people as 'Spurn').

Much of Spurn's low hummocky surface is composed of boulder clay; alluvium is deposited along its Humber Estuary shore (Figure 2.11). Spurn also comprises a series of sand and shingle banks now stabilised by colonies of marram grass and seabuckthorn. There is no rocky core to these banks, which are capped by sand dunes. **Saltmarsh** vegetation thrives on the sheltered estuarine **mudflats**.

In the last 2000 years, the Holderness coastline has retreated by almost 400 m. Since Roman times, more than 30 of its villages along the stretch of coast between Flamborough Head and Spurn Head have been overcome by the sea (Figure 2.12). An extreme example of the distress caused to local people by the disappearance of their villages occurred in the late eighteenth century, when a vicar and his sexton (assistant) actually fought each other on Owthorne Beach to decide who should take possession of some valuable lead coffins which were being washed away from their eroding churchyard!

Many measures have been taken to protect the Holderness coast and its economically important seaside holiday resorts. Such measures include sea walls at Bridlington, wooden **groynes** at Hornsea and Withernsea, and granite boulders to protect the base of the cliffs at Mappleton. Yorkshire Wildlife Trust (YWT) has repeatedly asked Holderness Borough Council to dismantle some of these defences so as to increase the southwards flow of beach material. Restricting the flow of sediment from the north is also threatening the stability of the mudflats in the lee of the spit. To date, the council's response has not been as positive as the YWT would like, but the existence of the local SMP may encourage council members to consider the wider implications of their local sea defence initiatives. Despite these measures, over one million cubic metres of beach material is transported annually along the Holderness coastline.

Spurn itself is protected by a series of Victorian defences which were maintained by the Ministry of Defence until the government sold the spit to YWT. These nineteenth century defences are now crumbling very rapidly and its eastern shore is retreating annually at an average rate of 2.75 m per year – twice that of any other coastal location in Britain.

FIGURE 2.12 The retreating coastline of Holderness

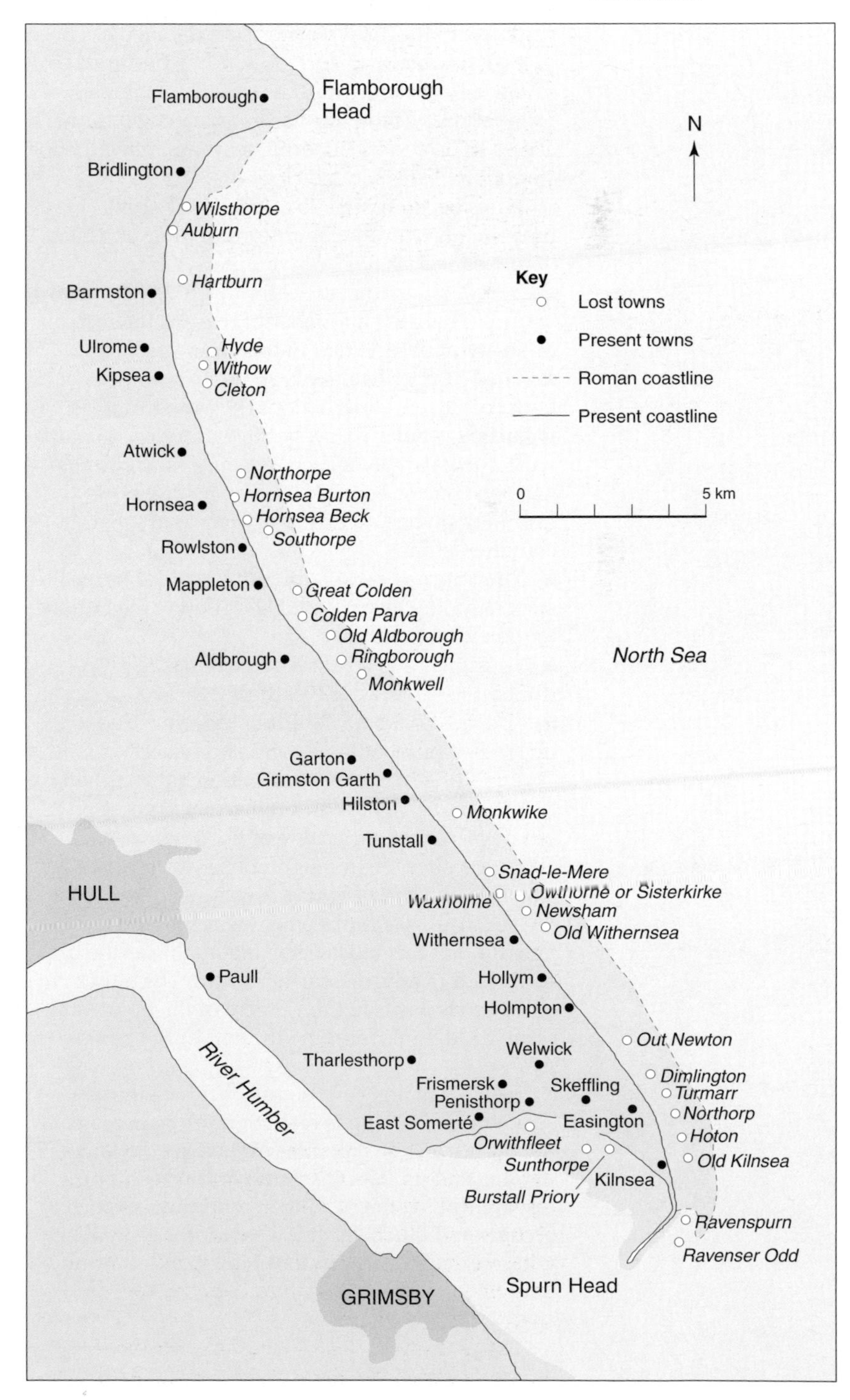

In the 1830s and 40s, a series of severe storms breached Spurn's increasingly narrow neck – a situation made worse by what used to be a common practice of whaling vessels taking shingle ballast from Spurn on leaving port. Up to 50 000 tonnes of shingle were removed annually in this way and dumped in the mid-Atlantic when the whalers' growing catch of fish provided enough weight to keep the vessels at a safe depth in the water. In February 1996 fierce storms almost severed Spurn from the mainland and a 300 m stretch of access road was washed away by gales and high tides. Storms of such exceptional strength have occurred at approximately 250 year intervals, and each storm has re-moulded Spurn's length and shape. One response to the most recent storm damage has been to lay a flexible road surface which is designed to follow any movements of its sand and shingle foundations. Doing this has guaranteed continued access to Britain's last permanently-manned lifeboat station, which is located close to its tip.

It has been possible to calculate the spit's movements over time with considerable accuracy, because the area has been monitored in much greater detail than is usual for this type of coastal feature. There are a number of reasons for such an intensity of observation in this case:

- The Humber Estuary is an important focus of both coastal and international shipping traffic – including regular passenger ferry services to north-west European ports such as Rotterdam. Grimsby has been one of Britain's leading fishing ports and Immingham has become a key terminal for container ships.
- The Spurn Peninsula presents a major hazard to such shipping because of its location at the entrance to the Humber Estuary.
- Spurn has a low profile, which makes it very difficult to see in poor weather.
- Spurn is constantly shifting in form – due to the unpredictability of local wind and sea conditions.

The siting of Spurn's navigation aids has had to be kept under review and replacements erected as old ones became untenable or no longer relevant. 100 years after the first lighthouse was built, at what was then the end of Spurn, it was swept away by the sea. In 1776 a replacement was built on the then new tip, but this too became unsuitable as the tip continued to advance further south. The only light on Spurn today is at the very tip of the point, and some fixed lights marking the end of the pilots' jetty.

Spurn remains very popular with visitors, although swimming is hazardous anywhere near the tip because strong tides often exceed 5 knots (6 km/hr) and its status as a fully protected nature reserve forbids the picking of wild flowers such as orchids and bluebells. It is a '**wilderness area**' within easy access of much of north and central England. The sea is rich in fish – particularly cod and whiting in winter, and skate and bass in the warmer months. Up to 200 species of bird have been recorded at Spurn, including some endangered species such as the rare Marmora's Warbler. Fossils abound because much of the beach material has been transported from Flamborough Head, whose chalk cliffs are rich in fossils. Rambling and riding are relatively easy, due to the absence of hills, steep gradients and outcrops of hard rock. Spurn Heritage Coast Visitors' Centre at Kilnsea provides up-to-date information for visitors and there is a bird observatory and second information centre at the entrance to Spurn Point Nature Reserve. Humber lifeboat and pilot stations, and the remains of discarded lighthouses, are also inevitable sources of interest to visitors.

ACTIVITIES

1 The following questions are based on the spit cross-section in Figure 2.11:
a) Does the topography of Spurn Head Spit follow a symmetrical pattern about the spit's centre point?
b) If not, what are the general characteristics of the topography of the spit?
c) Suggest possible reasons for the shape of the spit.
d) What are the general characteristics of particle size along the spit cross section from east to west?
e) Suggest reasons for the pattern which you have described in answer to d) above.
f) What links, if any, appear to exist between topography and particle size as you have described them?
g) Suggest reasons for your answer to f).

2 Use Figure 2.12 to answer the following questions about the physical and human circumstances on the Holderness peninsula:
a) What is the length along the Holderness coastline from Flamborough Head to Kilnsea?
b) What is the length along Spurn Head from Kilnsea to its tip?
c) How many Holderness villages have 'disappeared' since Roman times?
d) Is there any noticeable pattern to the distribution of village disappearance along the Holderness coast?
e) How does the pattern of existing settlements compare with that of the villages which have disappeared?
f) By what distances has the Holderness coastline retreated at Bridlington and Easington?
g) What do you deduce from your answer to f), in terms of changing rates of erosion and possible reasons for such changes?

Role-play Coastal protection – Holderness, Yorkshire

This activity takes the form of a role-play. Its objective is to examine the wide range of often conflicting physical and human issues which have to be taken into account when considering the future of environmentally sensitive areas such as Holderness. The role-play activity, which has four distinct phases, has been devised to debate the proposal: 'Sufficient finance should be made available to protect the Holderness coastline from the continuing effects of the present high rates of coastal erosion'.

Phase One: Preparation

All members of the group will be given one of two types of role. The local people will be required to think carefully about how the role-play proposal, if approved, would affect the lives of the community in which they live as well as the people who visit it. They will wish to put relevant questions to the experts during the course of Phase Two. It is therefore important that the experts undertake some detailed research ahead of the debate in Phase Two, and it is quite a good idea for them to meet to rehearse their answers to the questions which they are most likely to meet. The following role outlines are brief and students should seek advice as necessary from your teacher on how to undertake their roles. Possible roles are:

1 Parent with young children, who visit Holderness and Spurn Head every year for their annual holiday.
2 Local Geography Teacher, who is knowledgeable about longshore drift and spit formation.
3 Environmental officer of the local council, with responsibility for writing and implementing its Shoreline Management Plan.
4 Sewage disposal expert for the Holderness area.
5 Owner, Holderness Riding Stables.
6 Professor of Oceanography at Hull University, who is an expert on North Sea tidal patterns.
7 Chairperson of Yorkshire Wildlife Trust.
8 Member of Parliament, who is a member of the government's Agriculture Select Committee.
9 General Manager, Humber Dredging Company.
10 Harbour Master of Hull Docks, with responsibility for local ship movements and shipping pilotage services.
11 Trinity House Regional Officer for North-east England (Trinity House is responsible for lighthouses and other ship navigation aids).
12 Tourism and Publicity Officer for Bridlington.
13 Insurance consultant, who also operates a land-loss compensation agency.
14 Member of Greenpeace, with special responsibility for global warming matters.
15 Local farmer, who grazes sheep on salt marshes in the Humber Estuary.
16 Tenant of the Crown and Anchor public house at Kilnsea.
17 Local co-ordinator of the Royal National Lifeboat Institution (the R.N.L.I.).
18 Secretary, East Yorkshire Local Historical Society.

Additional parts may be suggested by your teacher or fellow students to cater for larger groups.

Phase Two: The debate

The purpose of this second phase is to give the local people an opportunity to put questions to expert witnesses. It is helpful for the room to be rearranged so as to create the atmosphere of a formal public enquiry and for each participant to have a card displaying his/her role. The chairperson (the teacher) will probably insist that every question is directed through him/her so as to ensure an orderly debate. This procedure is recommended so as to ensure that lively discussion does not degenerate into bitter or personal arguments. Before the start of the debate, the chairperson will inform members of the need to prepare themselves for the written tasks for Phase Four.

Phase Three: The vote

When all the relevant issues have been thoroughly debated, the chairperson will summarise the key points raised during Phase Two and instruct group members to disregard any previously held views. The members will then be invited to vote either for or against the proposal – basing their votes solely on the strength of the arguments made during the debate.

Phase Four: Follow-up work

Individual group members will undertake the following written tasks:

1 Prepare a summary of the key points discussed in the debate.
2 Record the vote reached in Phase Three.
3 State whether or not they agree with this whole-group vote and give reasons for his/her answer.

Study 2.4 Bars and tombolos – Loe Bar, Cornwall and Chesil Beach, Dorset

Spits such as Spurn Head (Study 2.3) may continue to grow until their tips join another stretch of the coast forming a **bar** – usually under the combined influence of the prevailing wind and the general direction of longshore drift. The initial effect of this growth is to impound a section of bay from the sea. If the bar is also able to continue to grow vertically, it will eventually cut off the trapped bay's supply of salt water and turn it into a fresh water lagoon fed only by streams flowing into it from the landward side. Ultimately, the sediments transported by these streams will in-fill the lagoon and transform it into wetland and then new land. Figures 2.13 and 2.14 provide information about the location and formation of Loe Bar, whose ridge of flint pebbles and sand created Cornwall's largest expanse of fresh water. In the past, the water in Loe Pool often threatened to flood the nearby village of Helston until a temporary channel could be dug through the bar to release its excess water.

Chesil Beach, 190 km further east, in Dorset, is a much larger structure and, as it links an island to the mainland, is more correctly called a **tombolo** (Figure 2.14). 98 per cent of the beach material is composed of erosion resistant pebbles of flint (eroded from the chalk cliffs to the west) and chert (from outcrops of greensand). Chesil Beach has protected a lagoon behind it and what is now Portland Harbour against all but the most severe storms. In 1590 the ship *Ebenezer* was lifted over the beach by waves which over-topped it and deposited the ship in more sheltered water to the north of Portland Island.

Chesil Beach was first formed as a bar of flint and other materials approximately 20 000 years ago. Rising sea levels at the end of the last Ice Age would have allowed storms to manoeuvre the mass of pebbles into the present-day 75 000 tonne, stable tombolo about 14 000 years later. The sorting of surface pebbles according to size continues, however, as it is a never-ending balancing act between the impacts of onshore winds and waves from many different directions.

ACTIVITIES

1 Contrast the locations, sizes and material compositions of Loe Bar and Chesil Beach.
2 Outline the likely impact of both features in this study on the lives of their local human populations.
3 Investigate at least one other example of each feature; possible choices are the tombolo on which Llandudno has been built and the many bars on the North European Baltic Sea coast.

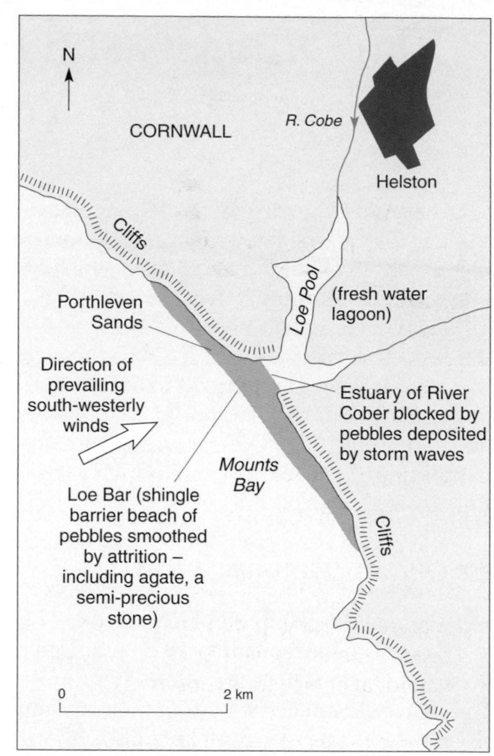

FIGURE 2.13 Loe Bar

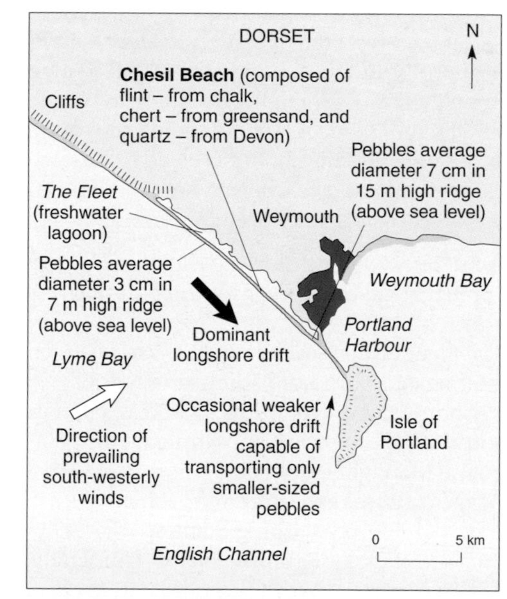

FIGURE 2.14 Chesil Beach

FIGURE 2.15 Loe Bar

FIGURE 2.16 Chesil Beach and the Fleet

Study 2.5 Psammosere ecosystem environments – Ainsdale, Merseyside

Psammosere ecosystems occupy coastal sand dune areas in many parts of Britain (Figure 2.17). Sand dune research has many potential geographical applications – human as well as physical – which makes it an ideal choice for synoptic A-Level based work.

Dunes tend to be in locations having:

- a large tidal range
- a low-gradient beach
- a relatively flat, low-lying strip of land behind the beach
- strong and predominantly onshore winds.

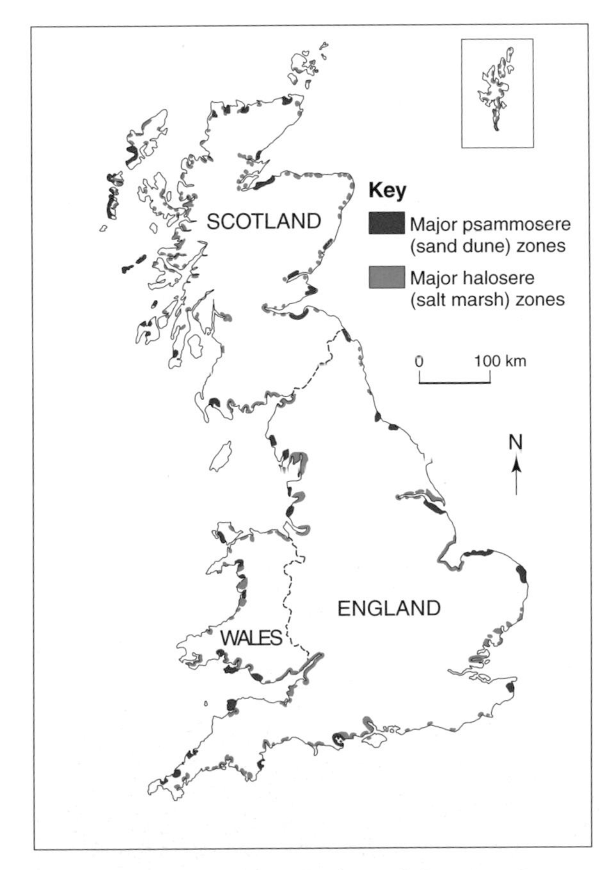

FIGURE 2.17 Sand dune and saltmarsh locations in Britain

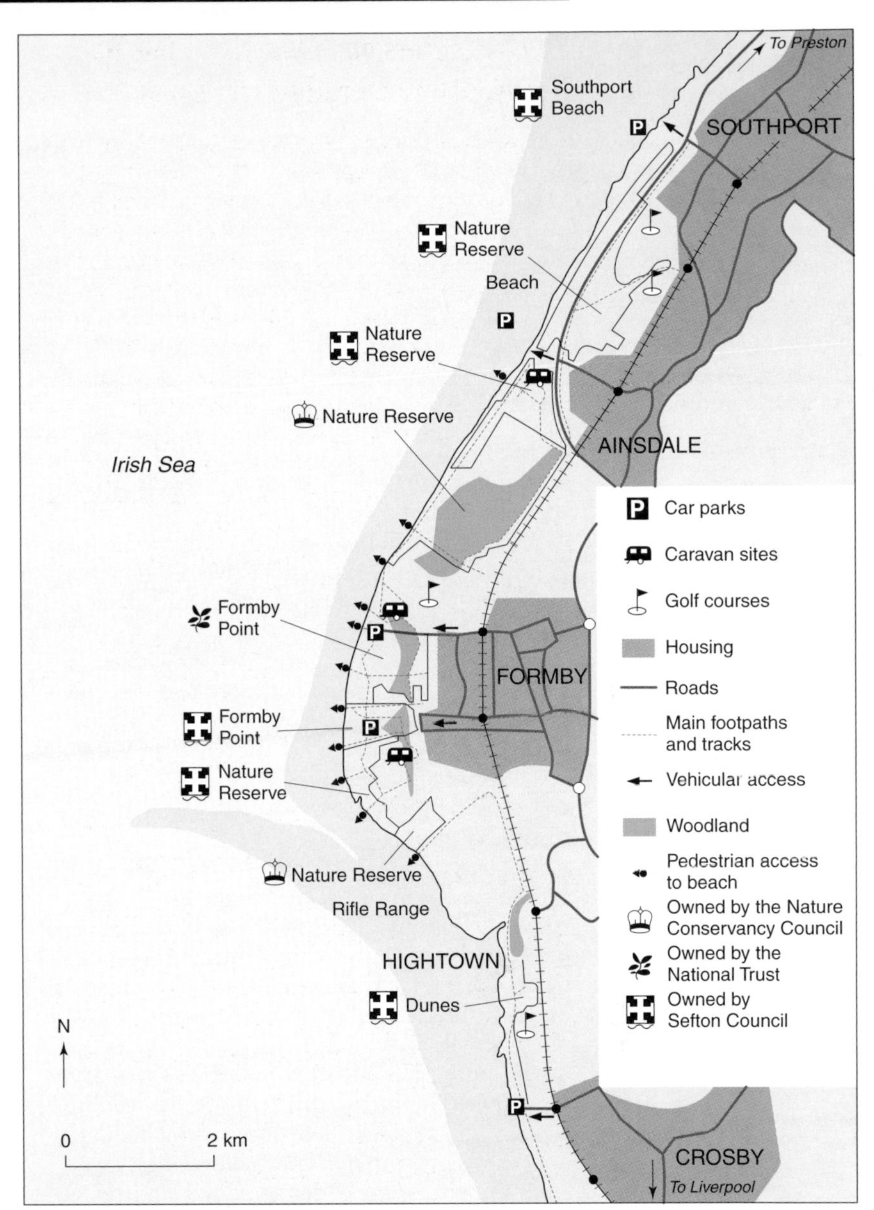

FIGURE 2.18 Ainsdale and surrounding area

FIGURE 2.19 Ainsdale: coastal dunes

FIGURE 2.20 Ainsdale: inland dunes

In common with many other coastal habitats, sand dunes are under increasing pressure from competing land uses and this important aspect is examined fully in the Ainsdale study which follows. Figure 2.21 (page 63) provides selected data for four of the most important types of coastal habitat in England.

The Sefton Coast (Figure 2.18), which lies between Liverpool and Southport, features one of the finest and most extensive dune systems in north-west England – if not the whole of Britain. It contains separate expanses of:

- exposed beach at low tide
- sand dunes
- Corsican pine woodland.

The coastal dunes nearest to the beach at Ainsdale (Figure 2.19) form the habitat of two particularly rare and legally protected creatures: the natterjack toad and the sand lizard. The more stable dunes further inland (Figure 2.20) support a wide range of wildlife which includes foxes, rabbits, moths, butterflies and dragonflies. The woodland was planted at the beginning of the nineteenth century and so does not form a natural habitat. It does, however, provide an ideal breeding environment for the endangered red squirrel. Over 450 species of flowering plants have been identified along the Sefton Coast and large numbers of wading birds feed on its shore during the autumn and winter months.

At Hightown, there is an exposed section of forest estimated to have been growing 4000 years ago before being submerged; this location also has stratigraphic evidence of 11 transgressive and regressive stages (coastline advances and retreats) during the past 9000 years. The entire Sefton Coast has now become an important area for both educational and research activities.

A number of local government and environmental organisations work closely within this fragile area, under an agreement called the Sefton Coast Management Scheme. This scheme requires these organisations to share knowledge, combine their individual resources whenever appropriate and ensure that their own initiatives are as compatible as possible with those of its other members. In 1965, a 508 ha area about halfway between Formby Point and Southport came under the protection of English Nature and was re-named the Ainsdale Sand Dunes National Nature Reserve. (English Nature is a Central Government-funded body having responsibility for the conservation of especially significant wildlife and natural features in England.) In 1980, a further 268 ha of land adjoining Ainsdale Sand Dunes NNR was declared a local nature reserve by Sefton Council, to be funded and managed by this local government authority. The National Trust property at Freshfield is a well-established Red Squirrel Reserve.

These organisations have undertaken many strategies to improve the management of the Ainsdale Sand Dunes NNR and other conservation areas along the Sefton Coast. Some of the more significant of these strategies are:

- Since 1977, several artificial ponds have been excavated to provide additional breeding habitats for natterjack toads.
- In 1993, a 'beach management plan' was put into action by Sefton Council. This enabled the foreshore at Ainsdale to be zoned into separate car parking and car-free areas. One effect of this strategy has been to increase the shore bird population within the car-free areas.
- The Sefton Coast projects have helped to pioneer the so-called 'dynamic approach' to conservation, which recognises that over-zealous dune stabilisation often puts rare habitats and species at risk. In line with this approach, large sections of the NNR were identified as being ideal for dune conservation and its natural processes should be left undisturbed.
- The removal of dune plantations is taking place at Ainsdale Sand Dunes NNR and dune habitats are being reclaimed. Small-scale sand mobility has been reactivated through sheep grazing and encouragement of the growth of the rabbit population, which had suffered depletion due to myxamotosis. Encouraging sustainable grazing has proved highly beneficial to the local farming community.
- Information of interest to environmental groups has been exchanged at regular workshop sessions and the general public has been made more aware of local initiatives by means of free leaflets and other publications.
- Some grassland areas have been mown to encourage the growth of a variety of plants.
- Woodland areas have been removed as part of an experiment to restore the original natural dune habitat. The remaining woodland is managed in such a way as to ensure that it remains healthy and will continue to support many trees of different species and ages. Species such as sea buckthorn, creeping willow and birch, which have particularly high rates of growth, have been rigidly controlled to allow other species to thrive.
- English Nature has created a network of visitor footpaths, marked with white-topped posts. Access to all other areas is by permit only.

The Sefton Coast has important recreational functions, particularly for families living in nearby Southport (population 90 000), Preston (135 000) and

Type of habitat	Current total area (in ha.)	Predicted loss over 20 years (in ha.)	Percentage habitat loss
Intertidal mudflats	250 000	10 000	4
Saltmarsh	34 375	2750	8
Sand dunes	12 000	240	2
Shingle	5000	200	4
Saline lagoons	1200	120	10

FIGURE 2.21 Coastal habitats in Britain

Skelmersdale (42 000). The city centre of Liverpool (4 615 000) is only 20 km from the National Nature Reserve. Communication links along the whole stretch of the Sefton Coast are excellent and frequent rail and bus services between Southport and Liverpool provide easy access to all of its open conservation areas; a network of footpaths (32 km total length) provides an even higher degree of access to the chief honeypot sites. Such access has, over many years, been a major factor in many of the coast's current environmental problems.

One recent threat has been the use of intensive, low-frequency soundings since the 1980s by companies seeking to exploit natural, onshore resources such as petroleum. As part of its exploration programme, Independent Energy UK has dynamited a number of sites since receiving planning consent to do this from the local authority in 1999. An RSBP site at Marshside, north-west of Southport town centre – which is a winter refuge for pink-footed geese, wigeon, black-tailed godwits and golden plovers – is believed to be at particular risk from the vibrations from this exploratory work. Sefton Council have now pledged that any applications to develop major onshore sites will be submitted to a rigorous public enquiry process.

Sand dunes are ideal locations in which to undertake small-scale research, for a number of reasons:

- There are well-developed dune systems within easy reach of most parts of Britain.
- They have relatively uncomplicated food webs and ecosystems.
- Their structure tends to follow a predictable sequence of zones – each of which provides an appropriate habitat for a small and distinct range of plant species (Figure 2.22).

ACTIVITIES

1 a) Describe the general distribution of sand dune locations shown in Figure 2.17.
b) Following group discussion, suggest reasons for the distribution pattern you have just described.
2 a) Complete a copy of the table in Figure 2.21. The formula to use for the last column is:

$$\text{per cent change} = \frac{\text{predicted loss of land}}{\text{total area of land}} \times 100$$

b) Choose two suitable graphs to display information from your completed table – one for the first column and one for the last column. Give reasons for each choice of graph you have made.
c) Use the information in the first and last data columns in the table to draw your two chosen types of graph.
d) Comment on the patterns of information illustrated by your completed graphs.
3 Using as much detailed information from the Ainsdale Sand Dunes study as necessary, give your own views on the policy of segregating widely different and often competing land-uses and activities within popular locations of environmental importance.

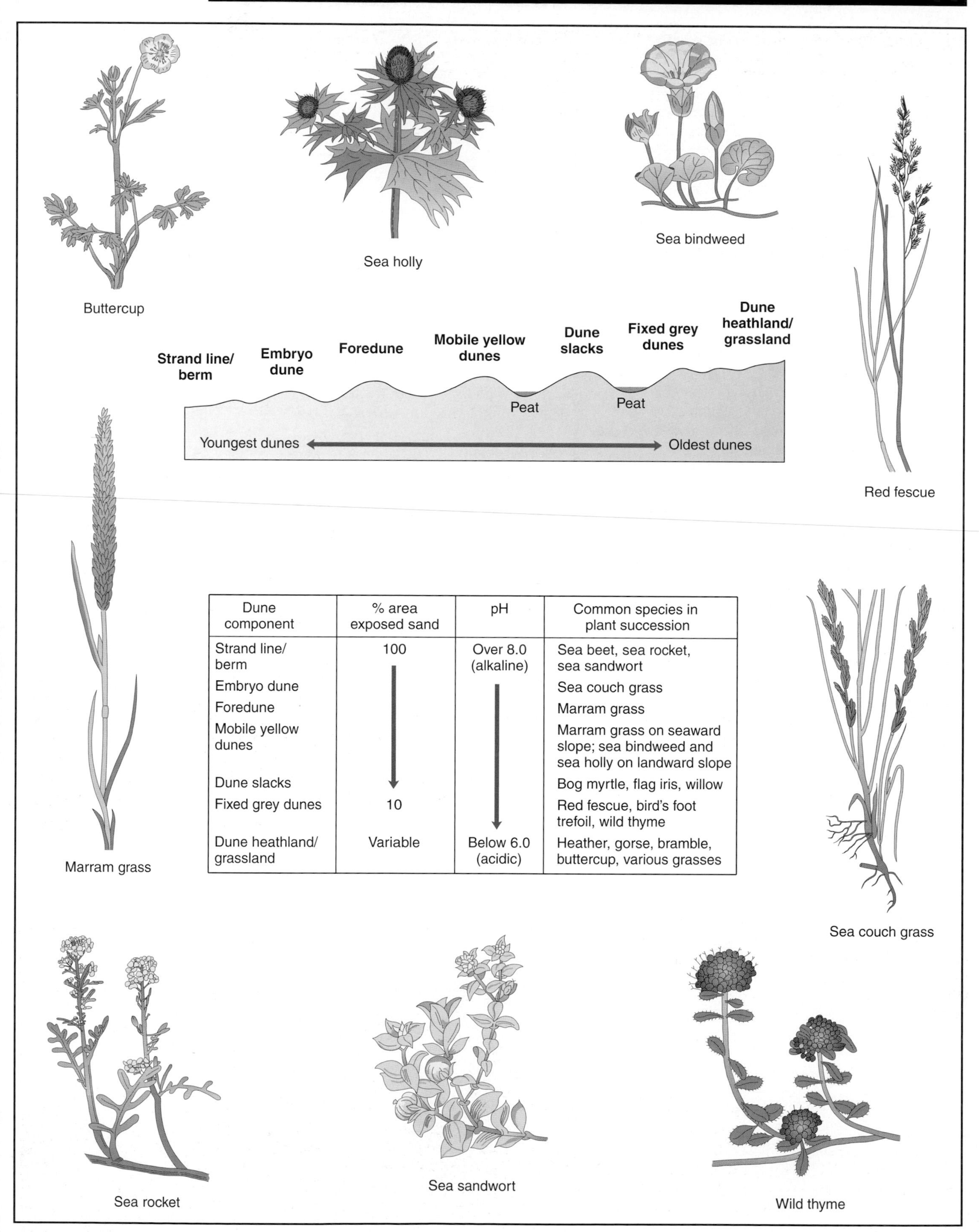

Dune component	% area exposed sand	pH	Common species in plant succession
Strand line/ berm	100	Over 8.0 (alkaline)	Sea beet, sea rocket, sea sandwort
Embryo dune	↓	↓	Sea couch grass
Foredune			Marram grass
Mobile yellow dunes			Marram grass on seaward slope; sea bindweed and sea holly on landward slope
Dune slacks			Bog myrtle, flag iris, willow
Fixed grey dunes	10		Red fescue, bird's foot trefoil, wild thyme
Dune heathland/ grassland	Variable	Below 6.0 (acidic)	Heather, gorse, bramble, buttercup, various grasses

FIGURE 2.22 Dune habitat zones

Study 2.6 Halosere ecosystem environments – Morecambe Bay, Lancashire

Psammoseres (Study 2.5) and haloseres are prime examples of plant succession within coastal environments. The crucial factors in determining which type of succession will develop on a particular stretch of coast are:

- distance from mean low water line
- height above mean low water line.

Both of these factors determine how sheltered the area will be from onshore winds. Additionally, individual plants within the **halophytes** range are tolerant of salt water to varying degrees; for such species, the duration of the twice-daily tidal submersion is especially crucial in determining their place within the saltmarsh and mudflat environment successions. Unlike sand dunes, both saltmarshes and mudflats are ideally suited to shorelines which are sheltered from high-energy wave action and the erosion which results from destructive wave sequences. Large bays, river estuaries and the landward shores of spits, bars and tombolos are ideal locations for saltmarsh development. Gently-shelving shorelines having high tidal ranges are also essential, as these increase the likelihood of initial deposition taking place. Most deposition actually takes place during the 'slack water' periods at high and low tides (on the higher marsh and lower mudflats respectively), when the tide is on the turn.

All shorelines exhibit irregularities to some degree – an excellent example being the gently undulating sands of Blackpool's central beaches. After submersion, the shallow depressions experience strong ebb currents, because the water which they have retained forms part of the tidal retreat. Being submerged to a lesser extent, the ridges between these depressions are less affected by such tidal scour and so are more likely to be covered by deposits of retreating sand and fine sediment. The twice-daily repetition of this process can, under favourable conditions, allow algae and pioneer, alkaline-tolerant plants such as eelgrass and marsh samphire to colonise the rising ridges. This is the crucial initial stage, as the roots, stems and foliage of these plants accelerate the deposition process and hence the upward development of the ridges. Although these hardy plants are quite small, they are able to trap seaweed, small pieces of driftwood and other forms of flotsam, which become entangled in their stems and so increase the sediment deposition rate.

The constant heightening of the ridges encourages the colonisation of less salt-tolerant (and less submergence-tolerant) halophytes such as rice grass, sea aster and spartina (cord grass). The availability of deeper mud allows these taller, longer-living and deeper-rooted species to thrive in conditions where submersion is reduced to only three to four hours in every twelve. The deposition rate then continues to increase, as the density of the vegetation cover increases.

FIGURE 2.23 Saltmarsh (a) and salt pans (b)

FIGURE 2.24 Mud flats and creeks

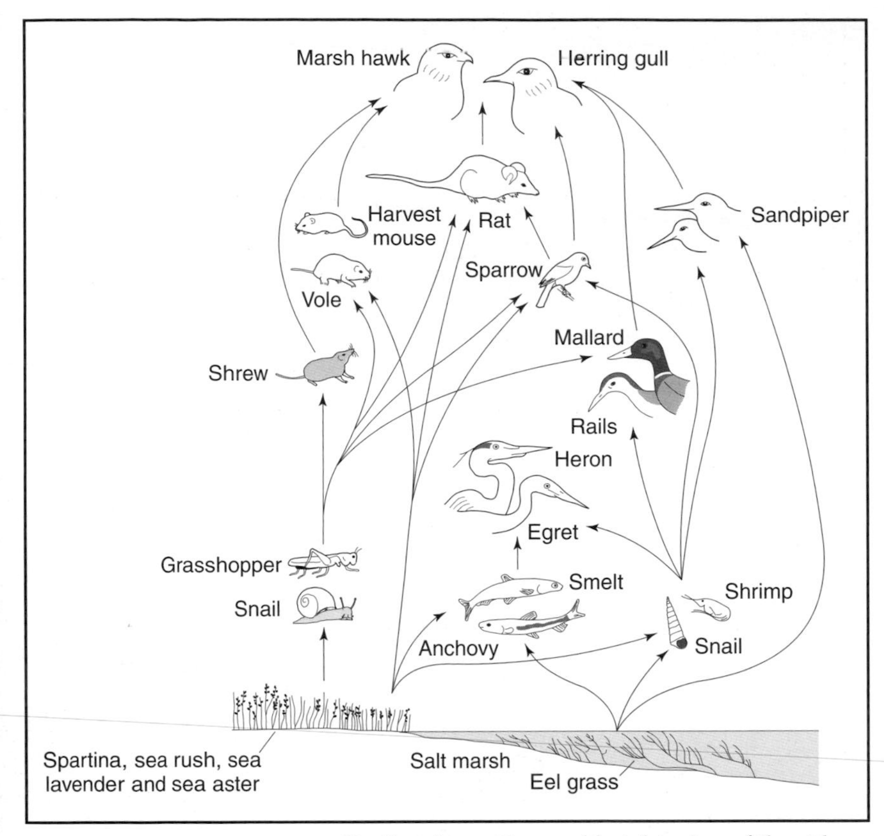

FIGURE 2.25 Saltmarsh ecosystem

Further deposition and heightening of the ridges encourage colonisation by grasses such as sea lavender, sea pink, sea blite, sea purslane and sand spurrey – all of which thrive best on briefly submerged stretches of mudflats. The widening range of plants within this succession sequence is proof that increasing **biodiversity** can take place when environmental conditions are less restrictive. The result is an almost continuous sward of saltmarsh which contains occasional **salt pans** (Figure 2.23). The high salt content of salt pans is largely due to evaporation during long periods of exposure, which makes them unattractive to less tolerant halophytes on the surrounding marsh turf.

As the marsh vegetation builds up and the area becomes ecologically more stable, any surviving depressions from the original undulating shore begin to stand out as muddy creeks with steep banks; these are the last saltmarsh components to be colonised (Figure 2.24). The creeks often form dendritic patterns, similar to river/tributary networks within drainage basins (Figure 1.8). Turf grasses such as fescue and junctus thrive on this penultimate phase of the succession, which can be reached only by the highest spring tides.

In the final zonation phase, the process of bioconstruction is complete, in that the land is now totally beyond the reach of all but the most severe, storm-enhanced spring tides. Salt-tolerance is no longer a major location factor and this allows meadow grasses, bushes and trees such as ash and alder to thrive in a newly-created dry environment.

Saltmarsh is a fertile and highly productive marine ecosystem (Figure 2.25), due to its high oxygen and nutrient content. For this reason, saltmarshes have provided cheap but highly nutritious grazing for cattle and sheep – especially when enclosed by walls or fences and underground drainage networks have been installed. Their coastal locations have, however, made them particularly vulnerable to a wide range of developments such as harbour extensions, oil refineries and other types of industrial complexes. Mature saltmarshes are now being increasingly recognised as crucial elements in the never-ending battle for supremacy between sea and shore. Because saltmarshes are highly efficient filtration systems, they are able to retain not only nutrients which aid plant growth but also industrial wastes such as heavy metals which retard plant growth. Offshore dredging is another hazard, as it greatly increases the volume of sediment in suspension and so modifies the established rate of saltmarsh development. However, marshes and mudflats do act as energy-depleting buffer zones along some vulnerable, low-lying stretches of coastline. Their ability to self-generate during intervals between major storms greatly enhances their ability to protect shorelines at risk from wave and flood damage.

ACTIVITIES

1 a) Describe the distribution of saltmarshes on Britain's coastline (see Figure 2.17).
b) In what ways is the saltmarsh distribution different from that of sand dunes?
c) Give reasons for the differences which you have highlighted.
2 Suggest reasons why saltmarshes are unlikely to develop on the beaches at Blackpool.
3 Devise a flow-diagram which identifies all of the following factors and ecosystem components at each phase of the saltmarsh succession:
- tidal submersion
- processes
- plant life.

4 Write an essay on the theme: 'Saltmarshes and mudflats are two of the most diverse and nutrient-rich of all coastal ecosystems, but are frequently at the greatest risk from development projects'.

Study 2.7 Barrier islands – East coast of North America

Barrier islands are so named because their offshore locations allow them to protect a mainland coast, by absorbing much of the force of oceanic storms. The world's chief barrier island systems fringe about 2 per cent of the total global coastline. Of these, the longest system is that which follows over 3200 km of the eastern seaboard of North America (Figure 2.26). The entire system was visible to the crew of the Apollo 9 spacecraft, 200 km above the Earth's surface, following its take-off from Cape Canaveral, Florida, on 12 March 1969.

Barrier islands usually occur where:

- The tidal range (the difference between high and low tides) is small; for example, the average tidal range off the North Carolina and Florida coasts is only 1 m – (compared with 10 m in the Bristol Channel).
- Offshore sea currents are weak, but the prevailing (dominant) onshore winds are strong enough to produce much local wave activity as well as longshore drift along the coast.
- Rising sea levels of about 100 m following the end of the last Ice Age swamped what used to be coastal lowlands, producing shallow areas of sea, edged by newly-created islands. Rising sea levels produced areas of water shallow enough to be exposed at low tide; in time, these also became islands, due to the processes of sand deposition.

Once established, barrier islands continue to evolve, under the influence of erosion, transportation and deposition which helped to create them in the first place. Over a period of time, most barrier islands tend to migrate towards the mainland shore. This migration is due partly to the ability of storm waves (and the winds which create them) to move beach material from one side of an island to the other; in effect, the islands are slowly but surely 'rolling over themselves' towards the mainland. Their migration is also due to the action of the Atlantic currents, which scour the barrier's outer beaches then transport this eroded material through narrow inlets between the islands and into the sheltered lagoons; there, deposition takes place as the rate of the current's flow decreases.

In spite of their instability and exposure to onshore winds up to hurricane force, the barrier islands have always been of great economic significance to the eastern states of the USA. Many Indian tribes (e.g. the Hatteras) established summer fishing camps on the islands; but they returned to their permanent village settlements on the mainland every autumn to escape the fierce winter storms. Most of the early European settlers, who later forced these tribes to migrate much further inland, lived on the more sheltered western sides of the islands. The barrier's recreational use was well-established long before the Second World War, by which time holiday resorts and privately-owned beach houses already covered 10 per cent of its total land area.

FIGURE 2.26 Barrier islands – east coast of North America

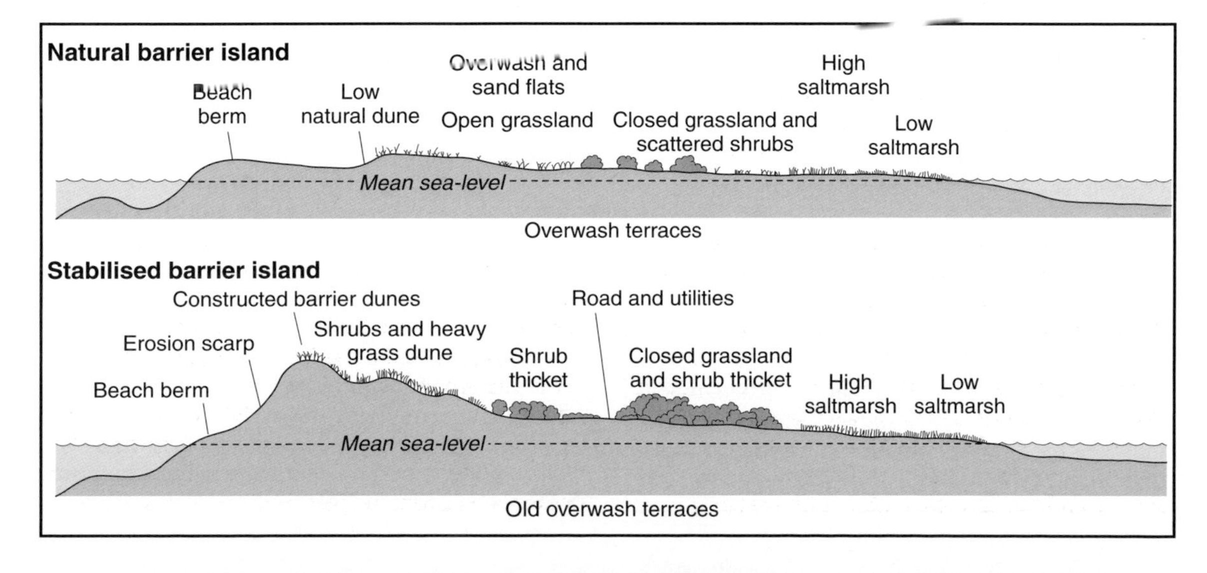

FIGURE 2.27 Comparison of natural and stabilised barrier islands

By the early 1980s, residential and leisure developments were absorbing more than 2500 ha of additional land every year. In 1982, the United States Congress became so concerned by the accelerating rate of barrier island urbanisation that it approved the Coastal Barrier Resources Act. This Act required 180 000 ha of as yet undeveloped barrier land to be permanently maintained in its natural state. The success of the 1982 Act led to further, more ambitious legislation in 1990 – the Coastal Barrier Improvement Act, which provided immediate protection for a further 280 000 ha. The popularity of the more southerly barrier islands continues to increase, however, and the largest honeypot centres such as Ocean City in Maryland attract over 8 million visitors annually.

Very large sums of money are now allocated to beach 'nourishment' programmes to counter the constant removal of sand by natural processes. The restoration cost of just one hectare of beach averages $2.5 million, and the whole process has to be repeated every decade or so. In spite of its high cost and short-term effectiveness, this form of 'soft' engineering is considered preferable and more natural than 'hard' engineering alternatives such as high concrete defensive sea walls. Figure 2.27 compares a typical stabilised barrier island in North Carolina with one in its natural state.

ACTIVITIES

1 a) Use atlas maps and/or other resources to discover the prevailing wind and ocean current patterns along the eastern American seaboard barrier island chain.
b) Add this information to an enlarged copy of Figure 2.26. Add further annotations to your map to name the main islands, headlands and barrier island cities within this map's area.
2 a) Devise a flow-chart to trace the formation and subsequent development of a typical American barrier island.
b) Suggest reasons for the methods used to stabilise barrier islands. When answering this question, bear in mind the need to provide some degree of protection from hurricane-force storms tracking westwards from the Atlantic Ocean.
3 a) Write a scene in which two characters debate how the American barrier islands should be allowed to develop in the immediate future. One character is a wealthy visitor; the other is a member of a family who has owned part of a typical, undeveloped island for many generations. The two people are certain to argue fiercely – because they will hold strong, opposing views – but do try to keep your debate as constructive and geographically-focused as possible.
b) Act out your play with a fellow student, then request your audience to suggest additional ways in which the dialogue between the two characters might have developed.

Barrier islands throughout the world are much more vulnerable to rising sea levels than most other coastal features. Along the Atlantic coast, the sea level is estimated to be rising at about 15 cm per century, but United Nations predictions for the future indicate a trend up to three times faster than this. Coastal retreat rates vary according to local conditions, but average about 1 m per year. An extreme case of coastal retreat is Timbalier Island, whose shore is eroding by more than 3 m annually.

Study 2.8 Coral reefs – Aqaba, Yemen Republic and the Great Barrier Reef, Australia

Coral reefs are the Earth's oldest form of life zone and one of its richest marine ecosystems – even though they cover only 0.2 per cent of the total ocean floor area. Figure 2.28 shows that coral reefs form in tropical regions, in locations with clear, warm, shallow seas.

Coral reefs are formed by massive colonies of tiny animals called polyps, which are close relatives of jellyfish. They slowly build reefs by secreting a protective crust of limestone (calcium carbonate) around their soft bodies. When they die, their empty crusts or outer skeletons remain as a platform for more reef growth. These coral builders work slowly, taking over two years to create only 1 cm of coral. The resulting tangled maze of cracks, crevices and caves formed by different types of coral provide sheltered niches for a huge variety of marine plants and animals (Figure 2.29).

Many species of reef fish maintain mutualistic relationships with other reef species, as coral reefs are a joint venture between the polyps and tiny, single-celled algae called zooxanthellae which live in the tissue of the polyps. In this mutualistic relationship, the microscopic algae provide the polyp with food and oxygen. The polyps in turn provide a well-protected home and make nutrients such as nitrogen and phosphorus available for the algae. The algae and other producers give corals most of their bright colours and provide plentiful food for a variety of marine life.

By forming limestone shells, coral polyps take up carbon dioxide as part of the carbon cycle. The reefs also act as natural barriers that help protect 15 per cent of the world's coastlines from wave and storm damage. In addition, the reefs build beaches, atolls and islands and they provide food, jobs and building materials for some of the world's poorest countries.

Marine biologists estimate that humans have directly or indirectly caused the 'death' of 10 per cent of the world's coral reefs (especially those in south-east Asia and the Caribbean). Another 30 per cent of the remaining reefs are in critical condition and 30 per cent more are threatened; only the remaining 30 per cent, therefore, are stable and healthy. If the current rates of destruction continue, another 60 per cent of reefs could be lost in the next

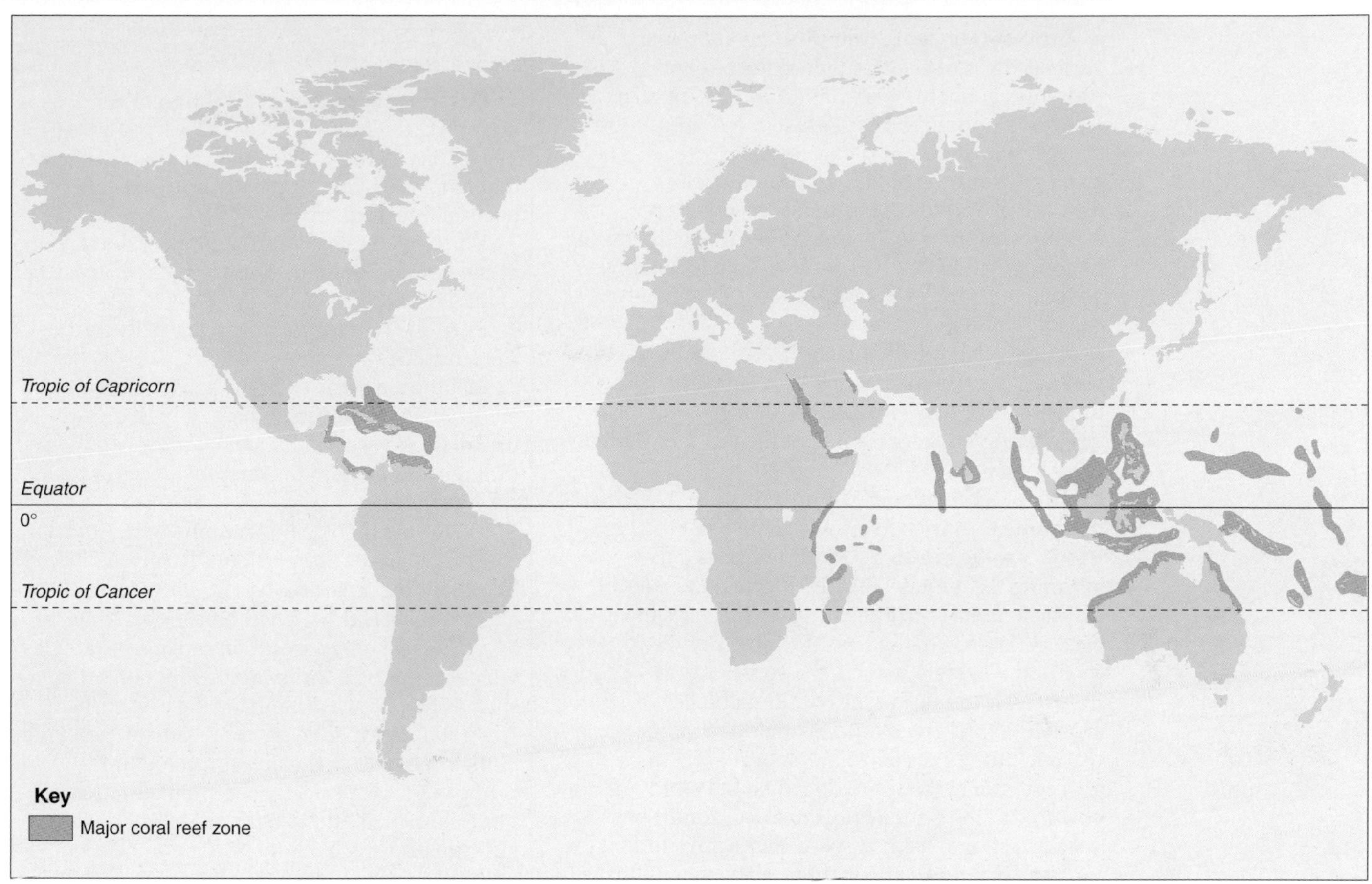

FIGURE 2.28 Global distribution of coral reefs

20–40 years. Some 300 coral reefs in 65 countries are protected as reserves or parks and another 600 have been recommended for similar protection.

Protecting reefs is difficult and expensive, however, and only half of the countries with coral reefs have set aside some of their reef areas as reserves. Biologists warn that, unless people act now to protect the world's diminishing coral reefs, another important part of Earth's vital biodiversity will be greatly depleted.

Coral reefs have always been exposed to severe natural disturbances such as typhoons, but their remarkable ability to recover from such short term threats has ensured their survival. Unfortunately, the reefs are now experiencing far greater and much longer term threats, for a wide range of reasons:

- The deposition of river-born materials such as mining waste and eroded soil smothers the polyps or blocks their sunlight.
- Deforestation in many tropical regions has resulted in streams transporting much greater quantities of eroded sediment to coastal areas.
- Offshore activities such as dredging have similar smothering and light-screening effects.
- The ever-increasing strength and usage of pesticides and insecticides significantly raises pollution levels in coastal areas.

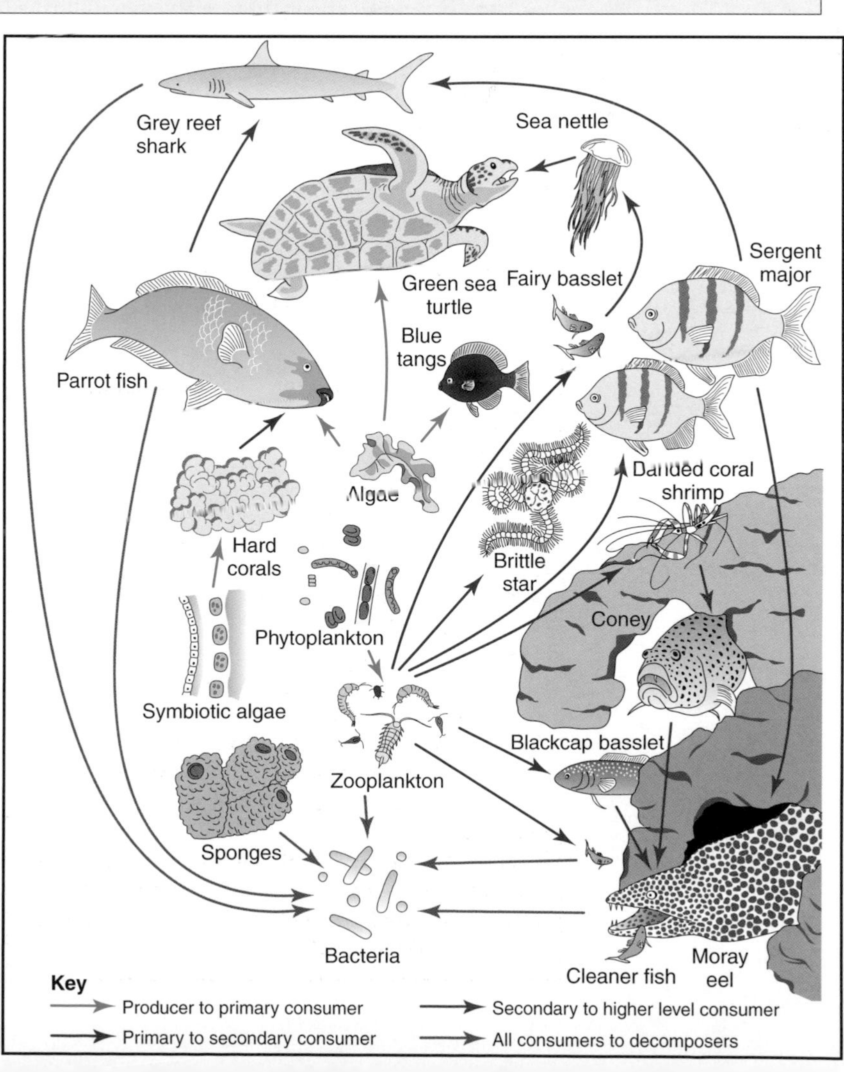

FIGURE 2.29 Coral reef ecosystem

- Increased offshore dumping of raw sewage enriches the nutrient content of the sea and promotes an increase in plankton. This, in turn, leads to a population explosion in crown-of-thorns starfish which graze on live coral.
- Coral is much sought after for aquarium decoration and for use as personal jewellery.
- Oil spills from sea-going tankers – whether due to accidental spillages or unlawful tank-cleaning – are one of the most potent sources of short-term coastal pollution.
- Ships' anchors, diving equipment and even bare feet may weaken or fracture sections of reef. The increased popularity of 'reef walking' is having a devastating impact on some of the more accessible coastal reefs.
- The practice of blast fishing – the use of dynamite to stun fish – is widely used in the Far East. Cyanide is also in common use as a fish-stunning technique. Blast fishing causes instant physical damage to a reef's outer surfaces and cyanide is toxic to the coral-forming polyps.
- Coral is widely used in the construction industry. It can be cut into building blocks or crushed to provide a crucial raw material for the manufacture of cement.
- Corals yield compounds which have proved effective in the fight against medical conditions such as asthma, heart disease, leukaemia and HIV viruses. This makes their protection even more urgent, but also makes reefs especially vulnerable to future exploitation.
- Long term rises in seawater temperature can lead to coral bleaching. This makes the coral very unattractive and its legal protection far less likely.

ACTIVITIES

1 Describe the global distribution of coral reefs, according to Figure 2.28.
2 List any items of information which indicate that coral reefs form a particularly delicate type of coastal feature.
3 With the help of Figure 2.29, note down some of the key components of the coral reef ecosystem and their linkages.
4 Read the two newspaper articles in Figures 2.30 and 2.31.
a) List the similarities in the physical states and environmental circumstances of both reefs.
b) Produce a second list which indicates the key differences between the circumstances in the two coral reef areas.
5 a) Outline the various reasons why Aqaba's coral reefs are under particularly serious threat.
b) What other reasons explain why the existence of coral reefs is threatened on a global scale?
6 a) Suggest an appropriate range of strategies which might be adopted to protect coral reefs such as that at Aqaba and off the east coast of Australia. Try to give an approximate indication of the timescale of the measures you are proposing; it is perfectly acceptable to use phrases such as 'short-term' and 'long-term' to indicate projects likely to be completed within/over five years. It will be almost impossible for you to give any accurate estimates of cost, but you could rank-order your measures according to their likely approximate and relative financial implications.
b) Now identify those two strategies from your selection in a) above which (in your opinion) are most likely to succeed.
c) Give reasons in support of the two decisions which you have just made.
d) Briefly present your chosen strategies and supportive arguments to a number of your fellow students, who should then take turns to do the same. Make appropriate summaries of any additional strategies proposed by your group which you believe to be of value.

Death knell for Jordan's coral reef paradise

Jonathan Cook on the perils facing Aqaba.

At the end of the long dusty drive south on Jordan's Desert Highway can be found an army of ageing lorries laying siege to the Red Sea resort of Aqaba. The thousands upon thousands of vehicles packed into the town's lorry park – a sea of rusting steel and iron stretching to the horizon on either side of the road – are evidence of the economic pressure that threatens to overwhelm the kingdom's 20-mile coastline.

The port has become an increasingly threadbare lifeline as Jordan competes for international business. Trucks loaded with salt, phosphates and potash – the only major natural resources – rumble towards it, returning with food, livestock and oil.

Hidden by the lorry park is Aqaba, a pale, low-rise reflection of Israel's popular resort of Eilat, whose towering hotels and apartment blocks on the other side of the narrow gulf seem designed to taunt its poor Arab neighbour.

Posters everywhere advertise the town's biggest tourist attraction: the brightly-coloured fish that live among the fingers of coral.

The tourists are here to swim over these reefs, the most northerly in the world and, until now, some of the most pristine. In the sheltered waters of this small gulf, a delicate ecosystem has evolved, supporting more than 1,000 species of fish and hundreds of types of soft and hard coral, many unique.

But the death knell is sounding for this underwater paradise. Akram Bederat, an instructor at the Royal Diving Centre, eight miles south of Aqaba, says the busy container port has wreaked havoc on the reef's flora. During the past two years there has been a dramatic fall in the number of fish, even at great depths.

"The coral is a living thing full of bacteria and tiny organisms; stand on it and an area of one square metre is killed. It is like a crate with one bad tomato in it – all the rest are soon contaminated," he says. "And once an area is damaged it needs 20 years to repair itself. That can be a death sentence for the fish that live on the corals."

He mostly blames the hundreds of tonnes of phosphates which are spilt each year due to the loading, showering the coral below with a fine dust that is suffocating the organisms. The export of Dead Sea salts has also raised the salinity of the coastal waters to dangerous levels.

Added to this, the gulf is being polluted by the increasing number of ships using the port. They flush out their tanks and, because the gulf waters take up to three years to make their way back into the Red Sea, the oil has longer to do its damage.

However, a new and bigger threat looms in the form of a free trade zone recently created to tempt international companies to a nascent industrial area close to the Saudi border. Petrochemical firms and fertiliser manufacturers are among the businesses queuing to do business. They are certain to add to the heavy burden on the gulf. Even before they arrive, the infrastructure to support the expansion is taking its toll.

Last year four power cables and two telecommunication lines were laid between Egypt and Aqaba, cutting directly through the reefs.

A seawater-cooled power station in the industrial zone is trebling its capacity to cope with the new business, prompting concerns about what temperature changes in the water will do to the corals.

Bruce Pollock, a British scientist based in Aqaba with the Global Environment Facility, a World Bank project which is trying to limit damage to Israel and Jordan's coasts, says Jordan is aware of the dangers. "It's a matter of balancing development with conservation, but for Jordan that's difficult because of the limited coastline."

Environmentalists are equally worried by the regional authority's land-use plan, which was recently revised to take account of Aqaba's other boom industry – tourism. An eight-km-long "marine peace park" will be dominated by a holiday resort, golf course, tourist village, camping area and a Disney-style theme park.

Philip Reichel, a Dutch volunteer working for the Royal Ecological Diving Society, says: "Along the 30-km coast there is not a metre of land they have not accounted for. How can the reefs survive when they will be damaged either by the port and free trade zone or by entrepreneurs exploiting them for the tourists?"

Because many of the reefs are close to the beach, holidaymakers can walk straight on to them to look at the fish, thereby killing swaths of coral. Mr Bederat says local by-laws mean visitors can be fined 20,000 dinar (£18,200) for damaging coral, but he has never heard of anyone being prosecuted. "Warnings are needed on the beaches where they would be seen," he says.

He has been diving in these waters since he was a boy. A few years ago he saw shoals of more than 100 barracuda. Now, he says, it is unusual to see one, even on a restaurant plate.

The Guardian, 24 June 1998

FIGURE 2.30 The threat to reefs in Aqaba

Great Barrier Reef going off-colour

More than 88 per cent of inshore coral on Australia's Great Barrier Reef is suffering from bleaching. Scientists are unsure how much of it will suffer long-term damage.

It is the worst degradation the reef has suffered. Details will be reported by the Great Barrier Reef Marine Park Authority to the Government next month. Bleached coral loses its colour, and its attractiveness for tourists, when the brownish algae which gives its colour are ejected, leaving the white skeleton visible through clear coral tissue. Bleaching can be natural, caused by heightened temperatures, low salinity or the voracious crown-of-thorns starfish, but man is also responsible for some of the damage.

"What we don't know is whether the natural event is being made worse by what people are doing," said Richard Kenchington, executive director of the authority. Finding the answer may mean the difference between life and death for the 20-million-year-old reeds, home to about 1,500 species of fish, 400 species of coral and 4,000 species of mollusc.

The report says: "During the 1998 event (March-April), over 88 per cent of inshore reefs exhibited some coral bleaching, with 25 per cent of all inshore reefs having more than half the corals affected. Although the level of mortality associated with the 1998 bleaching event has not yet been determined, up to 50 per cent of the corals died on some reefs in a previous severe episode in 1982."

The algae are the fuel that keep the coral growing in an environment both stable and delicate; so delicate that a tiny change can send the coral into shock. Australian reefs have not suffered the wholesale damage that has affected the coral of the Philippines and islands of the Pacific. Careful control has protected it from fishing with dynamite and the sale of coral.

Global warming is one phenomenon that can produce bleaching – just a one-degree increase in water temperature can trigger it, while the run-off of fertilisers and pesticides and industrial waste from Indonesia and Papua New Guinea may also be having an effect. Heightened temperatures have affected some of the inshore areas of the reef in the past, when the algae found temperatures unacceptable. The temperature of the reef waters has only to rise above 30C (86F) to stress the coral with increased solar radiation.

Serious study of the reefs was not done until the 1970s. The first bleaching was only observed ten years later. With such an ancient structure, it is hard for scientists to know what is normal cyclical occurrence and what is potentially disastrous.

"We've got a while to go before we know what is normal," said one expert. "But by the time the scientists find out, it may be too late. There is some evidence to suggest recovery can take place over an extended period. It's the faster-growing, smaller coral which are most at risk. The larger ones may last up to 150 years."

The third potential culprit is the crown-of-thorns starfish, which feeds off the coral and has in the past devastated up to 500 km of the reef. Man has all but wiped out the starfish's natural enemy, leaving the female, which lays up to 60 million eggs, free to produce a replacement rate for the starfish of about 1,000 to one. This time, however, the starfish is not top of the list of suspected causes.

The Times, 30 October 1998

FIGURE 2.31 The threat to the Great Barrier Reef

Study 2.9 Coastal mineral extraction – Druridge Bay, Northumberland

Druridge Bay sand dunes are located in Figure 2.17. The bay is in an isolated location, about half-way between Newbiggin and Alnwick on the Northumbrian coast. However, it is only 2 km from the A1068 and within a 25 minute drive of the Tyneside Conurbation.

The bay area includes an unbroken 7 km-long crescent of golden sand backed by sand dunes now owned by the National Trust (Figure 2.32). It sustains a rich plant life which includes purple vetch, yellow buttercups, crimson campions and scarlet poppies. Skylarks, oystercatchers and black-headed gulls are just three species within its varied bird population. All these factors should make Druridge Bay an ideal location for day-visitors in search of solitude and natural beauty, and it was for most of its recent history. Unfortunately, two problems (one potential, one actual) have threatened the attractiveness and ecological balance of the area in recent years.

The potential hazard first became clear in 1979, when the British Government announced plans to build up to three nuclear power stations there. The visual, noise and air pollution which accompanies such building projects would have seriously degraded the area both during and after the construction phase. Partly in response to a well-organised campaign over many years, but also because of growing concerns about safety issues in the nuclear power industry since the Chernobyl radiation leak in April 1986, these plans for nuclear

FIGURE 2.32 Druridge Bay

developments at Druridge were formally scrapped in December 1996.

The second problem did affect the area; large quantities of high-grade sand were extracted for many years from the point where Blakemoor Burn (a stream) crosses the beach and enters the North Sea. Planning approval for sand extraction was granted in the 1960s, in spite of fierce local opposition, permitting extraction until 2042 – even though the immediate area has since been granted SSSI status. Damage directly caused by the extraction of up to 40 000 tonnes of sand annually included entire sections of the beach surface disappearing and underlying expanses of stones, peat and clay from primeval forests becoming exposed. Five years of sustained public pressure so damaged Ready Mixed Concrete's image that it decided, in 1996, to cease extraction and to focus its activity on alternative sites. A plaque commemorates this environmental victory and is a constant reminder to today's visitors of the vulnerability of coastal ecosystems.

Druridge Bay's beach and dune areas now seem to be recovering, although local wind strengths and tidal heights are both crucial factors in their rate of recovery: milder, summer conditions encourage beach regeneration, whereas the strong, north-easterly winds in winter tend to scour the beach surface and delay its recovery.

ACTIVITIES

1 Put yourself in the position of one of the leading Druridge Bay environmental campaign organisers! You have been given the important task of preparing material for the use of other campaigners, who will be trying to convince people how important it is for the beach and sand dune areas to be protected from nuclear power station building and sand extraction work. It is entirely up to you how you present the facts and arguments which you have chosen to convince potential campaign supporters.

Study 2.10 Coastal interventions – The Ijsselmeer, Hiroshima and Newhaven

Coastal plains and river estuary banks are the locations of many of the world's most densely populated lowlands. Technological advances have made it possible to enhance the economic, residential and recreational potential of such locations, whilst also reducing some of the hazards caused by erosion, silting and flooding.

The examples of human coastal intervention in Figure 2.33 demonstrate contrasting ways in which communities have modified their local coastal environments. Some of these interventions have, in due course, caused considerable difficulties for neighbouring communities.

ACTIVITIES

1 a) With reference to a range of named examples taken from different parts of the world, suggest reasons why many coastal plains and river estuary banks have become major urban areas.

b) Use atlas maps to locate and then suggest reasons why some examples of coastal plains and river estuary banks have remained sparsely populated.

2 For each of the examples in Figure 2.33:

a) Note down its name, then briefly describe its location.

b) State the chief purpose for the coastal modification and make a reasoned judgement as to whether it was successful.

c) Outline any problems which the modification created, then make a reasoned judgement as to whether these problems could have been foreseen before work started on the modification.

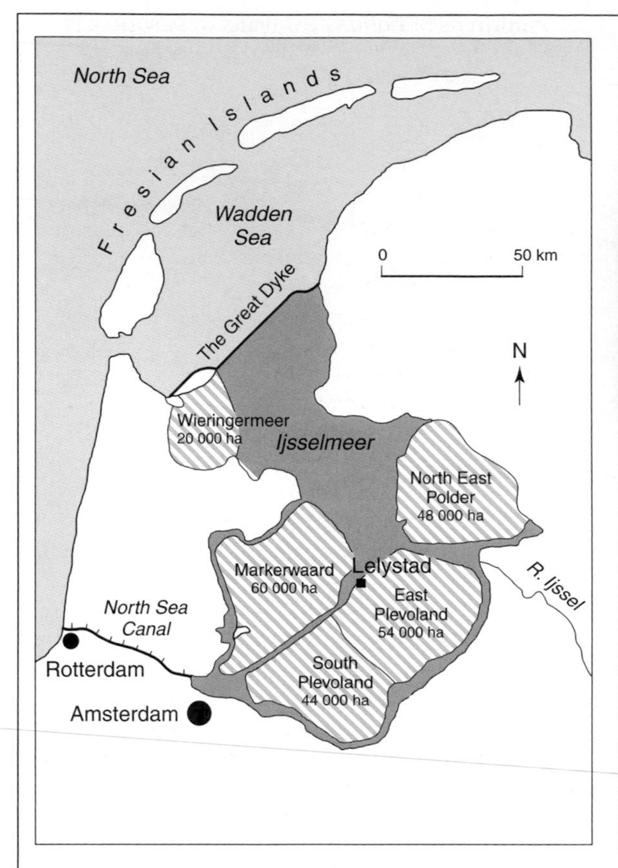

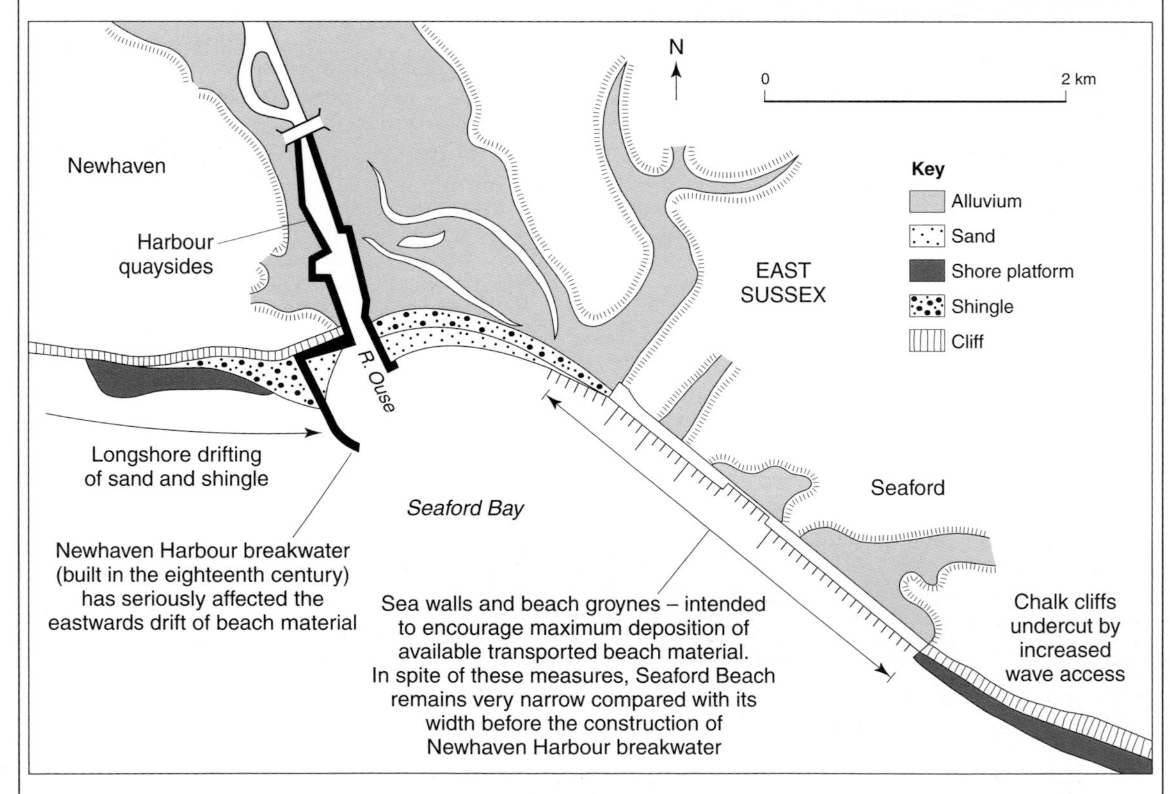

FIGURE 2.33 Ijsselmeer, Hiroshima and Newhaven

Study 2.11 Coastal flooding – Towyn, North Wales

On 26 February 1990 the Towyn stretch of the North Wales coast suffered its worst storms for over 20 years. A severe depression, whose easterly track across Britain was much more southerly than usual for that time of year, had created onshore winds of 110–145 km/hr; these winds coincided with a series of exceptionally high spring tides.

On 26 February, the sea finally breached the already much weakened defence wall in a number of places – the worst affected area being Towyn, where 467 m of the sea wall crumbled and allowed seawater to flood 10 km^2 of low-lying land immediately behind it. For two hours, huge quantities of beach material such as shingle and pebbles were either swept over the protective wall like a waterfall or through its breached sections. The swirling mass of water and debris uprooted trees which, in turn, greatly added to the damage caused directly by both wind and water. Flooded septic tanks and damaged sewers provoked fears of outbreaks of diseases such as cholera. As many as 2800 individual properties were flooded and 5000 people had to be evacuated.

Hundreds of holiday caravans were seriously damaged along one of the most popular caravanning locations in Britain. Towyn is widely acknowledged as having the highest concentration of static caravans in the whole of Europe and it is for this reason that its peak summer season population is seven times greater than that in winter. It required a massive collective effort to prepare the area for its next influx of visitors. Even so, many valuable Easter bookings were lost and a number of visitors who had planned to stay there the following summer suddenly changed their minds and decided to go elsewhere for their annual holiday. Insurance premiums have since increased sharply for both caravan and house properties along all the low-lying stretches of the North Wales coast.

The A55 North Wales coastal road has been protected by boulder defences (a hard engineering option) along its most vulnerable stretches. For much of its length, the A55 and the nearby railway route to Holyhead pass very close to the shore (due to the proximity of the Snowdonia foothills) and there is very little flat land available for use as washlands. Caravan parks and their associated tourist infrastructure already occupy any patches of lowland which might otherwise have been utilised for soft engineering alternatives.

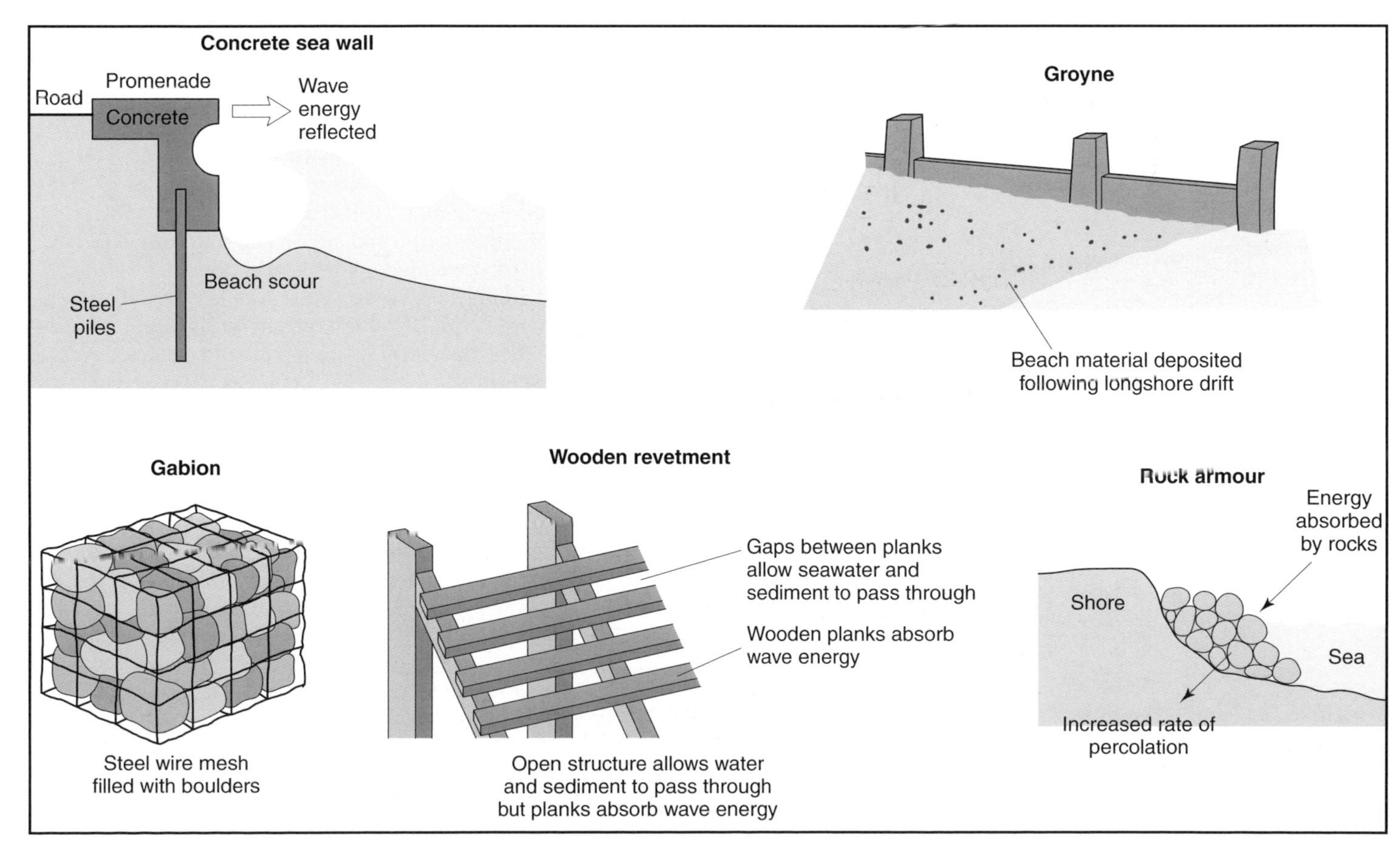

FIGURE 2.34 Hard and soft engineering coastal defences

ACTIVITIES

1 Figure 2.34 illustrates the most common forms of 'hard' and 'soft' engineering techniques. Use the information in these drawings (and your own judgement) to prepare a discussion entitled: 'Recommended engineering options to avoid future coastal flooding at Towyn, North Wales'. Your teacher may recommend group discussion of the relevant issues and options (e.g. visual pollution v. tourism interests; engineering efficiency v. economic restraints in an area of high unemployment) before you start to draft your discussion contribution.

2 The discussion prepared for Activity **1** should be presented to the whole class and/or displayed to generate further discussion of the issues surrounding the hard and soft engineering options in coastal flood prevention. Make brief notes of any innovative suggestions generated by this exchange of ideas.

FIGURE 2.35 The Ganges Delta Area

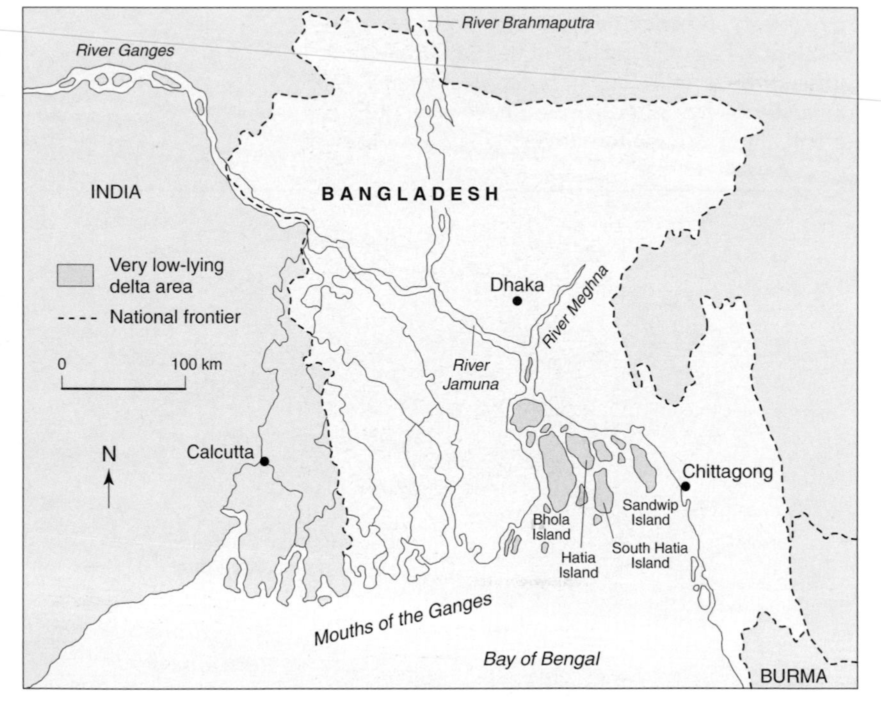

Study 2.12 Storm surges in LEDCs – Bangladesh

Bangladesh has a long history of coastal disasters linked to the relief, hydrology and climatology of the northern and eastern regions of the Indian sub-continent. Much of its central and southern regions are occupied by the world's largest delta (The Ganges Delta) 75 000 km^2 in area (Figure 2.35). The following summaries of some of the most devastating storm surge events within the last 40 years indicate the scale of the problem:

- In 1737, a 12 m storm surge killed 300 000 people in the Bay of Bengal region.
- In 1962, 25 000 Bangladeshis drowned in coastal floods.
- On 12 November 1970 (a day still called 'Black Thursday' by local people) a tropical cyclone created wind speeds exceeding 200 km/hr which produced a storm surge 8 m high. 26 000 km^2 of coastal lowland was flooded where the River Ganges and the Brahamaputra River flow into the Bay of Bengal. Over 300 000 people lost their lives in the flooding, but lack of food, medical care and shelter later increased the death toll to 1.5 million people and 2.5 million were made homeless. At least half a million farm animals were drowned and about 10 000 fishing boats (65 per cent of the entire coastal fishing fleet) were dashed to pieces. Saltwater ruined vital rice-growing areas and 80 per cent of the rice crop itself was made inedible. Entire villages were devastated and countless items of farm equipment, from modern tractors to basic hand-hoes, were swept away by the floodwaters. International help included 5000 tonnes of wheat from Australia, 210 000 doses of cholera and typhoid vaccine and an interest-free loan of £22 million provided by the World Bank.
- In May 1985 over 11 000 people were killed by a storm surge 12 m high caused by 180 km/hr winds which extended 150 km inland and occurred soon after the same area was in the process of recovering from recent disastrous river floods. 6000 people were swept out to sea – some whilst still asleep – and many survivors were observed clinging to bamboo rafts and floating roof tops to evade attacks by sharks and crocodiles. The rice crop was devastated and it was predicted that the sea salt would not be flushed out of the topsoil until the following cycle of monsoon rains.
- In May 1991 a 200 km/hr cyclone resulted in a 6 m storm surge and well over 150 000 deaths – mainly in the south-east of Bangladesh. 10 million people were rendered homeless, property valued at £2 billion was destroyed, and large areas lost their electricity supplies.

Deaths and reductions in the '**quality of life**' caused by storm surges are especially serious on the new land formed as coastal islands by silt deposition (almost two billion tonnes of silt are generated in the Himalayas each year and transported downstream by the Ganges and the Brahamaputra alone). Much of this silt is deposited within the delta area to form rich alluvial islands known as **chars** (Figure 2.36). Chars rarely rise more than 1 m above the high tide level and are so vulnerable to the storm surges triggered by cyclones that many of them have been encircled by 7 m high protective embankments. River channels can shift up to 600 m of sediment in a single day but even this constant threat does not seem to deter the habitation and cultivation of even the most precarious sandbanks! Rice is the dominant crop for the chars' subsistence farmers and the consequences of crop failure are inevitably catastrophic for them.

More than half of the char population is considered landless and at least 30 per cent of its people are either unemployed or under-employed agricultural workers. Char island literacy levels are only 20 per cent of the population, which can lead to serious delays in responding to disaster warnings. Many children do not attend school – simply because they are required to help their parents with farm work. In a favourable year, a whole day's work picking vegetables such as lentils may just be sufficient to provide three or four meals for the family. Most char roads are made of dirt and quickly become unusable in heavy rain or flooding; the islands themselves are almost impossible to evacuate when their surrounding waters are turbulent.

Following the 1970 cyclone, a Cyclone Preparedness Programme was initiated in Bangladesh and this included efforts to improve the country's advance cyclone forecasting systems. Satellite images of cyclone tracks are now monitored, and warnings are transmitted from a central base to regional stations such as Chittagong and then to local field stations, from which groups of volunteers can issue warnings to the people in their communities. The Preparedness Programme also involved the construction of cyclone shelters (Figure 2.37). These are two-storey concrete structures, which can accommodate up to 2000 people, and have water storage tanks to meet their basic drinking needs. When not required for emergency use, the shelters serve as mosques, schools and health clinics. There are insufficient shelters, however, and over-crowding has led to casualties and health problems.

The prognosis (predicted future) for Bangladesh is not encouraging. Sea level rises due to global warming of 10–30 cm are anticipated for the Bay of Bengal and it is expected that monsoon rain will increase by 10–15 per cent. The country's population is almost certain to double to 220 million by 2025, putting additional demands on the economy which will further reduce the already limited ability of Bangladesh to respond to the storm surge threat.

FIGURE 2.36 A typical char island, Bangladesh

FIGURE 2.37 A cyclone shelter in Bangladesh

ACTIVITIES

1 List some of the main difficulties faced by the delta people of central and southern Bangladesh, then link these problems to their causes.

2 a) Make a record of the strategies which have been taken so far to alleviate the problems experienced by the Bangladeshi people due to storm surges. When doing this, attempt to make some general assessment of their likely effectiveness.

b) What other measures might be taken to alleviate storm surge related-problems, bearing in mind the limitations imposed by a country which experiences chronic economic, unemployment, educational, technical and administrative difficulties?

Unit 3
NATURAL ENVIRONMENTS 1

Biomes

Biomes are the largest terrestrial ecosystem unit. Whilst some writers also apply this term to water-based units, these are more accurately known as **aquatic life zones**. Much of the information in this section is equally applicable to both environments – except for the significance of climatic factors. In water, it is salinity which holds the crucial key to plant adaptations, although temperature, light and nutrients also play important roles. Biomes are delimited by a shared, characteristic plant community which is adapted to the specific environmental conditions of the region. There is a high degree of correlation between biome distribution and world climatic zones, simply because climate is the most influential factor in natural vegetation distribution (Figures 3.1 and 3.2).

Precipitation is the chief single determinant of vegetation type – be it forest, grassland or desert. Precipitation, temperature and soil type influence whether a region's vegetation is tropical, temperate or polar, forest, grassland or desert. Figure 3.2 shows that there are nine biome divisions on Earth. At more advanced levels of study, these nine basic divisions are often enhanced with various sub-categories.

Maps tend to suggest that biomes are fixed regions with inflexible boundaries which mark dramatic changes between adjacent plant communities. They can also give the impression that totally uniform vegetation communities cover huge expanses of the planet. Nothing could be further from reality. Biomes are merely convenient (and extremely broad) classifications of dominant vegetation type – each characterised by its variations as well as its similarities. Any sizeable forest, for example, will consist of a range of species of trees, shrubs and flowering plants reflecting local variations in environmental conditions. Continental areas sharing the same biome classification may support different vegetation species and (even more commonly) will host different animal species. Additionally, both within and between continents, soil types vary considerably according to parent rock, glacial histories and local micro-climates.

Within any one biome, a wide variety of ecosystems is likely to develop in response to local conditions and the degree to which intervention by humans has taken place. Surprisingly rich and dense **ecotones** often mark the transition between neighbouring ecosystems; likewise, individual biomes are linked with others through a similar series of ecotones. As a result, the Earth's surface is not covered by uniform blankets of identical vegetation communities but by 'patchwork quilts' of ecosystems/ecotones, with each combination exhibiting its own distinctive locality-based variations.

At the global scale, four distinct factors inter-relate to produce the biome distribution; these are climate, topography, soils and biotic factors.

Climate

The single most important aspect of climate (in defining a biome) is **precipitation;** its influence extends far beyond annual totals of rain, snow, hail etc. Especially important is the distribution of precipitation throughout the year – but associated temperatures, the nature of the precipitation and its availability during the **growing season** are also important considerations. For example, large quantities of rainfall in one season, followed by prolonged drought will favour one type of mature (climax) community, whereas precisely the same total annual rainfall distributed evenly throughout the year will lead inevitably to the dominance of a totally different plant community. Where rain falls all year round, forest growth is possible. In tropical areas, heavy rain all year can off-set the high evapotranspiration demands of the high temperatures. Low precipitation totals in cold, high latitudes can also support forests – because the evapotranspiration demands on available moisture are significantly less in such locations. Where prolonged summer droughts coincide with high rates of evapotranspiration – as in **Mediterranean** climate regions – plants must be **xerophytic** in order to survive. Clearly, the key to the relationship between precipitation and biomes is not quantity but effectiveness. **Effective moisture** is the term used to describe the net soil moisture which is available for the vegetation to use as and when it is required.

Temperatures also affect plant adaptation and survival. All plants have maximum and minimum temperature tolerances within which normal growth may take place, and this tolerance range varies quite widely between plant species. Within this maximum/minimum range, however, is a much narrower temperature band which actively facilitates growth; this smaller temperature range is known as the plant's **ecological optimum condition**. Most plants cease to function (i.e. produce chlorophyll) and so become dormant when the air temperature falls below 6 °C; ideally, temperatures should exceed 10 °C for effective photosynthesis to take place. Where temperatures exceed 15 °C throughout the year, there is the potential for a continuous growing season, although plants often exhibit distinct signs of stress in temperatures above 35 °C. For most plants, the optimum mean annual temperature for growth is 25 °C. Higher mean (and actual) temperatures inevitably increase plants' water requirements. Wherever winter temperatures fall below 6 °C for up to five months of the year, trees adapt by shedding their leaves to protect themselves against frost damage. Where temperatures fall below 6 °C for more than six months of the year, tree adaptation is to retain leaves – in order to maximise photosynthesis as soon as temperatures rise above the critical temperature. The influence of temperature upon vegetation is not usually experienced in isolation, but operates in conjunction with other factors such as humidity, precipitation and light intensity. For example, well-established olive groves have the ability to survive periods of extreme cold – provided that the weather remains dry; yet fairly short spells of cold and wet conditions can be extremely damaging however long the olive trees have been established.

The availability of light determines the rate at which photosynthesis takes place. Light availability and its intensity vary considerably between places and at different times. Both factors are influenced by latitude, season, local relief, climate (especially cloud cover) and the proximity of nearby plants. As light decreases, fewer plants are able to exist and so they become more widely dispersed. Light quality is also an important factor and high ultra-violet light levels appear to be one reason why the range of plant species is greatly diminished in mountainous areas.

Atmospheric movements also affect ecosystem and biome development. Wind direction and strength influence both evapo-transpiration rates and air temperatures. In colder climates, the wind-chill factor can lower the effective ambient temperature by many degrees. Wind can also hasten the drying-out of soil in exposed locations, leading to a reduction in effective soil moisture and a marked increase in the potential for soil erosion. Permanently windy habitats invariably support grassland rather than tree cover – partly because grasses have adaptations which allow them to bend easily without damage to their internal structures.

Topography

Topography (or relief) can influence vegetation successions in several ways. Increases in altitude are very closely correlated to decreases in temperature and consequently increased altitude invariably results in more stunted plant growth, fewer plant species and a reduction in protective soil cover. Relief may also increase an area's exposure to (or, indeed, protect it from) heavy rainfall and strong winds. Gradient or angle of slope also directly affects soil depth, texture, acidity and drainage, with most steep slopes having thinner but less waterlogged soils.

Aspect is usually only of local importance. Within the Northern Hemisphere, south-facing slopes are much more favourable locations for plant growth as their orientation increases access to sunlight, leads to higher temperatures and reduces their susceptibility to frosts.

Edaphic factors

The equilibrium between soils and vegetation is extremely delicate. The availability of organic and inorganic matter within the soil is vitally important for the continued well-being of the entire biome. Whilst biomes are generally characterised by certain specific soil types, considerable variations in vegetation can occur across a single biome as a result of local differences in soil and/or underlying parent rock. Britain provides an excellent example of such variation; although the regional classification is 'temperate deciduous forest', the mature vegetation across large tracts of the British countryside is predominantly grassland and heathland – due to underlying chalk and limestone which give rise to local variations in soil type. Deciduous forest did occur very widely on the clay-based soils which cover much of England, Wales and southern Scotland, but these forests are now much reduced due to long term human interference. Vegetation cover is also greatly affected by local variations in **edaphic factors** such as soil texture, structure, acidity and depth as well as water, oxygen and nutrient content.

Biotic factors

The most important biotic factor affecting biomes is inter-plant competition for light, root space and water; such competition is especially fierce in the lower latitudes where vegetation cover is particularly dense and greater numbers of species are involved. However, **grazers** and **predators** also

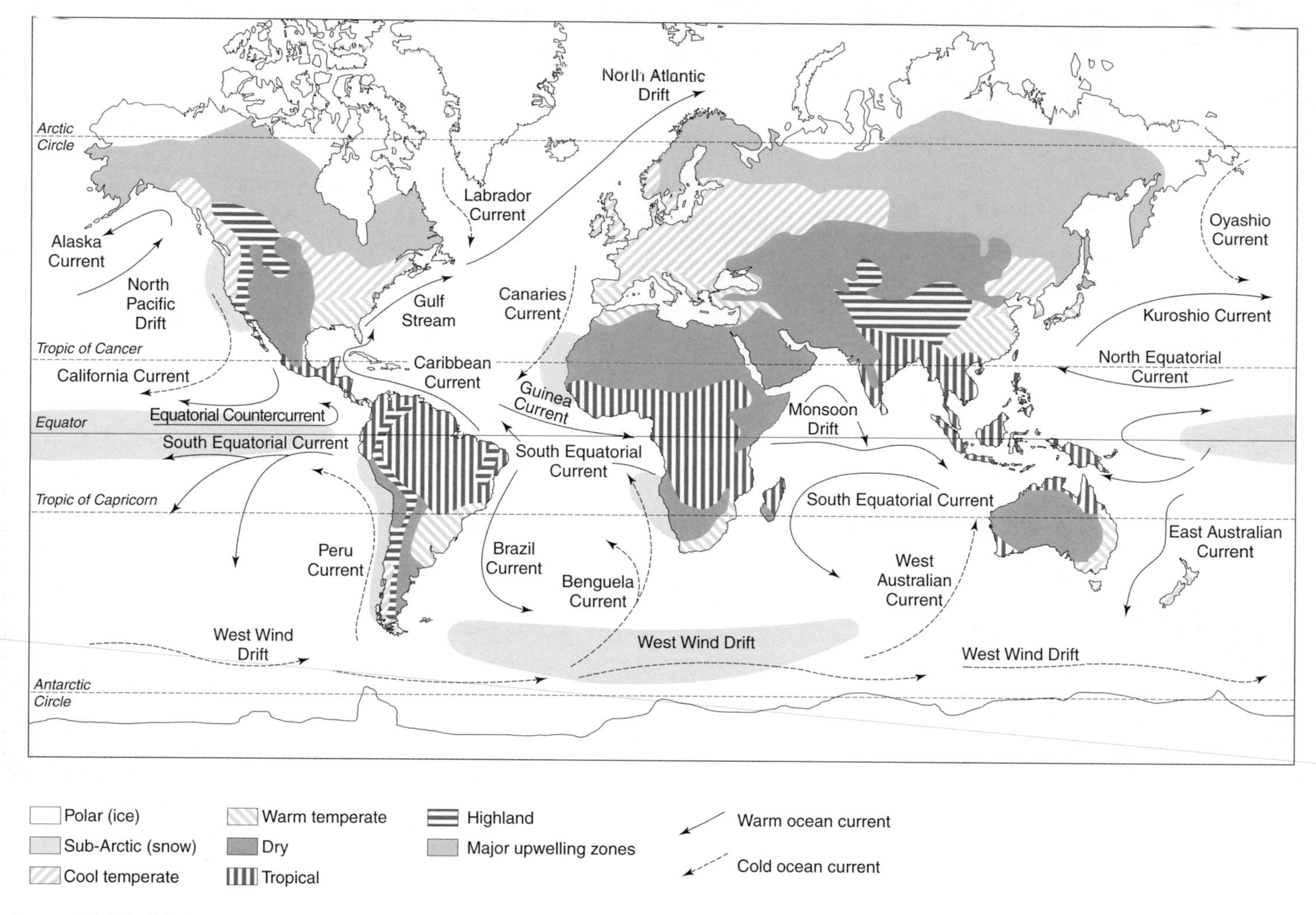

FIGURE 3.1 World climate regions and ocean currents

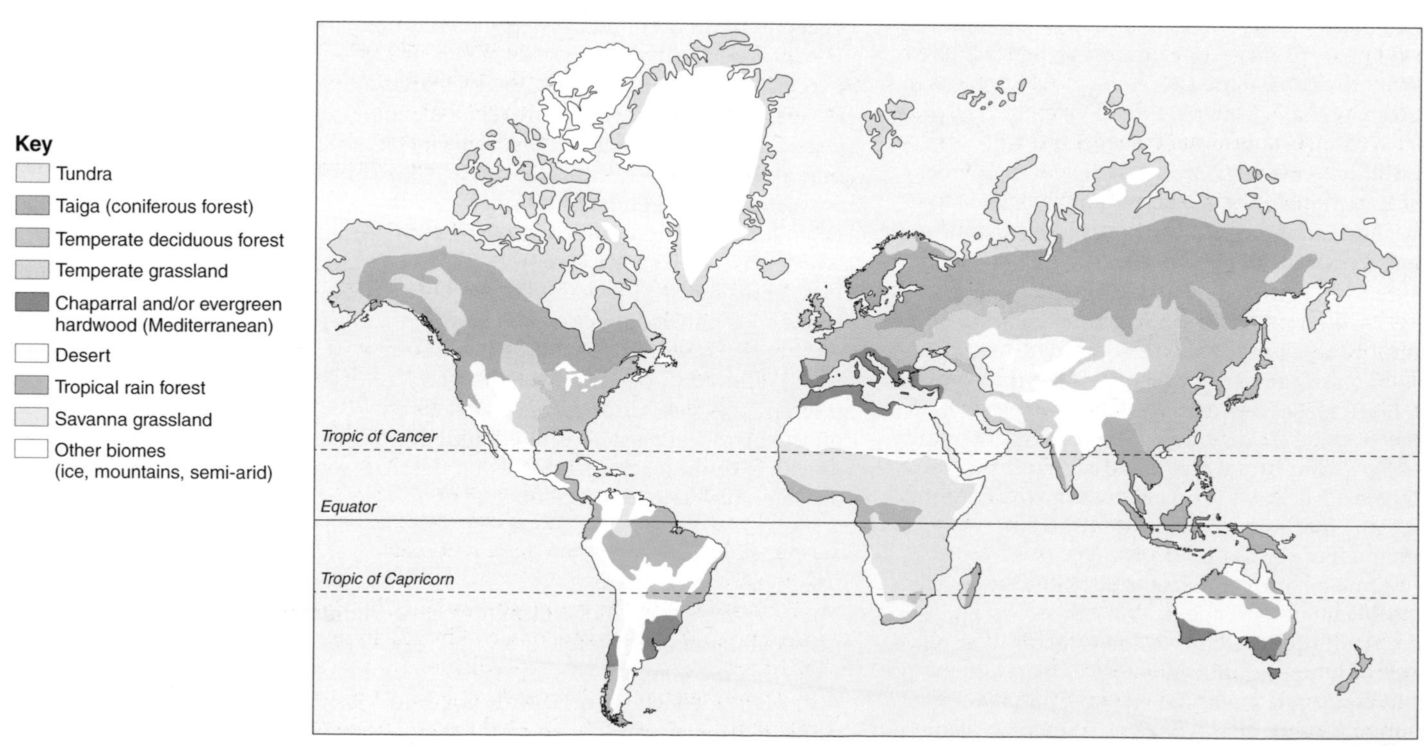

FIGURE 3.2 World biomes

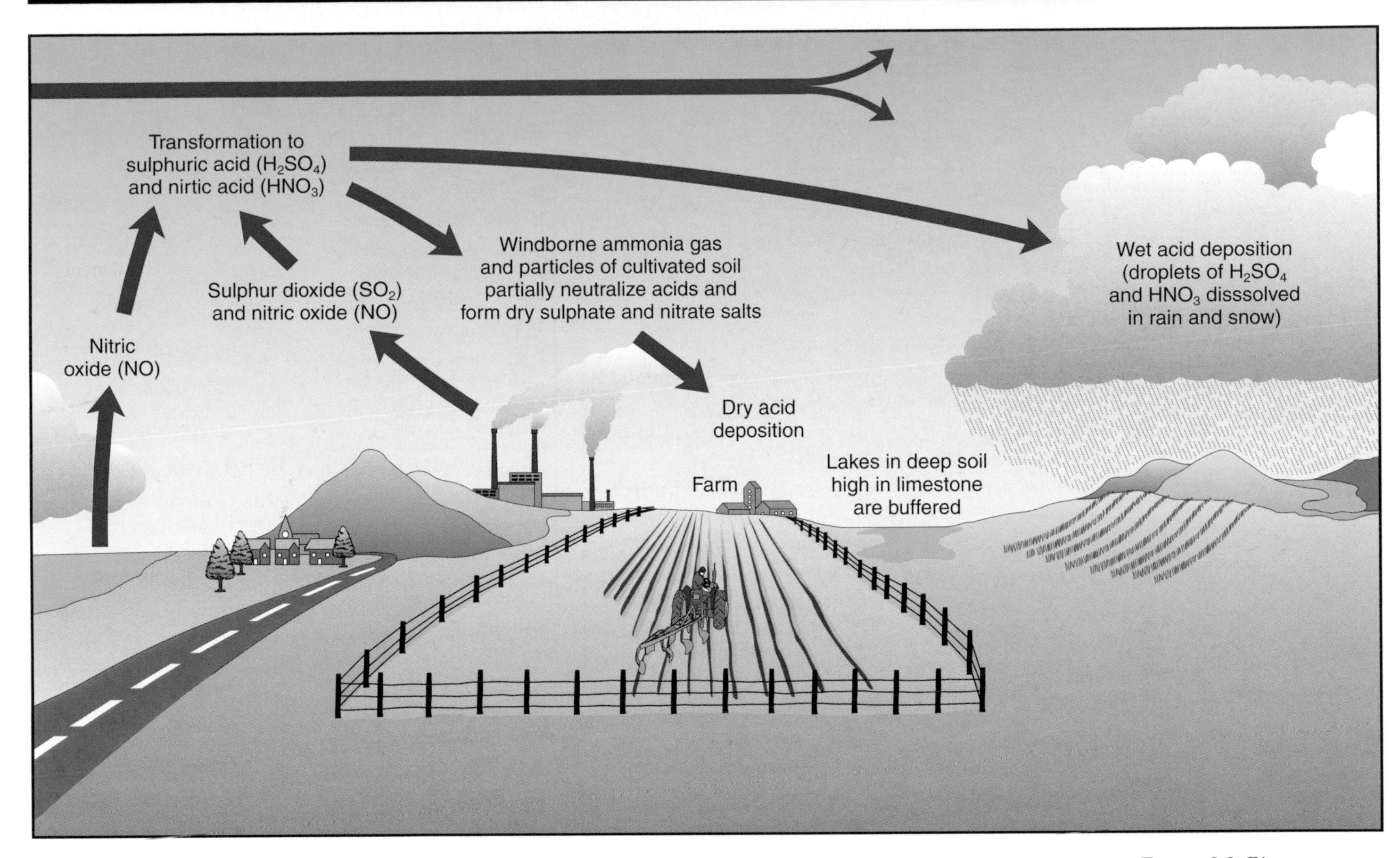

FIGURE 3.3 Biome damage caused by acid rain

exert a considerable influence on vegetation development. Herbivores and primary consumers are frequently responsible for pollination and seed dispersal, as well as the close-cropping of the vegetation itself. Second and higher-order predators are influential by controlling the populations of herbivores and primary predators such as insects. Animal diseases are an additional factor – as was clearly demonstrated by the unintentional introduction of myxamatosis to Britain in the 1950s. Myxamatosis is a viral infection of rabbits endemic only in South America. However, it was deliberately introduced into the rabbit population of Australia in the 1950s as a means of dramatically reducing their numbers. This was seen as a necessary measure because rabbits (which are not indigenous to Australia) were devastating precious arable and grazing areas across the country. Inadvertently, the virus was transmitted to Britain at about the same time and decimated the rabbit population. Unexpectedly, this rabbit cull led to record rates of grass growth, increased crop yields and successful sapling establishment.

In addition to the above natural influences, the role of human activity should not be underestimated. Very few areas in the world support 'undisturbed' mature (climax) communities or biomes and in many places – particularly the major urban conurbations – the natural biome has been either totally destroyed or very comprehensively manipulated in order to meet the perceived needs of the local human population. Human interventions in the natural world may be categorised as being of three distinct types:

1 **In the atmosphere**, the delicate balances of carbon dioxide and ozone within the lower and upper atmospheres respectively are being destroyed – leading to temperature rises and increased ultra-violet radiation. The current rates of sulphur dioxide and nitrogen oxide emissions are devastating large areas of forest biomes through the effects of **acid rain** (Figure 3.3).

2 **On land**, our pursuit of economic activities has led to rapidly increasing urban sprawl as well as widespread stripping of natural vegetation to provide additional land for both building and large-scale agriculture. Subsidence, soil exhaustion, topsoil erosion, deforestation, afforestation, fire and flooding are just some of the many factors currently responsible for both biome destruction and modification.

3 **Disturbance of the ecological balance** has taken place due to the widespread use of increasingly potent fertilisers, pesticides, herbicides and fungicides. Such disturbance, combined with **over-grazing** by animal herds, has led to changes in the inputs and outputs of the natural systems and

created new, unnatural, feedback loops. Often, this feedback is positive feedback, resulting in increasingly rapid movement away from equilibrium and accelerated rates of change.

ACTIVITIES

1 a) What do you understand by the term 'biome'?
b) In what ways are ecosystems and biomes different?
c) What are 'aquatic life zones'?
2 Compare the two maps in Figures 3.1 and 3.2.
a) Identify and name six separate regions of the world where there is a high degree of correlation between the two maps.
b) For each of these regions you have identified, name their climatic type and biome cover.
c) Identify and name three separate locations where an extensive climatic region comprises more than one of the major global biome types.
d) Identify the climatic type in each of the three regions you named in c) as well as all of the biomes which occur within each region.
3 Write notes to show that you understand the concepts linked to the following key terms:
- ecological optimum conditions
- edaphic factors
- effective moisture
- growing season
- photosynthesis.

4 a) Describe briefly how light quality may affect plant growth.
b) Explain why increased ultra-violet radiation due to ozone depletion is currently affecting the Earth's biome cover.
5 Undertake sufficient research of your own to enable you to write, in some detail, about the ways in which pollutants affect vegetation growth and hence have a knock-on effect on whole-biome survival.
6 Produce an account of how human initiatives have led to a wide range of adverse effects on the natural environment within your own local area.

Forest biomes

Forest tends to be the natural vegetation response in most areas having moderate to high annual precipitation. Forest is rarely species-specific, although small areas of woodland are often referred to by their dominant tree species – as with the 'oak woodlands' of Britain. The vegetation within any particular forest biome is, however, usually quite distinctive in its appearance and/or character and is often described as being deciduous, coniferous, evergreen etc.

We also tend to classify forests according to their general global location, within the three broad categories of tropical, temperate and polar. Although superficially vague, these generalisations do give some indication of the temperature and precipitation regimes within each location – and hence the likely vegetative response. Tropical forest exhibits the greatest range of **biodiversity** (i.e. the species, genetic and ecological diversity within any ecosystem) in both its flora and fauna; its NPP (approximately 9000 kcal/m^2/yr) is the joint-highest of all biome regimes. This reflects the fact that both biodiversity and NPP tend to decrease with temperature reduction. Whilst trees are the dominant form of flora within any forest, it is usual for them to co-exist with populations of much smaller-scale vegetation such as bushes, shrubs and flowering plants.

Evergreen coniferous forests

Evergreen coniferous forests are also known as **boreal forests** or **taiga**. Such forest exists (in its natural state) as an almost continuous belt of evergreen coniferous trees across North America and Eurasia, covering about 11 per cent of Earth's total surface area. It is restricted almost exclusively to the Northern Hemisphere and is located in habitats which are climatically sub-Arctic and cold-continental (Figures 3.1 and 3.2). Such areas are characterised by low precipitation (40–50 cm/year), but with a distinct summer maximum. Long, dark winters (with up to six months below freezing point) and short, mild summers result in a very modest annual total of between 50 and 100 frost-free days (Figure 3.4). Extremes of temperature are a particular feature of such areas with Verkhoyansk (in the Russian Federation) recording temperatures with variations as wide as ± 36 °C. Winters are dark, with only a few hours of daylight in December and January; yet in summer, there may be unbroken sunshine for up to 20 hours every day. Although precipitation is low, so too are evapotranspiration rates and the climate is technically classified as being humid. Most of these areas have been glaciated relatively recently and in places are still underlain by a thick layer of **permafrost** (see Unit 4).

Boreal forest is exceptionally sensitive to local variations in climate as well as edaphic conditions and this sensitivity is reflected in the successions and plagioclimax communities which we see today. As might be expected, trees are the dominant plants of the taiga; indeed, trees are the only major form of vegetation which has successfully adapted to survival in such extreme conditions. Four main types of coniferous tree occur – each consisting of only a few, common species. Although there is little variety of species, there are huge numbers of trees! Spruce, fir and pine are all evergreen conifers; larch is the exception, being a deciduous conifer. In North America, fir and spruce dominate (with two or three species of each being the norm), whilst in Eurasia, the Scots Pine is the single most common species. Although evergreen conifers are the dominant mature vegetation, broad-leafed deciduous trees and shrubs such as alder and birch are part of the early stages of both primary and secondary successions. Because of the rigorous environmental conditions, the natural vegetation has had to adapt to:

- intense cold
- extreme variations in temperature
- an effective winter drought caused by the soil being frozen
- a short growing season.

The ways in which evergreen conifers have made the necessary adaptations are illustrated in Figure 3.5. Most boreal forests are not homogeneous units; instead, they tend to be zoned within broad latitudinal bands. Taking a north-south transect through a typical forest is likely to reveal the following four distinct vegetation zones:

- **A tundra/taiga ecotone** This represents the northern-most extent of the forest and generally follows the meanderings of the 10 °C summer isotherm. This isotherm frequently corresponds to the **tree-line** – the biome boundary that follows the initial isotherm along which temperatures fall below those at which trees can survive. In some places there is no boundary ecotone whatsoever; in such places, tree cover ceases abruptly, to be replaced immediately by true tundra biome. In other places, the ecotone may extend for hundreds of kilometres, providing a rich mix of flora and fauna from both biomes and creating an expanse of relatively open ground punctuated by increasingly sparse and stunted trees.
- **An open evergreen coniferous forest** This is the true taiga. It exhibits underbush and undergrowth vegetation and ground-hugging plants such as those to be found in classic tundra conditions.
- **A closed canopy evergreen boreal forest** This has tall stands of trees underlain by a thick cover of leaf litter but little undergrowth.
- Either: **Mixed evergreen/deciduous forest ecotone** This marks the transition to temperate, broad-leafed deciduous forest in the south, or: **Open evergreen, coniferous forest with grassland** This marks the transition to continental-interior, temperate grasslands.

In most boreal forests, there is a far greater variety of fauna than flora. Animal life includes **seed-eaters** (e.g. squirrels and nutcrackers), insects and mammalian herbivores (such as hares), together with large **browsers** (i.e. animals which nibble leaves and twigs) such as elk and moose. Beaver are also associated with the aquatic/terrestrial ecotone of such regions. Predators include the weasel family

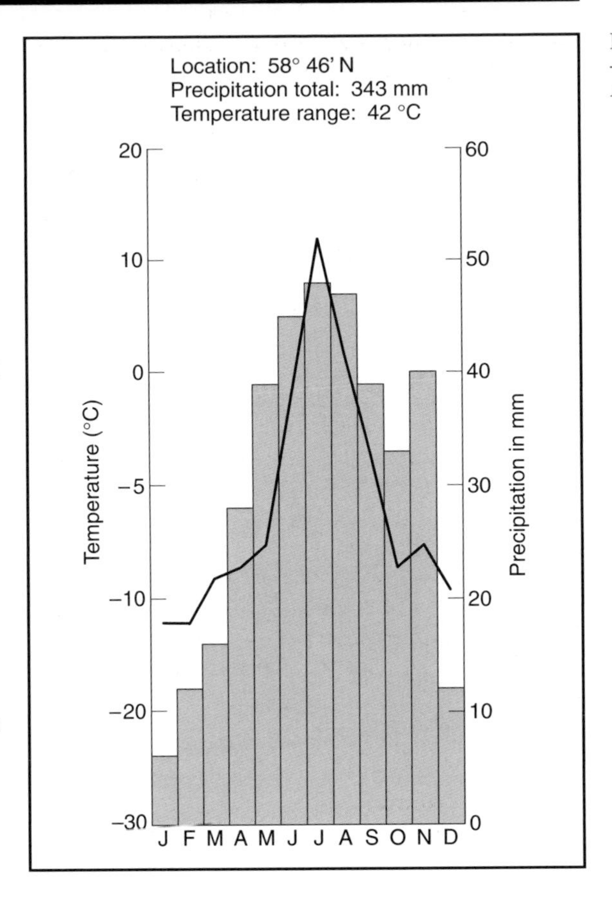

FIGURE 3.4 Climate graph – Fort Chipewyan, Alberta, Canada

(pine martin, mink, ermine), lynx, bears and wolves. In North America, the caribou migrate annually from summer grazing in the Arctic tundra to winter shelter within the forest biome. In summer, thousands of insect-eating birds (especially warblers) in-migrate to the forests to feed on flies, mosquitoes and caterpillars. Seed-eating birds such as sparrows, together with omnivores such as ravens, reside permanently in these habitats.

Beneath the trees is a deep layer of leaf litter which is only ever partially decomposed; decomposition is particularly slow because of:

- low temperatures
- the waxy coating on the leaves
- the high acidity of the leaves.

As the leaves (or needles) begin to decompose, they cause the already thin, nutrient-deficient soil to become even more acidic. The physical barrier which they create prevents other plants from colonising the forest floor and so is very effective in reducing **interference competition**. The overall effect is to produce huge stands of tall trees with little or no undergrowth/underbush. These vegetation conditions, combined with the prevailing low temperatures, give rise to a distinctive zonal soil known as **podsol**. Zonal soils such as podsols result from the combined effects of climate and local vegetation. Podsols were first named by Russian peasants who noticed a distinctive white **horizon** (layer) below the surface leaf litter and an equally distinct dark (almost black) layer at a greater depth (Figure 3.6). They assumed that this black layer was

Shape: The triangular shape of the trees, together with their short, springy, down-sloping branches, protect the tree from damage by the weight of lying snow in winter. Their shape assists in snow-shedding.

Evergreen: Retention of foliage allows plants to recommence photosynthesis immediately temperatures rise sufficiently in spring. The growing season is too short to grow new leaves from bud.

Leaf shedding: Leaves are shed only when they are old and no longer functional.

Colour: Spruce and fir particularly have very dark green foliage which helps to absorb the maximum amount of heat from the Sun – allowing photosynthesis to recommence as soon as possible.

Roots: Although the trees are often exceptionally tall, they are also extremely shallow-rooted. This means that they can survive above both hard-pan and permafrost.

Leaf shape: Leaves are needle-like; this reduces the surface area available for transpiration.

Leaf coating: Needles have thick waxy coating which is waterproof and windproof; this protects the stomata from drying winds – again reducing transpiration.

Cones: Seeds are protected from the intense cold within the cone. These open-out to release the seed only when conditions are favourable.

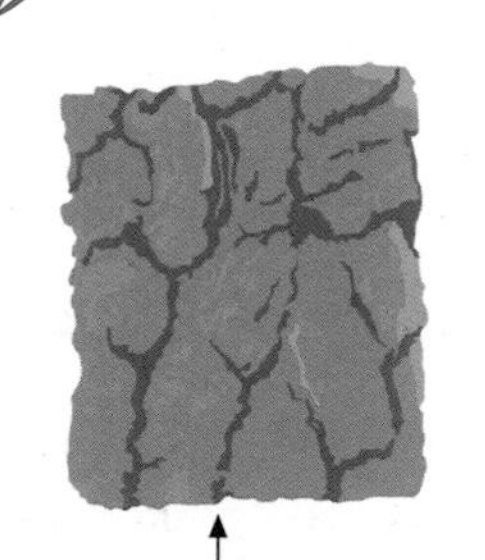

Barks range in colour from grey → red → brown. They are often ridged or scaled to add greater protection to inner vesicles of tree

Larch: Deciduous, coniferous (birch and aspen) predominate in permafrost areas where the winter temperatures fall so low that even waxy, needle-like leaves are insufficient protection against freezing.

FIGURE 3.5 Evergreen coniferous vegetation adaptations

charcoal, a residue from previous forest fires and concluded that that the upper white layer was the ash remains from such fires. Thus they gave these forest soils the name podsol – which literally translates as 'ash-soil'.

Podsols are most often found in cool/cold climates, beneath coniferous forests. In such areas, precipitation usually exceeds evapotranspiration, and often, these areas coincide with extensive glacial outwash plains and/or regions of acidic parent rock (e.g. sandstone, granite or acidic metamorphic rocks). Whilst these latter conditions are not critical for the formation of podsolic soils, they do further increase the acidity of the overlying soil and perpetuate its continued existence. The process by which podsols are formed is called **podsolisation** (often defined as 'intensive **leaching**') but is not confined to boreal forest locations; it is also found in tropical areas – where it gives rise to ferruginous soils.

As noted above, the needle-like leaves of coniferous trees have thick, waxy coatings and decompose very slowly. The humus which they provide is very acidic and is known as **mor**. As there are few earthworms (because they cannot tolerate such cold, acidic conditions), there is little mixing of materials between the soil layers, so allowing very distinctive horizons to develop. The organic breakdown products released from the leaf litter include **fulvic acid** and **humic acid**. Humic aid aids the conversion of iron, aluminium and silica minerals into more soluble forms. Downward percolation of water through the soil profile transports and precipitates these **sesquioxides** and clays to the lower sub-surface layers. This process is particularly active during spring, when snowmelt produces greatly increased amounts of percolating water. This intense **leaching** process produces the white (or pale) upper horizon; the removal of sesquioxides and clay is known as **eluviation** (or out-washing).

The later precipitation of these minerals within the 'B' horizon is known as **illuviation** (or in-washing). This process creates the distinctive dark layering beneath the surface. Often, the organic material which is deposited creates a black layer, whilst the deposition of iron creates a dark red/orange (rusty) coloured layer capable of forming **hard-pan**. Hard-pan can, over time, become an impermeable layer which prevents both the downward movement of percolating water and the penetration of tree roots. Ultimately, it can cause widespread and long term waterlogging of the upper soil horizons and lead to the development of **gleyed** (or **gleyic**) **podsol**. Such gleying refers to the situation in which waterlogged soil provides anaerobic conditions under which the iron minerals are unable to oxidise – leaving them grey in colour rather than their usual 'rusty' red colour. Where permafrost occurs within 200 cm of the surface, a variety of podsol known as **gelic podsol** develops.

Podsols are generally low in nutrient content and are not naturally fertile. They can be improved by the application of lime (to neutralise acidity) and fertilisers. However, this is not a common practice – due to the fact that few plants other than evergreen conifers and larches can survive in such adverse climatic conditions, and that this type of vegetation merely perpetuates the acidic state of the soil.

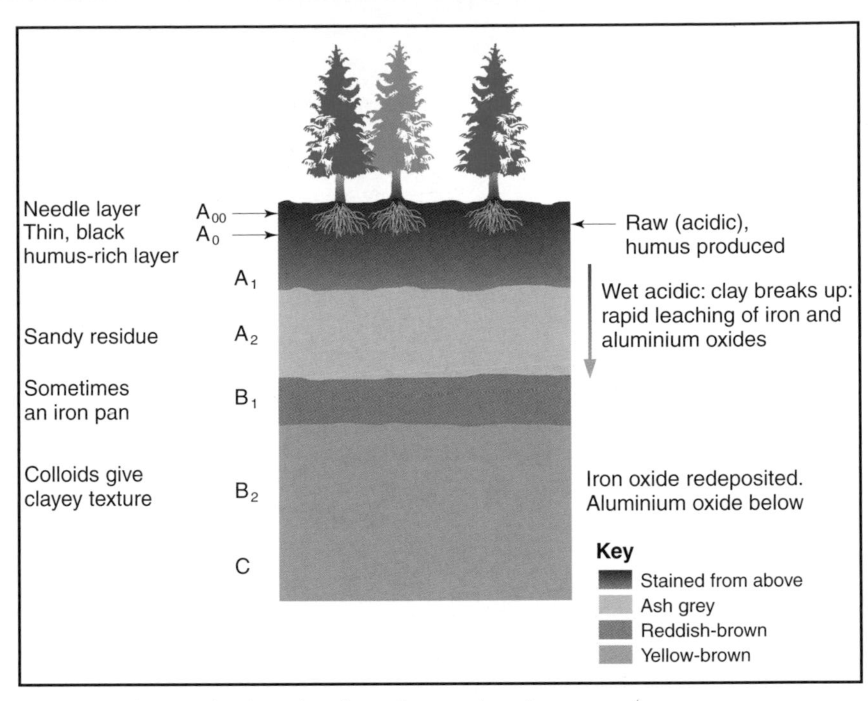

FIGURE 3.6 Podsol soil profile

Edaphic factors can cause extensive areas of plagioclimax vegetation. The three most common variations within spruce/fir-dominated ecosystems are:

- **Bog** (or **muskeg**), commonly found in poorly-drained hollows and depressions. The ponds of trapped water support thick layers of sphagnum moss upon which tundra species such as cottongrass and heathland shrubs thrive.
- **Pine forest**, which is common, particularly in North America, glacial **outwash plains** and former dunes. The sandy sub-strata significantly reduce effective soil moisture, creating drought-like conditions and low-nutrient soils.
- **Larch forest**, which inhabits intensely cold areas underlain by permafrost where the usual range of evergreen adaptations to extreme climatic conditions are ineffective. In such areas, whilst larch is the dominant species, deciduous birch and aspen are also often found. Such forests are more open than evergreen boreal forest and so have underbush vegetation comprising shrubs, mosses and lichen.

Current research on the taiga biome suggests that, whilst spruce and fir are the dominant species at the present time, they actually represent only one sere within a 200-year vegetation cycle. This conceptualised cycle includes a sequence of changes from the present vegetation type to an aspen-dominated climax before the cycle returns to spruce/fir domination. Figure 3.7 provides additional information about the operation of this cycle.

In the sparsely populated North American boreal forests, farmers, ranchers and hunters have

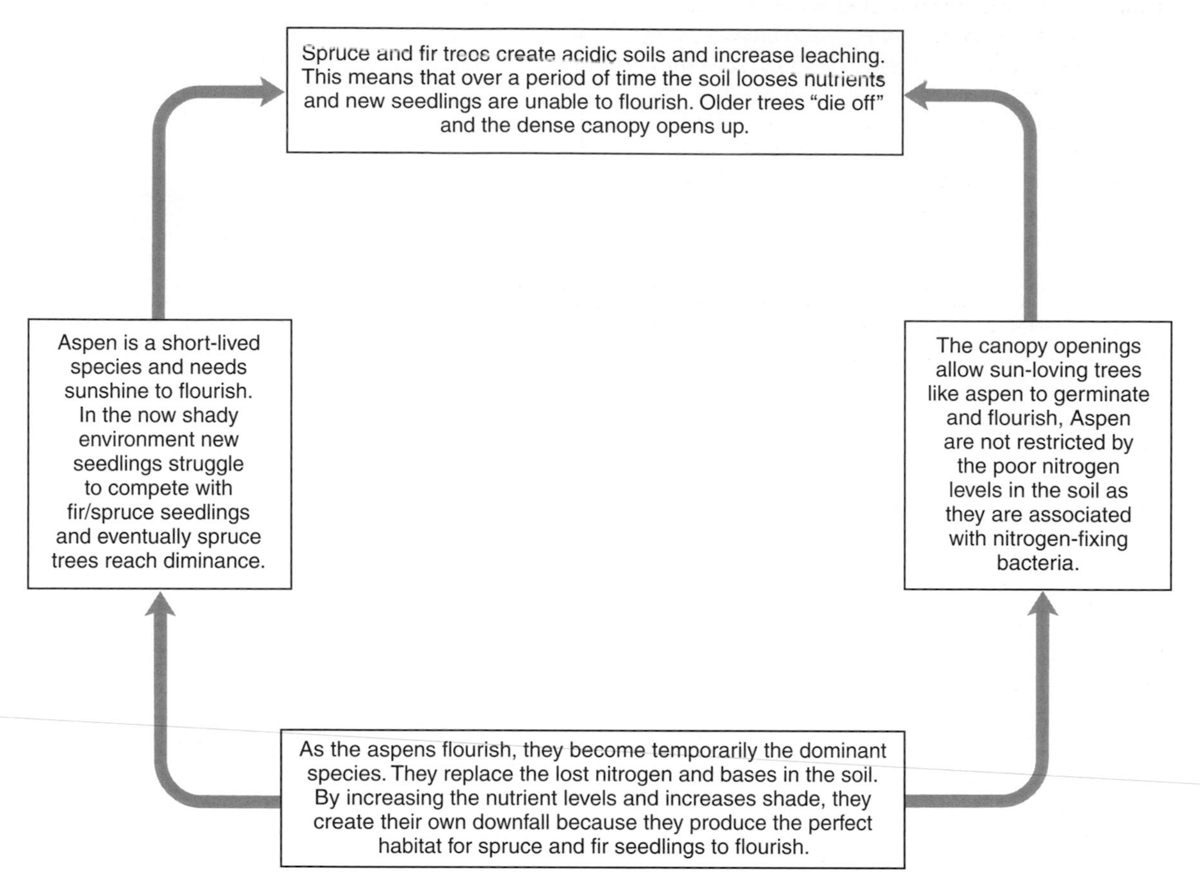

FIGURE 3.7 The spruce-aspen cycle

eliminated the large carnivore predators which preyed on their livestock and game animals. Timber production has involved the deforestation of enormous swathes of forest and many of the remaining stands are currently under threat as our demand for softwoods continues to increase at an almost exponential rate. In particular, Japan is the major importer of Canadian timber. In Europe most natural forests have disappeared – although many European countries introduced effective re-afforestation programmes during the latter half of the twentieth century. In Siberia there is an ever-increasing demand to export timber in order to boost ailing regional and national economies. Unfortunately, wood regeneration is a long-term process in high latitudes, because the intense cold means that newly planted saplings may take a century or even longer to reach maturity. In contrast, coniferous plantations in more temperate regions such as Britain reach maturity within 30 years or so. Extensive mining is a further threat to taiga biomes – especially in Russia, where iron ore, gold, diamonds and other valuable natural reserves occur. The economic pressure to recover and export these valuable resources far outweighs the conservationists' arguments to protect the forest biome. In Europe and Canada, boreal forests have suffered serious damage due to air pollution and acid rain since the late 1950s. In addition, their naturally acidic soils are unable to neutralise the effects of acidic water percolating through them – further compounding the problems of these areas.

Finally, extensive areas of boreal forest have been submerged as a result of HEP schemes such as the James Bay project in Canada. This major project harnesses the potential energy of 19 large rivers which flow into James Bay and Hudson Bay. Stage One of the project was completed in the early 1990s. If and when the second construction phase is undertaken, no fewer than 600 dams will reverse or modify the flow of all 19 rivers – flooding a total area equivalent to that of Germany. In addition, it will permanently displace thousands of indigenous Cree Indians and Inuit who continue to support themselves by traditional hunting, fishing and trapping activities. In 1994, Stage Two of the project was deferred indefinitely – partly because of sustained opposition from environmental organisations and the local populations, but chiefly because the US State of New York decided to withdraw from its contract to purchase the power which would have been generated by the project.

Whilst true boreal forests are restricted to the most northerly latitudes, isolated but substantial pockets of locally-adapted boreal forest occur within mountainous regions in other parts of the world. Boreal adaptations within the Western Cordillera of North America include the giant

Douglas Firs, Sequoia and Redwoods. Timber operations threaten these habitats – but the conservation lobby within the USA is more active than its counterparts in Canada and Eurasia, and it has so far been more successful in preserving these impressive giants than have been the attempts to protect the more humble spruces, firs and pines of the true boreal forest biome.

ACTIVITIES

1 Explain clearly what you understand by the terms:
- eluviation
- gleyic podsol
- hard-pan
- illuviation
- interference competition
- mor
- muskeg
- primary productivity
- podsolisation
- sesquioxide
- soil horizon
- spruce-aspen cycle

2 a) Working alone or with a partner, research a specific named area of boreal forest/taiga. Use the Internet to provide up-to-date information as necessary. You are advised to focus your research activities upon:
- specific climate data for this location
- local flora and fauna adaptations
- local edaphic conditions
- economic activities (both on-going and planned for the future)
- the effects of human activities, at a wide range of scales.

b) Present the results of your research, using a number of different, but appropriate, techniques.

3 You must respond to this question in essay form: 'Explain fully what you understand by the concept of a forest biome and indicate the range of ways in which such biomes are currently under threat.' Enrich your essay with sketch maps and other illustrative material as an integral part of your writing. You should refer to a minimum of two named, contrasting forest biomes.

Grassland biomes

Most grasslands are located in continental interiors. Grassland biomes are usually a direct response to precipitation scarcity and/or unreliability, which means that the **effective soil moisture** is insufficient to support forest vegetation. However, in numerous locations around the world, grassland is more directly a response to edaphic factors or ecological disturbance, which is discussed later in this section. Usually, the habitats which grasses colonise experience at least one season of drought which, in addition to inhibiting forest succession, also greatly increases the likelihood of fire damage. Spontaneous (and, increasingly, managed) fires inhibit invasion by colonising tree saplings – and this process is often further controlled by the **browsing** of large herbivores. Whilst tree saplings are irreversibly damaged by both browsers and fire, grasses are perennial and renewable. There are three types of grassland, each determined by temperature as well as moisture considerations:

1 **Tropical grasslands** (often called **savanna**) These are located where temperatures are high, precipitation is low to moderate and drought is a seasonal occurrence.

2 **Temperate grasslands** These are located where temperatures are lower (and often extreme), precipitation is medium–low, evapotranspiration rates are high and wind strength is also high.

FIGURE 3.8 Climate statistics – Harare, Zimbabwe

Total precipitation 828.0 mm
Mean temperature 18.3 °C
Temperature range 7.2 °C

Latitude 17° 50' s

	J	F	M	A	M	J	J	A	S	O	N	D
Precipitation in mm	195.6	177.8	116.8	27.9	12.7	2.6	0.0	2.5	5.1	27.9	96.5	162.6
Temperature in °C	20.6	20.6	20.0	18.9	16.7	13.9	13.9	15.6	18.9	21.1	21.1	20.6

FIGURE 3.9 Longitudinal transect through a savanna biome

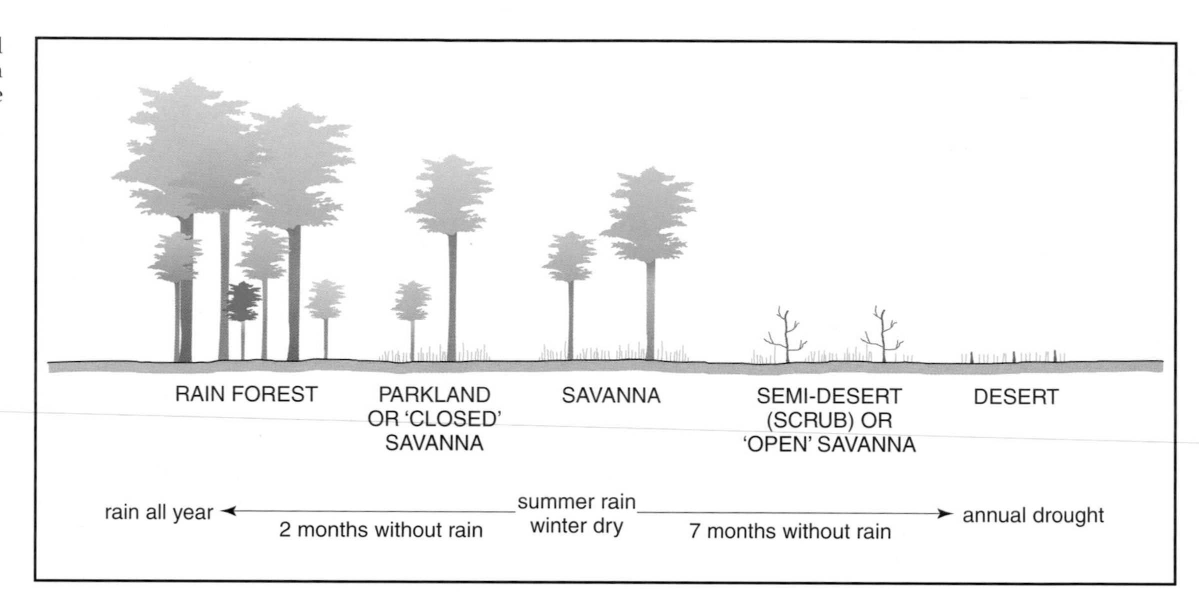

3 **Polar** (or **Arctic**) **grasslands** (often called **tundra**) These are located where temperatures are extremely low, precipitation is also very low and one season is almost without daylight; any precipitation usually falls as snow.

Tropical grasslands – savanna

Savanna grasslands are located in a broad band about 5–15° north and south of the Equator – between the tropical rain forest belt of the equatorial regions and the hot deserts of the tropics. They are associated with tropical continental climates, covering about 12 per cent of Earth's surface – with by far the greatest extent being located within Africa (Figure 3.2). Such geographic areas exhibit high mean temperatures throughout the year (>20 °C), low to moderate annual precipitation (600–1250 mm) and a prolonged dry season of at least five months' drought (Figure 3.8). This dry season usually corresponds to the cooler season which is known as the **low-sun period** (i.e. the season when the Sun is not directly overhead) but, as temperatures are still relatively high, evapotranspiration demands remain high. Rain is frequently abundant in the wet season – so much so that the entire annual rainfall may be achieved in only two or three weeks, although it is more usually spread over a few months, and is often associated with intense electric storms which can be very destructive. Strong, dry trade winds dominate these areas during the dry season.

Although such biomes are classified as grasslands, this does not mean that trees are totally absent from their landscapes. Traditionally, savanna grasslands have been considered as a transitional biome, between those of tropical rain forest and desert. At the forest margins, trees are usually abundant, frequently in very extensive stands separated by more open country which has isolated trees within extensive grassland; such areas are often referred to as **savanna parkland**. The parkland trees tend to be shorter than those of the forest and, as annual rainfall totals diminish, they become progressively shorter and more 'gnarled'. At the desert margins, grasses occur only in sparse, isolated tufts separated by increasingly extensive unvegetated and hence exposed surface areas (Figure 3.9).

Within the archetypal grassland environment, the vegetation comprises wind-pollinated grasses, small broad-leafed deciduous plants pollinated by insects and open-canopy, xerophytic (drought resistant) trees – the most notable of which are **acacia** and **baobab** in Africa and the **eucalypt** in Australia. Sometimes, this open-canopy layer may be replaced by an open-shrub layer. Small stands of trees do occur – usually along river courses or in other localities where the soil moisture content is higher. All forms of savanna vegetation have adapted to the dry season by remaining dormant

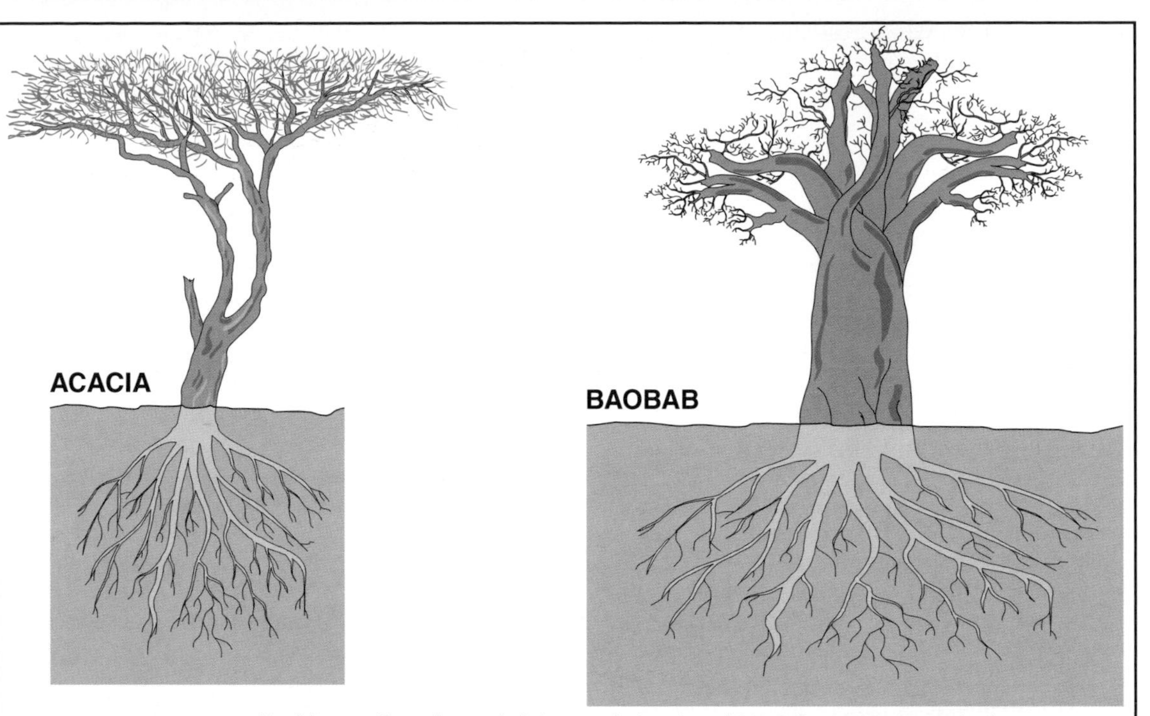

Deciduous: Trees loose their leaves during drought to reduce evapotranspiration.

Leaves: Leaves are often waxy and needle-like – again to reduce evapotranspiration. Needle-like thorns also deter browsers.

Roots: Trees such as acacia have root systems which extend both laterally and deep underground so that the tree can tap all available moisture. The roots may reach the water table.

Spacing: In addition to having wide root networks, trees are also widely spaced to reduce the competition for scarce water resources.

Bark: Thick bark helps trees to reduce moisture loss by insulating their internal structures from the intense heat.

Water storage: In order to survive the drought, some species of tree (notably the Baobab) store water in the trunk creating huge, bulbous trunks.

Fire-resistance: Fire resistant trees are described as being porphyritic. The best example is the Palm tree – which is a monocot, i.e. its internal vascular bundles are scattered throughout the stem, so that if it is scorched, the tree does not loose its vascular function and die.

Seeds: Seeds are drought-resistant.

Life cycle: Vegetation dies back in drought.

Structure: Grasses more flexible – will move in wind without damaging their vascular networks.

FIGURE 3.10 African savanna adaptations to seasonal drought

throughout the drought period. Some of the more common tree adaptations are displayed in Figure 3.10. Savanna grasses lie dormant until the summer rains, then grow rapidly in large tufts – often attaining heights of three to five metres, which means that they dwarf the more stunted trees around them. They tend to be yellow-coloured, in stark contrast to the lush greens of the fertilised fields in Britain. As the dry season advances, the grasses turn a pale straw colour and die back, to remain once again in their dormant state until the next rains. Most grasses are **perennial** (live for several years) and their seeds are so drought-resistant that they are able to lie dormant for many years. This capability is now increasingly important in the many savanna areas where rainfall is becoming less reliable. By the end of the dry season, land surfaces are dry and exposed, making them vulnerable to soil erosion caused by the strong winds which are prevalent during the dry season and by the torrential rains which often herald the start of the rainy season.

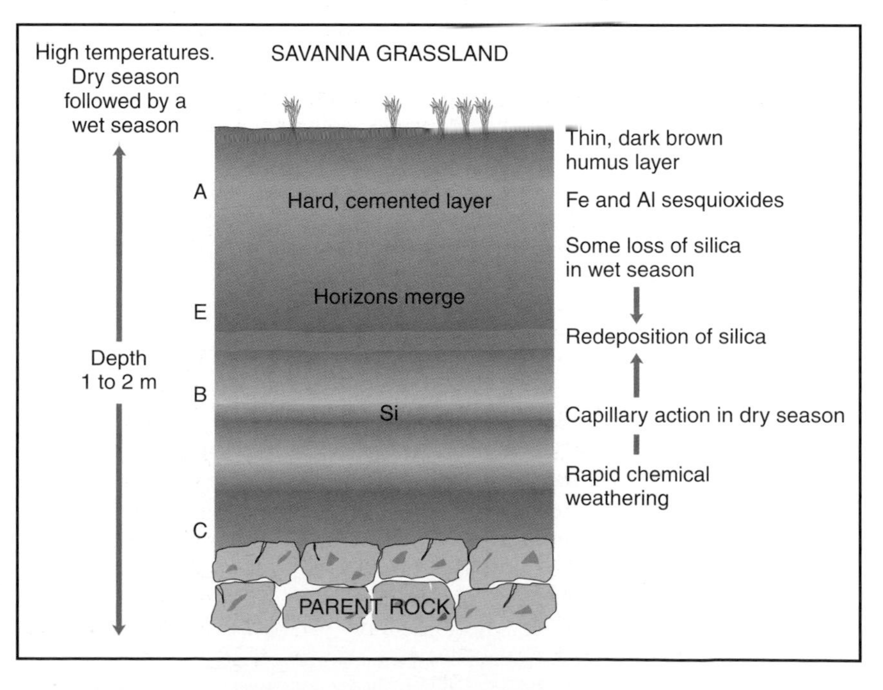

FIGURE 3.11 Laterite soil profile

Although savannas are essentially grassland areas, their individual names often incorporate the names of dominant tree species within their habitats. Examples of this curious nomenclature (naming system) include palm savanna, pine savanna and acacia savanna for south-east Asia, Central America and Africa respectively.

Net primary productivity in savanna areas (2–20 tonnes/ha) is much lower than in neighbouring tropical rain forests and is generally very low for tropical locations; within 'the tropics', primary productivity is usually higher because of the high temperatures. Low productivity in the savanna is due to:

- fewer species and/or layers of vegetation in the grasslands.
- Decomposition rates significantly lower in zones subject to prolonged drought.

Savannas show clear variations from continent to continent. In Britain, our mental image of savanna lands tends to be of acacia savanna – because we have become so familiar with pictures of safari holidays and wildlife programmes based on the East African region. However, the **llanos** of South America is typified by grasslands which are maintained not by drought but by the annual flooding of the Orinoco River, following which long periods of soil saturation inhibit tree growth. The **cerrado** of Brazil is the most species-rich of all the savanna areas and is typified by open woodland rather than grassland; the flora of this particular area has specific adaptations to cope with unusually high aluminium levels in the local soils. Specialised species of trees inhabit the African savannas, whereas the vegetation of other savanna biomes merely exhibits specialised adaptations to survive in the localised edaphic/environmental conditions. In Central America, the pine savannas are the consequence of edaphic, rather than climatic variation; soils are sandy and therefore have poor water retaining capacity. Under such conditions, pine trees are one of the few species which can survive. Australia exhibits flora and fauna responses to the savanna habitat based upon adaptations of the eucalypt vegetation which is unique to that continent.

Fire is both a major natural and managed event in savanna areas. Savanna vegetation is, for the most part, well adapted to survive this fire risk; many trees are **pyrophytic**, whilst grass roots (which is where the plant grows from) are rarely damaged by fire – and so the grass simply regrows after the fire has passed. Fire can be a positive input within grassland ecosystems, as well as being a potential hazard for those who live there. About 50 per cent of the northern savanna regions of Australia are destroyed by fire annually, and fire is now believed to be responsible for the existence of the pine savannas of south-east Asia. Palms are **monocots**, which means that their vascular bundles are scattered throughout the stem; consequently, scorching of the trunk does not kill the plant.

It is now generally believed that tropical savanna grasslands are not the climax (mature) vegetation of such large expanses of tropical continental areas. Certainly, in the lower latitudes, precipitation is sufficient to sustain much denser forest vegetation than exists at present. We have already noted above that in the Americas and south-east Asia, factors other than precipitation contribute to the existence of savanna biomes. Evidence suggests that much current savanna grassland is, in fact, a plagioclimax response to edaphic factors, fire and overgrazing. Overgrazing in this context refers as much to natural overgrazing as to environmentally-damaging farming practices. Particularly in Africa, large herbivores play a significant role in keeping the incursion of tree populations in check by their browsing and trampling activities. Elephants especially are well known to be responsible for opening up forest margins to grassland succession by toppling trees along the forest/grassland periphery.

Savanna grasslands provide ideal habitats for a wide range of insects, but amongst larger animals, the number of individual species is relatively low (whilst their actual populations are large). Significantly, savanna food chains tend to be short, often stopping at the secondary consumer level. Such regions are famous for their large herbivore species, especially elephants, giraffes, antelope and zebra (in Africa) and koala and kangaroo (in Australia). The savannas of south-east Asia and the Americas do not have specialised indigenous fauna. Typical predators (again in Africa) include cheetahs, lions, hyenas, eagles and hawks. In Australia, the large herbivores have no predators except humans. The herbivore populations of the savanna (particularly in Africa) have evolved highly-specialised eating patterns. They may be divided into two groups:

- browsers, which nibble twigs and leaves
- grazers, which eat grasses and herbs.

These herbivores can also be categorised by the height at which they eat, for example:

- gazelles and wildebeests eat short grasses
- zebras eat longer grass and stems
- elephants eat leaves and the low branches of trees
- giraffes eat leaves and shoots at the top of trees.

Resources are also divided temporally – with animals often eating at different times of day (or night) and at different times of the year. Combined, space and time differentiation minimises inter-species competition for food and is known as **resource-partitioning**. The animal population responds to drought in one of two ways. Smaller animals hibernate (in just the same way that the vegetation becomes dormant) or modify their diet by existing on dormant seeds instead of grazing. Larger animals migrate to wetter and more productive areas.

The savanna fauna is rich in scavengers and decomposers. Termites are especially abundant and play a significant role in the maintenance of the biome as a whole. They feast on abandoned food debris and prepare it for decomposition by other organisms. In areas where decomposition is halted during the long dry season, such acceleration of the natural decomposition process is vital in ensuring that sufficient nutrients are recycled to support the system. The grasslands themselves provide a high input of potential organic matter, but much remains on the surface as **leaf litter** during the dry season. Matter which is recycled tends to be processed quickly; warmth and moisture encourage the speedy breakdown of organic matter and the high temperatures also accelerate chemical processes within the soil. Nutrients which are returned to the soil tend to remain there until utilised. Leaching (the loss of surface nutrients due to solution following heavy rainfall) is not a widespread problem – at least where the soils are clay-based. Soil erosion, on the other hand, is a major problem in continental interior regions, and is the most usual cause of nutrient loss from any locality.

Soils vary between the continents and within specific areas. However, laterisation is the dominant soil-forming process. The lateritic soil is typified by a hard, impermeable layer (Figure 3.11). In the wet season, this can create pools of standing water which may be present for several months and which prevent the establishment of most tree species. During the dry season, the **hard-pan** prevents root penetration and this too inhibits tree growth. Palms and course grasses are best able to tolerate such adverse growing conditions. Sub-strata of sand also inhibit tree growth as a result of their low moisture-retaining properties; such soils are often also nutrient-deficient. Both the pine savannas of Central America and the treeless plains of the East African Serengeti are the natural vegetation's response to edaphic factors such as these.

The most widespread human impact on tropical savannas has been due to overgrazing and trampling by cattle. Some 50 million people in Africa and Asia alone raise livestock in such areas and their numbers are increasing. This practice leads to soil damage, increased erosion and decreased fertility. Within the grassland/desert ecotones particularly, this has also led to extensive **desertification** and the loss of thousands of hectares of potentially productive land. Traditionally, pastoralism has been nomadic, with herdsmen following the natural migration routes of wild animals. However, the twentieth century witnessed a widespread shift from nomadism towards more settled patterns of farming. This change has accelerated many problems – particularly where the soils are driest and cattle have been allowed to overgraze the available pasture. Added to this, savanna soils are not particularly well suited to agriculture, as they contain few nutrients; in the past, attempts to cultivate such areas rapidly led to soil erosion and desertification. With careful, scientific (and, inevitably, more expensive) farm management/irrigation techniques, long-term agriculture may indeed be possible, but few populations in the savanna areas are able to practice such farming techniques without intensive external support. Some scientists believe that the increased preference for arable farming poses a major threat to world climates. Burning the grasslands followed by ploughing has the potential to release huge quantities of CO_2 into the atmosphere. This will contribute far greater quantities of CO_2 to the **greenhouse effect** than clearing and burning the tropical rain forests – the effects of which have been widely publicised in recent years. Perhaps the greatest threat the to savanna habitats comes from the increased rate of **urbanisation**. In 1957, Kenya's population was approximately 7 million, compared with almost 30 million at the end of the twentieth century; such a high rate of population increase combined with the global trend towards urbanisation have resulted in many grassland habitats being overtaken by urban sprawl. So great are the current pressures on rural-urban fringe land that the traditional savanna wildlife in some areas is now restricted to isolated pockets of natural vegetation.

ACTIVITIES

1 Write definitions of the following key terms:
- browsers and grazers
- effective soil moisture
- greenhouse gases
- leaching
- leaf litter
- monocot
- overgrazing
- perennial vegetation
- pyrophytic vegetation
- resource partitioning
- soil erosion
- transitional biomes
- xerophytic vegetation

2 a) Draw temperature and precipitation graphs to represent the data shown in Figure 3.8.
b) Describe any trends which are highlighted by your completed graphs.

3 Undertake a piece of research into a savanna region of your own choice (from a continent other than Africa). When doing this, investigate all eight of the aspects listed below. Your findings should then be presented in note-form and be supplemented with a range of illustrative and statistical techniques.
- economic activities
- edaphic influences
- local flora and fauna
- local human activities
- population/urban growth
- specific climatic factors
- the role of fire
- vegetation adaptations.

4 Answer the question at the top of page 92, in essay form, using the notes which you made in response to Activity **3** as well as any other resources which you judge to be necessary. You may extend the area of your study to include examples from Africa. Your essay should be supported by – and include references to – any sketch maps and other relevant illustrative material.

'With specific reference to one or more named locations, state what you understand by the concept of a tropical grassland biome. Explain why such biomes are currently under increasing threat and the possible consequence(s) of their destruction'.

Study 3.1 Cool temperate maritime climate – British Isles

The British Isles and its adjacent coastal areas on mainland Europe are described as experiencing a 'cool temperate maritime climate'. The following explanations give reasons for the choice of words used to describe this particular climate type:

The British Isles are located in the Northern Hemisphere's **cool temperate maritime climate zone** which lies in the middle latitudes between the sub-tropical and sub-Arctic zones. The temperate zone's boundaries are usually taken to be the Tropic of Cancer (latitude 23.5 °N) and the Arctic Circle (66.5 °N). A similar zone exists in the Southern Hemisphere, between the Tropic of Capricorn and the Antarctic Circle. Figure 3.12 illustrates the significance of these imaginary lines and why it is impossible for anywhere within either of the two temperate climate zones to witness the Sun directly overhead.

Regions with a cool temperate maritime climate experience generally cool air temperatures throughout the year, although there are seasonal and frequent much shorter term variations; daytime temperatures rarely extend into either the 'hot' or 'very cold' temperature ranges (Figure 3.13) Their location on the western side of continental land masses is very significant, as east coast locations, even at the same latitudes, experience much more extreme seasonal temperature variations (Figure 3.14).

Temperature	Description
Below −10 °C	Very cold
−10 °C to 0 °C	Cold
0 °C to 10 °C	Cool
11 °C to 20 °C	Warm
21 °C to 30 °C	Hot
Over 30 °C	Very hot

FIGURE 3.13 Heat descriptions defined by temperature range

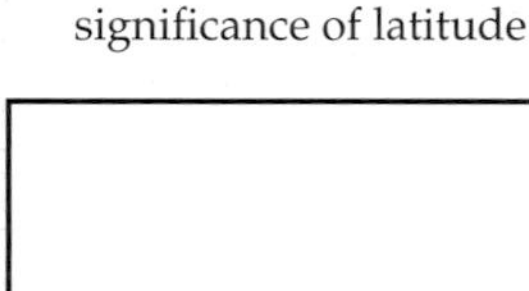

FIGURE 3.12 The significance of latitude

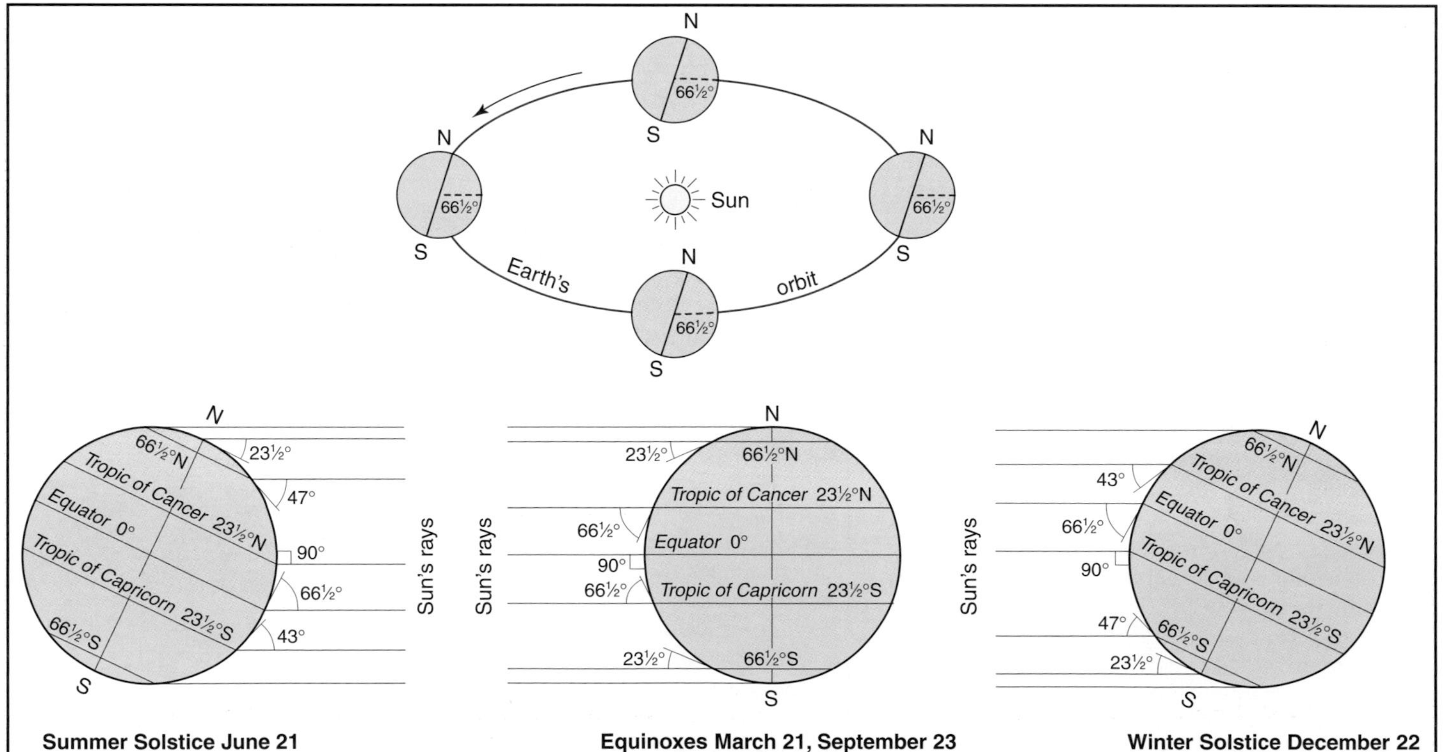

FIGURE 3.14 Typical west and Atlantic Ocean coastal comparison

City	Country	Latitude	Average monthly temperature (°C)												Average total annual precipitation (mm)
			J	F	M	A	M	J	J	A	S	O	N	D	
Brest	France	48° 24' N	7	7	8	10	13	16	17	18	16	13	9	8	866
Halifax	Canada	44° 38' N	−5	−5	−1	4	10	14	18	18	14	9	4	−2	1412
Lisbon	Portugal	38° 42' N	11	11	13	14	17	19	22	22	21	17	14	11	686
New York	USA	40° 45' N	−1	−1	3	10	16	20	23	23	21	15	7	1	1092

The inclusion of the word maritime recognises the dominance of those onshore winds which blow chiefly from the western and south-western sectors. The term 'prevailing south-westerly winds' is one with which you need to become familiar with. Figure 3.15 shows why these winds dominate the British Isles' air stream pattern throughout the year. Although these generally westerly winds are described as 'prevailing', they are quite frequently displaced by air movements from other directions. Figure 3.16 identifies the four major air mass movements, whose interactions are highly influential in determining the weather which the British Isles experiences at any particular time.

Wind direction is only one of the many factors which dictate the British weather on a day-to-day basis. Latitudinal position, nearness to the Continent of Europe, the impact of ocean currents, air pressure and the influence of relief (Figure 3.17) all interact to produce the regional temperature and rainfall variations shown in Figure 3.18.

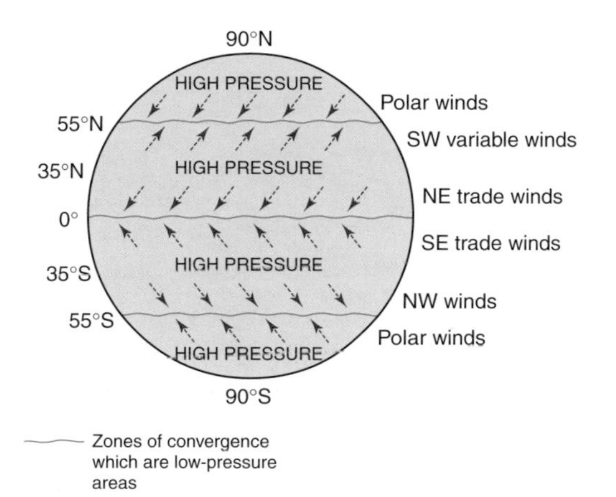

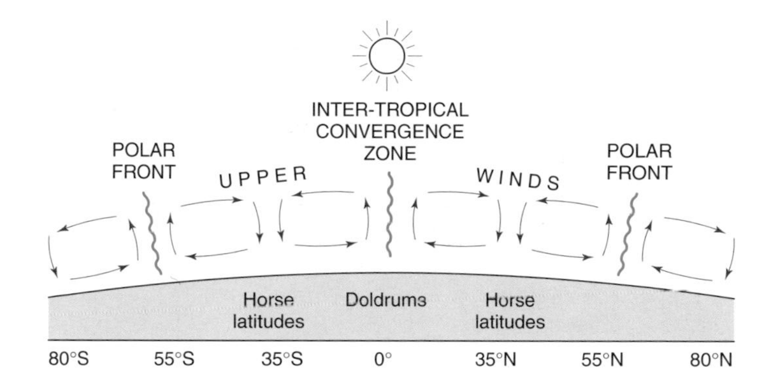

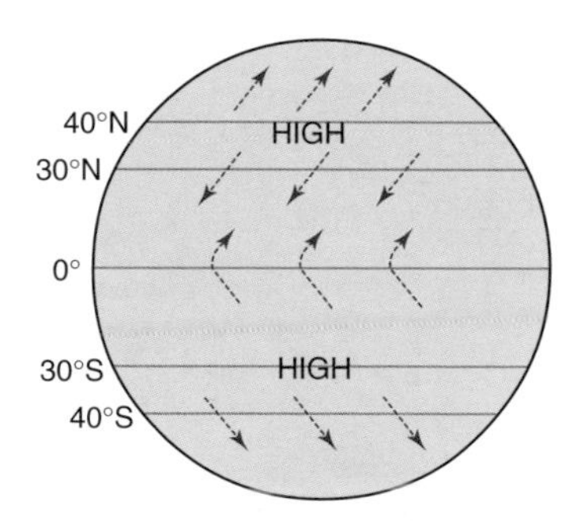

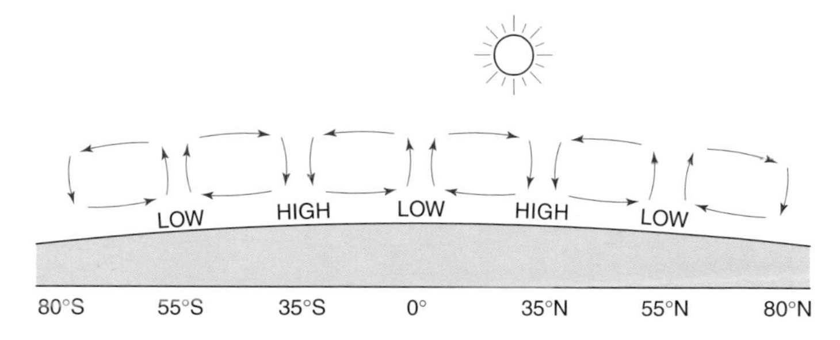

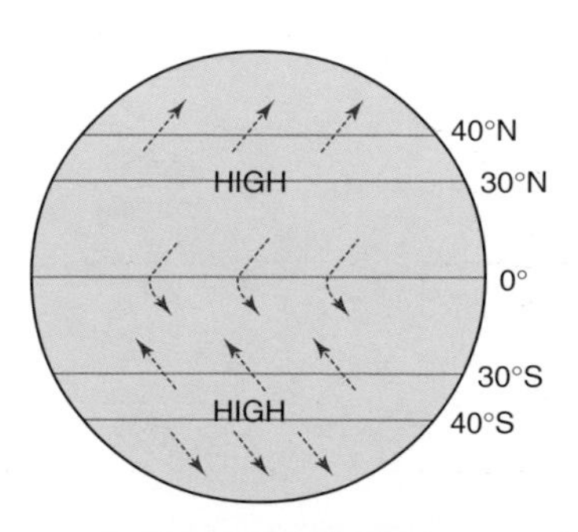

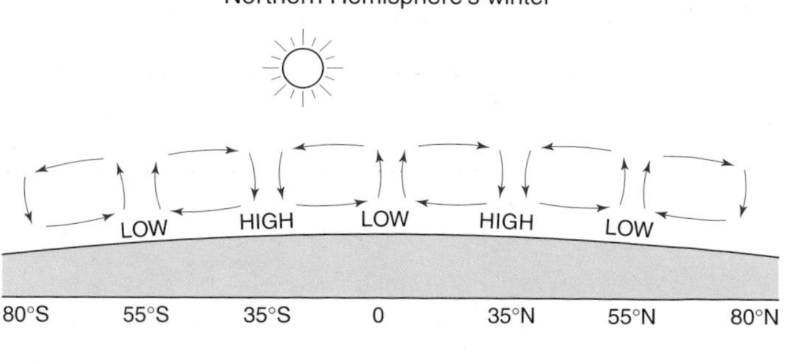

FIGURE 3.15 Global seasonal wind belt variations

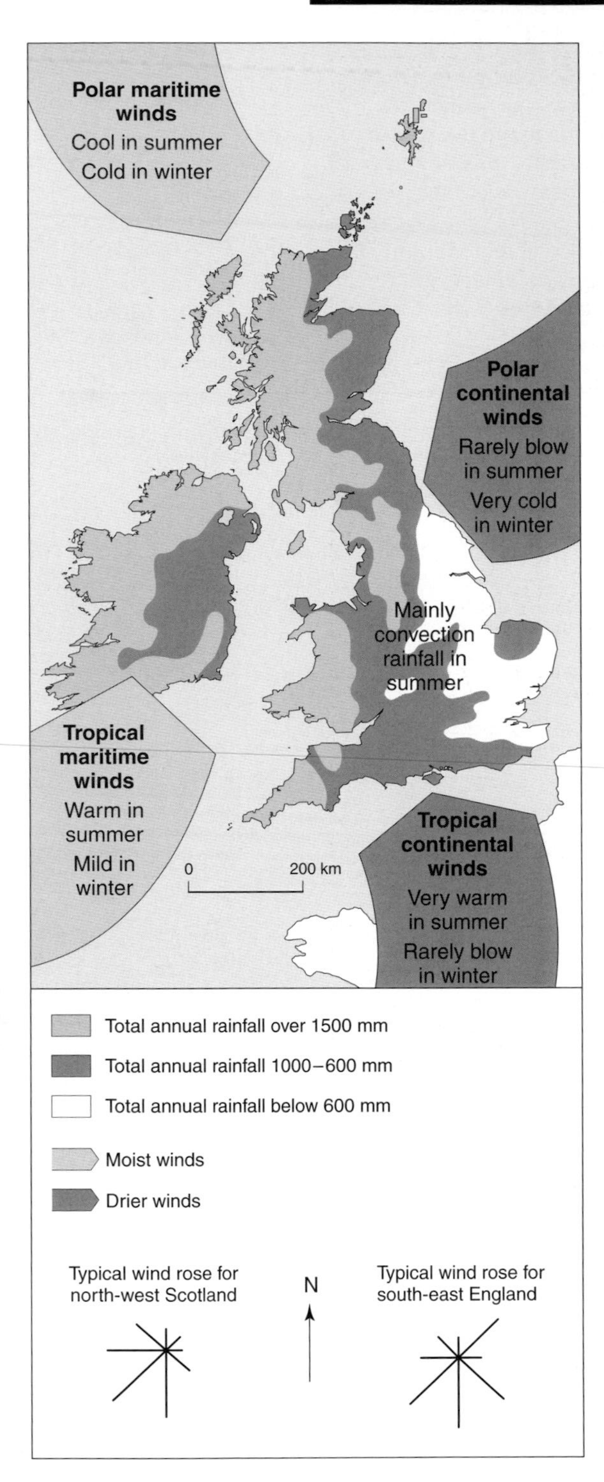

FIGURE 3.16 The four air masses influencing British climate

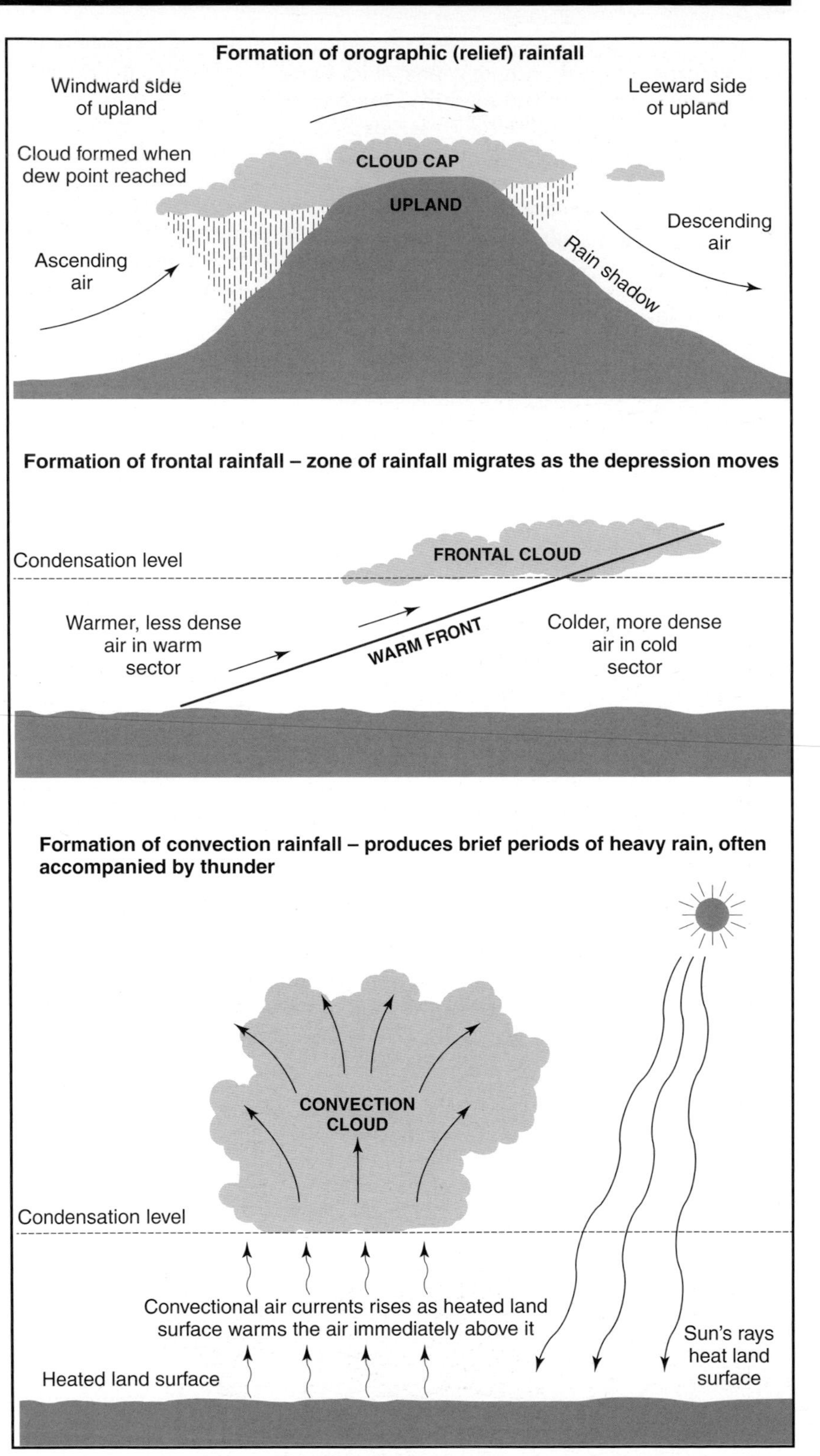

FIGURE 3.17 Factors influencing temperature and rainfall patterns

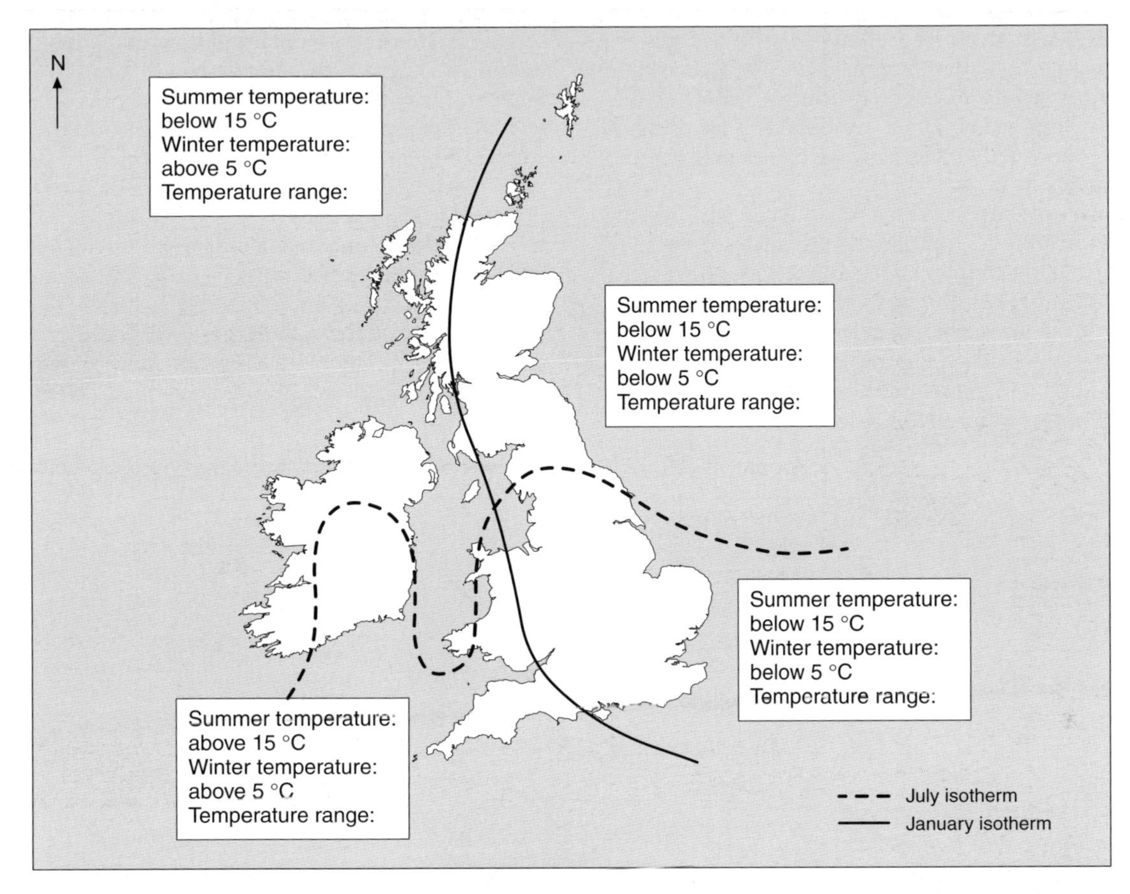

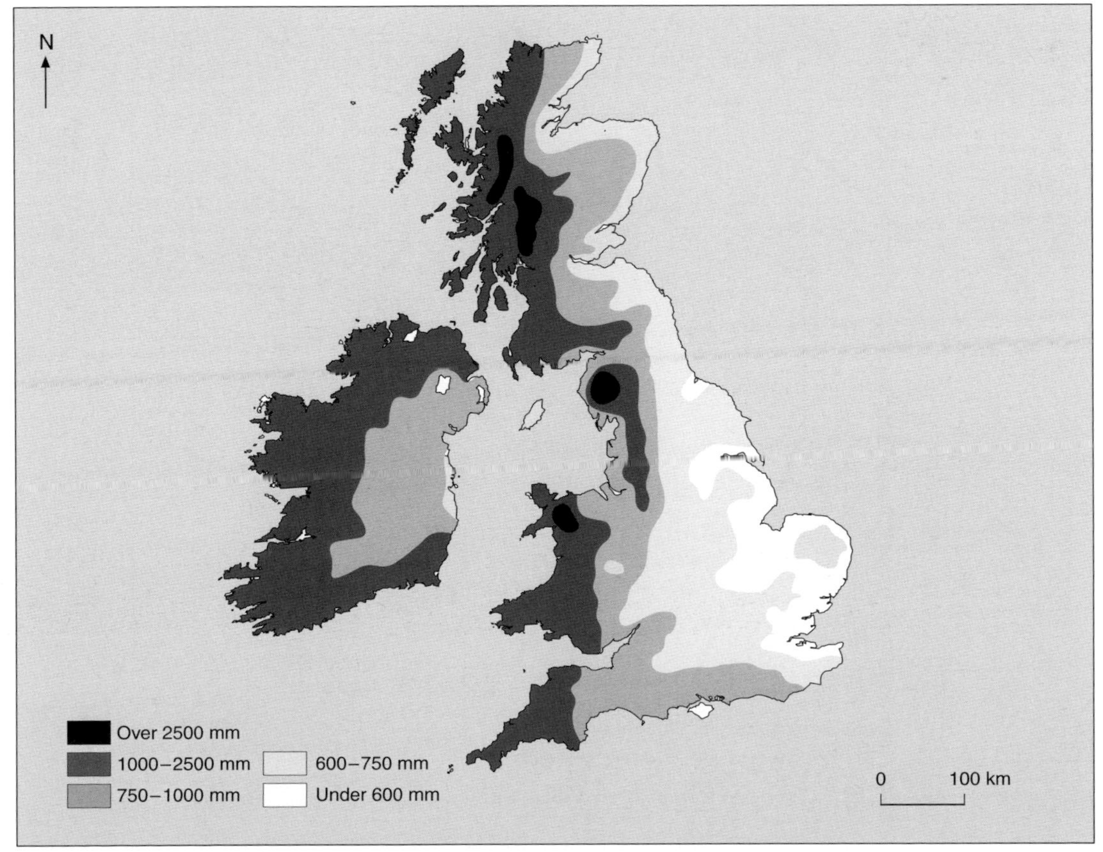

FIGURE 3.18 British temperature and rainfall variations

Air pressure is often just as changeable as the weather induced by it. Anyone who has followed a series of weather forecasts for a few days is almost certain to have encountered some of the following terms which are linked to air pressure and the weather sequences which result from changes in it:

■ **Depression** – an area in which the air pressure is significantly lower than the pressure around it. Depressions bring changeable weather having alternating wet, dry, warmer and cooler intervals. Depressions move in a generally easterly direction across the British Isles and are particularly influential in the winter months. Figure 3.19 explains how and where depressions are formed, how they develop into distinctive cold and warm sectors separated by fronts, and how they often become occluded before finally petering out over mainland Europe when an air pressure differential no longer exists between the depression and the surrounding areas. It is depressions which make it particularly difficult to forecast the weather for the British Isles (Figure 3.20).

■ **Anticyclone** – an area of air pressure which is higher than that surrounding it. Anticyclones are less mobile and less complicated structurally than depressions and, for both of these reasons, the weather which they induce is much more stable (Figure 3.21).

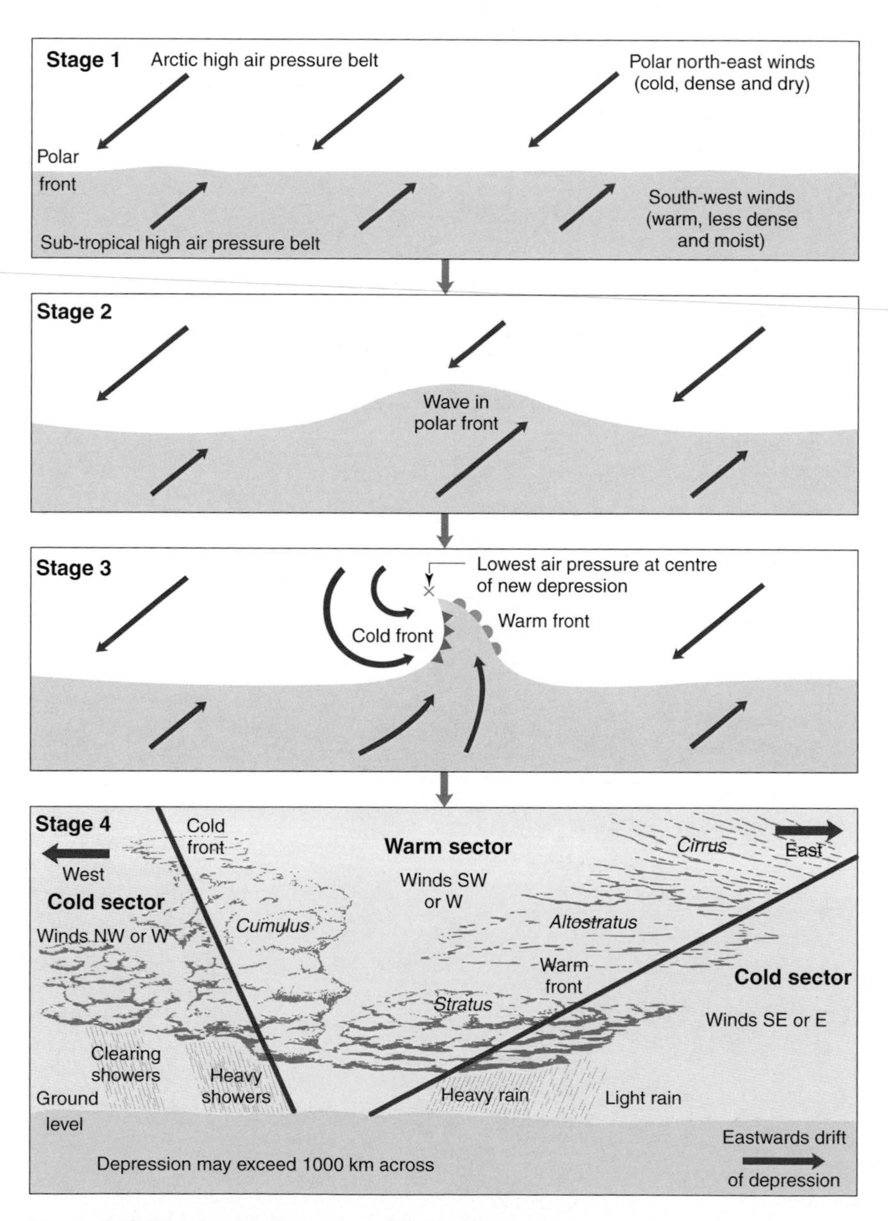

FIGURE 3.19 Depression formative and development

FIGURE 3.20 Synoptic chart showing a typical depression

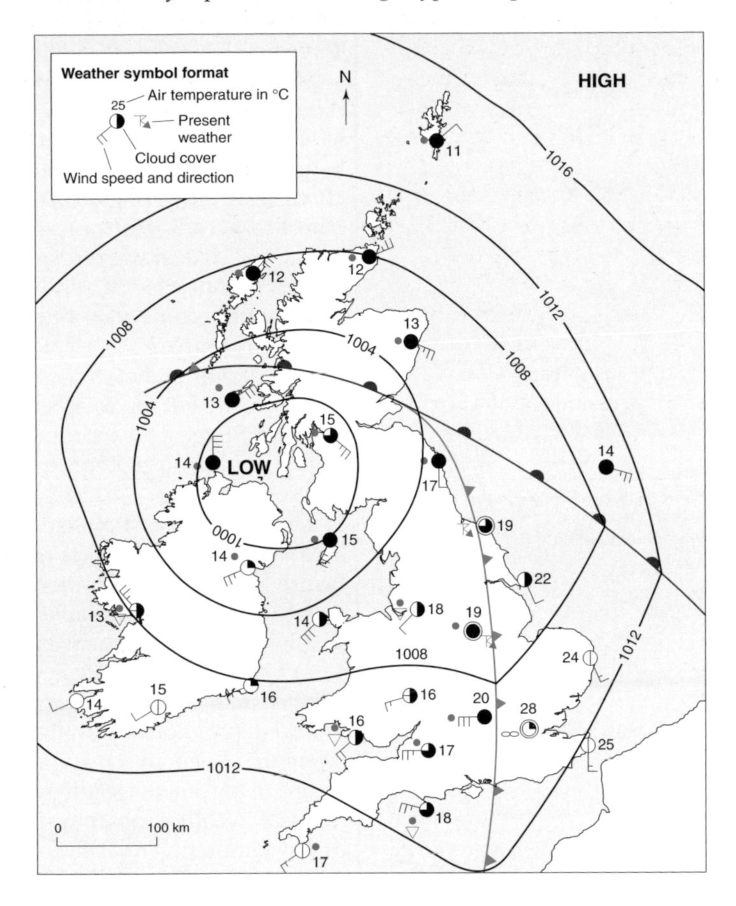

FIGURE 3.21 Synoptic chart showing a typical anticyclone

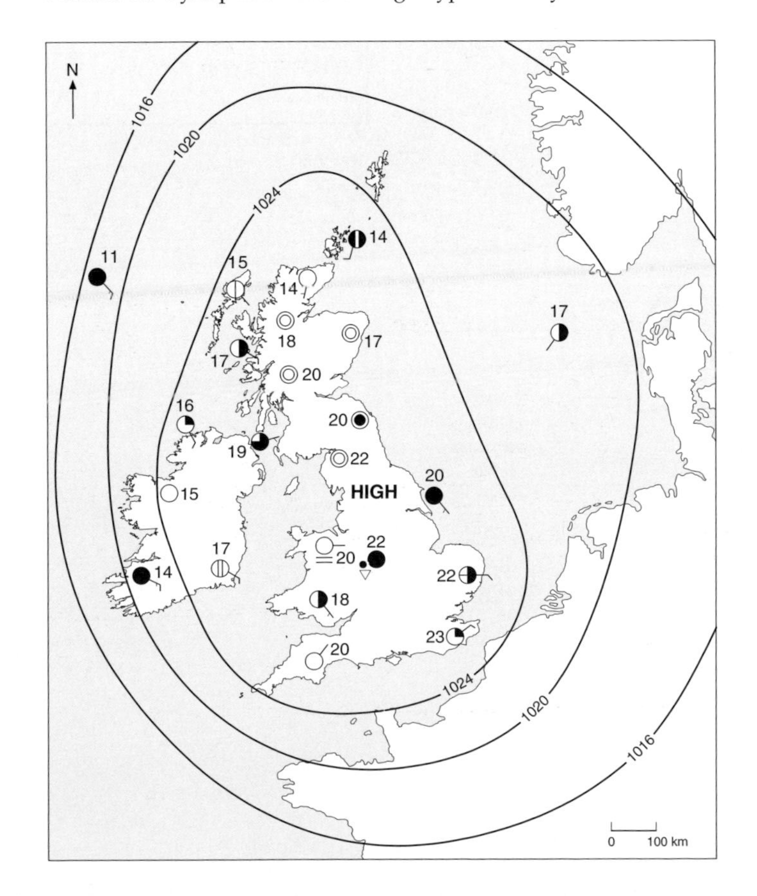

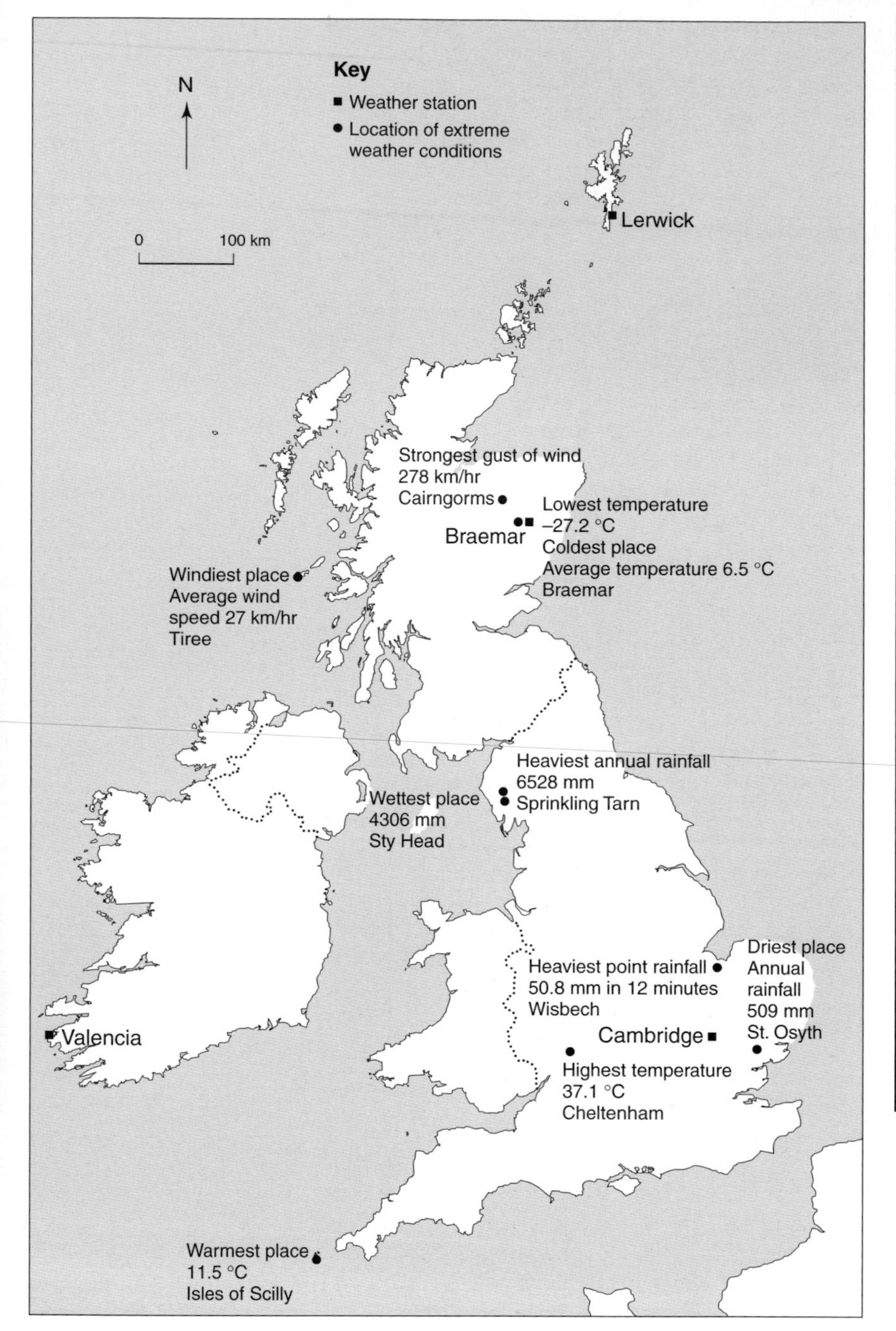

Weather station	Height above sea in mm	Average monthly temperature in °C												Total annual precipitation in mm
		J	F	M	A	M	J	J	A	S	O	N	D	
1	12	3	4	6	9	12	15	17	17	14	10	7	4	545
2	339	1	1	3	5	8	11	13	13	10	7	4	2	923
3	82	3	3	4	6	8	10	12	12	11	8	6	4	1028
4	9	7	7	8	9	12	14	15	15	14	12	9	8	1419

FIGURE 3.22 British climate statistics

ACTIVITIES

1 With reference to the many illustrations in this study produce a series of statements about the nature and causes of 'The British Climate'. It is up to you to decide whether these statements should be sub-divided under headings such as 'Chief characteristics', 'Long-term causal factors', 'Short-term causal factors' and 'Regional variations'. It is, however, very important that you include precise details so that your statements become a worthwhile examination revision resource.

2 The climate statistics shown in Figure 3.22 were recorded at the weather stations shown on the map. Suggest an appropriate location for each station, adding arguments for each of your location decisions.

3 Figure 3.22 also shows some of the record-breaking weather statistics recorded by the British Meteorological Office. Following group debate, note down plausible reasons for the location of each extreme weather event.

4 a) Use a series of daily newspaper weather forecast map cuttings to trace the progress of a typical depression across the British Isles. Whilst doing this, keep your own (at least twice-daily) record of barometer readings and an appropriate range of 'weather' observations.

b) On completion of a), make a detailed comparison of the predicted and recorded weather patterns – noting down any significant variations between the two.

c) With the help of your teacher as necessary, suggest reasons which help to explain the variations you identified in your response to b) above.

Study 3.2 British weather extremes 1 – The drought of 1976

Britain rarely experiences extremes of climate. Few British summers are hot by global standards and prolonged, heavy falls of snow in lowland areas are so infrequent that they soon cause widespread disruption. We tend to be quick to dramatise weather conditions which are somewhat out of the ordinary. For example, autumns which are much milder and drier than usual are nick-named 'Indian summers' and cold spells below −5 °C are described as 'Siberian' even though Siberia experiences many months below −30 °C every year. The British climate rarely justifies exaggerations such as these, but the drought of 1976 was one of the most notable exceptions.

Both the summer and winter of 1975 had been much drier than usual and this situation continued throughout the first half of 1976 (Figure 3.23); evaporation was an important additional reason for reducing water levels in reservoirs. From 25 June, temperatures exceeded 32 °C somewhere in Britain every day for two weeks, reaching a maximum of 35 °C. A state of crisis was officially declared on 15 July. The Drought Bill passed by Parliament on that day gave local authorities the power to fine people caught wasting water, but such measures failed to stop some reservoirs drying up completely during July and August (Figure 3.24).

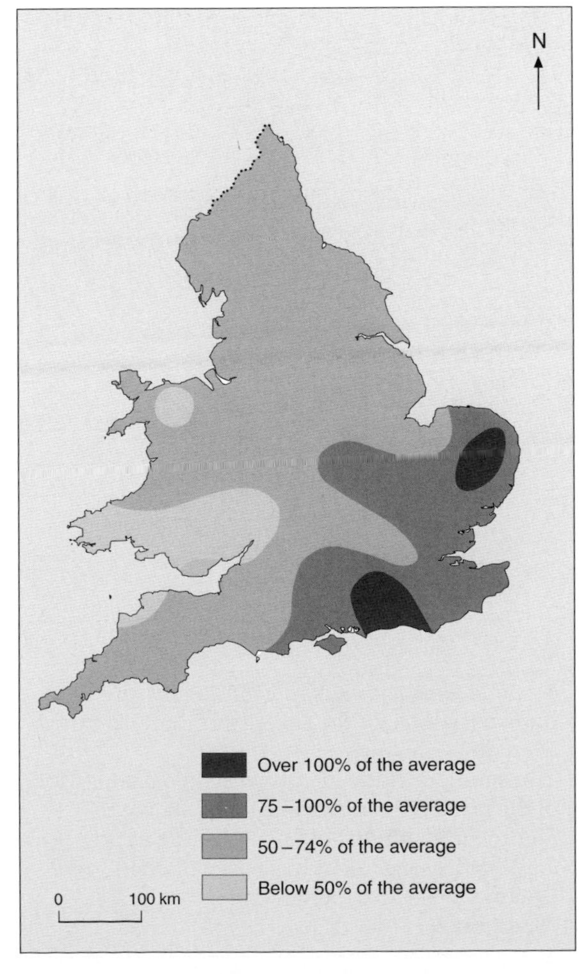

Unfortunately, Britain did not possess a national water grid which could have piped water to the regions in greatest need. Instead, fleets of road tankers took emergency supplies to remote farms and villages. Standpipes became a common sight and were used by more than one million people nationwide, even in cities which had previously invested very large sums of money to secure 'reliable' water supplies.

As the hot, dry summer continued, elderly people suffered strokes and heart attacks brought on by stifling daytime temperatures and uncomfortable nights. Farmers were the most seriously affected commercially and lost pre-harvest crops worth some £500 million. As the soil became dehydrated, root crops suffered particularly badly and the price of home-grown vegetables rose sharply. Cattle suffered great discomfort as their pastures dried out and many thousands of animals were slaughtered to avoid further distress; milk yields were significantly lower than usual. Supplies of many food items had to be imported to make up deficiencies, the additional cost involved in purchasing food abroad having a detrimental effect on Britain's annual balance-of-payments.

Trees were deprived of essential water and extended their roots deeper into the soil, causing damage to the foundations of nearby buildings. The shortage of water reduced the strength of older trees and made them more vulnerable to the spread Dutch Elm Disease some years later. Thousands of hectares of forest were lost to fire, which caused estimated damage of £1 million in Devon alone.

The reasons for the 1976 drought are easy to understand. Fewer depressions than normal passed over the British Isles during 1975–76 and it is depressions (whose moving air masses generate frontal precipitation) which are responsible for much of the annual precipitation and variable weather patterns which are a major feature of the British climate. Instead of being under the influence of precipitation-inducing depressions, the British Isles region was dominated by high air pressure anticyclones – which bring much calmer, stable and dry conditions. The depressions which Britain should have been experiencing were forced around the edges of this fixed area of high pressure – a blocking anticyclone – causing unusually heavy rain to the normally dry and sunny Mediterranean tourist areas (Figure 3.25).

By late August, many parts of England and Wales had experienced 35–40 consecutive days without rain. Then, as Britain prepared for the August bank holiday, the high pressure system which had so dominated the summer's weather finally began to weaken and retreat. As it did so, depressions on its edge moved swiftly inwards, bringing heavy rain and flooding to areas where the topsoil had become too dry and hard to allow that amount of surface water to infiltrate it.

FIGURE 3.23 Rainfall in England and Wales (February–July 1976) as a percentage of the normal total for these months

FIGURE 3.24 Reservoir at the height of the 1976 drought

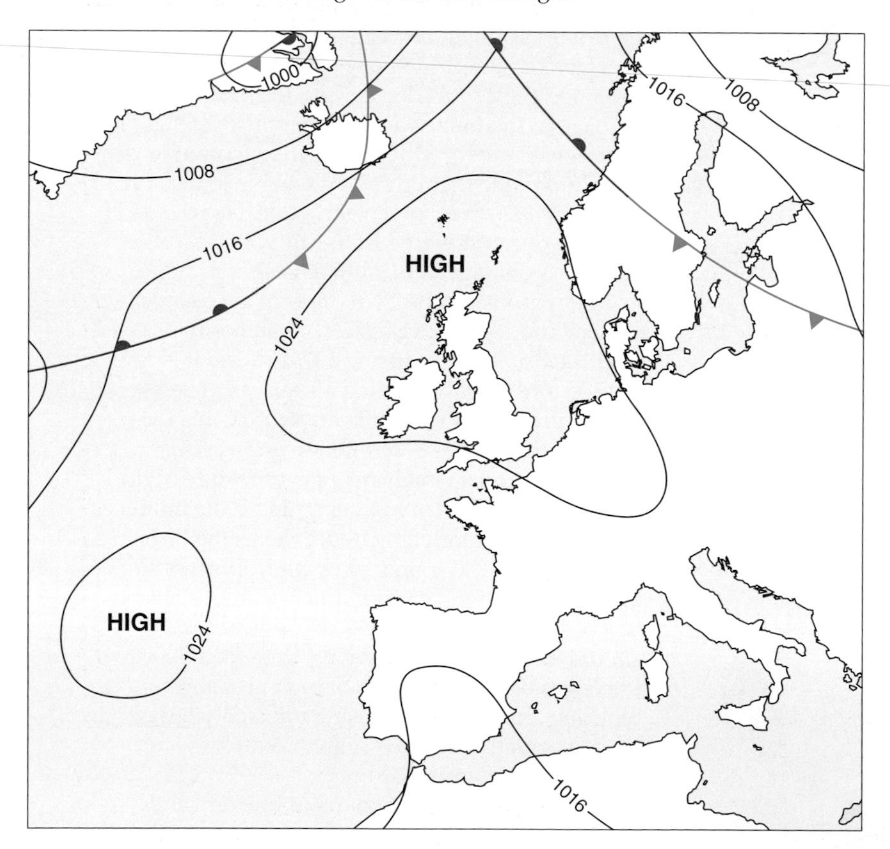

FIGURE 3.25 The blocking anticyclone which caused the 1976 drought

ACTIVITIES

1 What were the chief meteorological and human reasons for the severity of the 1976 British drought?
You should include in your answer a comparison of the rainfall distribution pattern shown in Figure 3.23 with that of an average year.
2 What lessons might have been learned from Britain's 1975–76 experience of drought conditions?

Study 3.3 British weather extremes 2 – The cold winter of 1962–3

The four coldest winters in Britain this century were 1916–17, 1939–40, 1946–7 and 1962–3 (the coldest of the four). The 1962–3 event witnessed snow lying continuously on the ground in south-east England, as well as higher areas throughout Britain, from Boxing Day 1962 until the following April (Figure 3.26). Levels of national economic activity dropped by about 7 per cent, unemployment increased due to the laying-off of 160 000 workers and at least 49 people died as a direct result of the severe weather.

Such extreme winter conditions are the result of prolonged, calm conditions in which cloudless night skies allow above average losses of ground heat due to it being radiated into the atmosphere; cloud cover acts as a protective blanket, trapping heat gained by daytime insolation.

FIGURE 3.26 Severe weather conditions – winter 1962–3

ACTIVITIES

1 Interview members of your family, neighbours and other adults who experienced the harsh winter of 1962–3. Find out what conditions were actually like and how people managed to cope with the resultant difficulties.

2 Suggest reasons why the economic life of Britain was so seriously affected by the 1962–3 winter conditions.

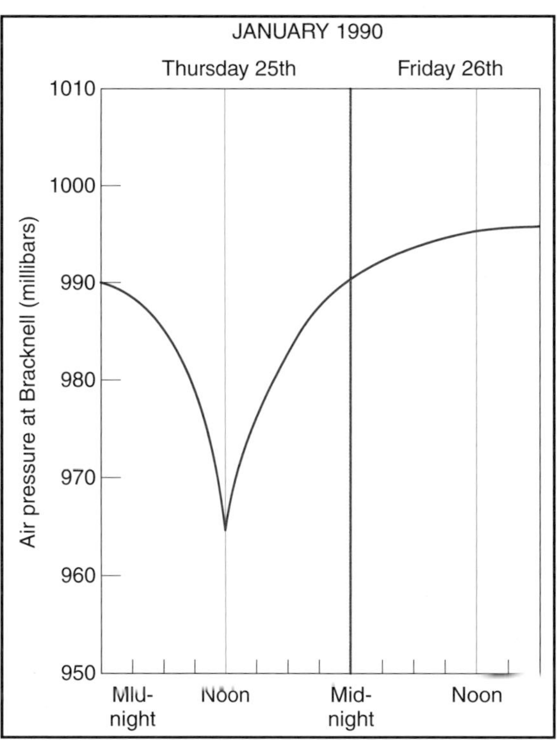

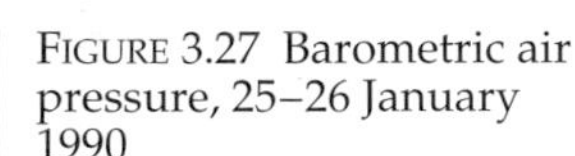
FIGURE 3.27 Barometric air pressure, 25–26 January 1990

Study 3.4 British weather extremes 3 – The storms of 1987 and 1990

On 25 January 1990 much of Britain experienced hurricane-force winds caused by a deep depression which, at its most intense phase, recorded an unusually low air pressure of 968 millibars (Figure 3.27). **Hurricanes** are defined as wind speeds exceeding 115 km/hr. Fortunately, the British Meteorological Office had predicted an event of that nature some five days previously; it had also stated that such conditions usually affect only the more sparsely-populated north of Scotland, where freak winds of 225 km/hr have been recoded but affected relatively few people. The seriousness of the damage caused on that particular occasion is clear from these statements:

- 38 people were killed – mainly due to the passage of the hurricane across southern Britain during the middle of the day.
- More than one million homes lost their electricity supply.
- Transport networks were widely disrupted – mainly due to fallen trees; many airline schedules were halted and one ferry broke down in the middle of the English Channel.
- Schools in the most affected regions closed much earlier than usual.

The newspaper article in Figure 3.28 describes an even more intense and potentially dangerous depression. During this event, the lowest recorded air pressure was just below 960 millibars (Figure 3.29) during the night of 16–17 October 1987. Although it became known as 'the Great Storm', this event actually caused fewer deaths than that of January 1990. It was also nick-named 'Hurricane Ethelred' (after Ethelred the Unready – the King of England from 979–1016 AD, who earned a reputation for never being sufficiently prepared for events, simply because he couldn't be bothered to seek appropriate advice before they happened).

Storm of the century lashes England

October 16. Yesterday evening, a viewer called the BBC and asked the weatherman if there was going to be a hurricane. He laughed off the suggestion. Within hours, south-east England was being battered by winds gusting up to 110mph causing greater havoc than any other storm this century. Northern France was also lashed by the hurricane-force winds.

In England the storm has killed at least 17 people and left a £300 million trail of destruction from Cornwall to East Anglia. Hotels and houses collapsed. Railway lines and roads are blocked by thousands of fallen trees, with no services running south of Rugby or Peterborough. A 6,000-tonne ship was washed on to the beach in Sussex; and most of south-east England was without electricity for at least some time. Many areas are likely to be without power for several days to come.

In London, the fire brigade dealt with a record 6,000 emergency calls in 24 hours. Kew Gardens lost a third of its trees. Sevenoaks in Kent lost six of the oaks which gave it its name; and casualty wards in every London hospital were filled with casualties from flying slates and other debris.

The Meteorological Office is under fire from both the public and politicians for its failure to predict the hurricane. A computer misreading is being blamed for the poor forecast. "We could have got it better," the Meteorological Office spokesman said.

Chronicle of the Twentieth Century, Longman p1289

FIGURE 3.28 The Great Storm, 1987

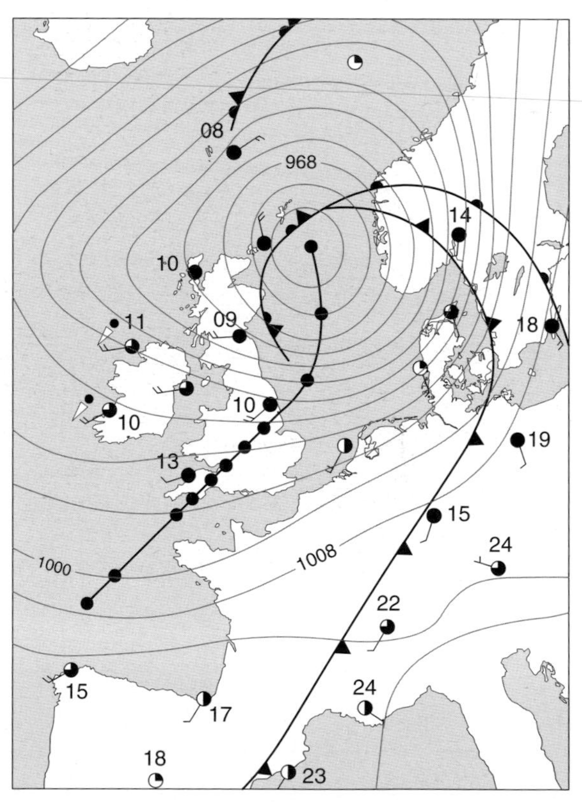

FIGURE 3.29 Synoptic chart for 16–17 October 1987

Additional pieces of information about the 1987 storm are:

- 19 people lost their lives
- 19 million trees were blown down
- almost 5000 km of telephone lines were brought down
- many schools were shut
- Chichester Cathedral suffered serious damage
- the London Stock Exchange ceased trading
- a Channel ferry ran aground on the English coast
- an oil rig in the southern part of the North Sea broke free of its moorings
- insurance claims totalling £1.5 billion were made.

ACTIVITIES

1 a) Account for the difference in number of fatalities between the 1987 and 1990 storms.
b) Note down the most common types of damage and disruption caused by depression winds of hurricane intensity over Britain, providing additional factual evidence as appropriate.

Study 3.5 Granite landscapes – Dartmoor, Devon and south Cornwall

Tors are hill-top exposures of resistant, impermeable rock. They may form isolated, tightly-clustered groups – as in the Pennine millstone grit outcrop at Brimham Rocks (Figure 3.30) or occur as more widely dispersed features such as the granite **tors** of south-west England (Figure 3.31).

The tors of Devon and Cornwall are merely the exposed tips of extensive **batholith** formations, connected underground as shown in Figure 3.32. Batholiths are the solidified remnants of huge magma chambers beneath extinct volcanoes. The cooling of the molten rock took place well below the surface, and this very prolonged solidification process resulted in the formation of granite – a rock composed of quartz, feldspar and mica. Most British granites were formed 290–270 million years ago. Subsequent erosion has exposed the tors to increased erosion and weathering which, together, have created a highly distinctive and deeply-etched type of upland feature. Even though at quite modest heights, the tors are so dominant in the local landscape that some eighteenth century writers thought them to be artificial features, erected by the Druids for use in religious ceremonies! Whilst more recent theories quickly disposed of such fanciful ideas, there remains considerable uncertainty as to the precise reasons for their formation and, particularly, their very weathered appearance. The matter is complicated by the fact that some tors are formed of solid bedrock, whilst others are groups of detached blocks.

One possible theory, proposed by David Linton in 1955, suggests that much of the chemical weathering of the tor structures took place during more humid (warm and damp) conditions during previous geological periods – most probably the Tertiary. The rate of weathering would have been greatest in the densely-jointed sections, whilst the more intact zones of rock between them would be so resistant to such weathering that they could form tors.

Granite is very durable, which makes it difficult to cut and shape, but this property also makes it a highly desirable building material, where the ability to resist storms and the ravages of weathering is far more important than cost. Many castles, coastal churches and older public buildings in south-west England have exterior granite walls for that very reason. Aberdeen, on Scotland's exposed north-eastern coast, has so many granite-construction buildings that it became known as 'the granite city'.

Although granite is very hard and durable it can, under certain conditions, break down into its component elements. The process of conversion from hard rock into a soft matrix is known as kaolinisation; the quartz and mica components remain relatively unchanged, but the feldspar is transformed into kaolinite. This process occurs naturally in underground, funnel-shaped deposits up to 300 m deep, which narrow with increasing depth. The most dense concentration of kaolinite pits in Britain is in the St. Austell district of south Cornwall.

Kaolinite usually consists of up to 15 per cent china clay, 10 per cent mica and at least 75 per cent quartz. Ten tonnes of material is quarried for every net tonne of clay needed! It is quarried by first removing the overburden layer of unwanted surface material, which may be 1–15 m thick, then dislodging the exposed kaolinite with high-pressure water hoses. The slurry produced by this unusual form of **hydraulic quarrying** passes through sedimentation, filtration and drying processes until a pure, china clay powder is obtained.

Almost 90 per cent of all the china clay produced in Cornwall is exported and the industry injects over £130 million annually into the local economy. Although the kaolinite quarrying landscape is barren and almost lunar in appearance (Figure 3.33), the Cornish accept that disfigured landscapes are a small price to pay for an industrial activity which employs approximately 3000 people, in a region where the collapse of fishing and tin mining has made it almost totally dependent upon income generated by holiday visitors and retired incomers.

The most important industrial uses of china clay are paper-making and the pottery industry, particularly in the manufacture of tableware, toilet fittings, tiles and porcelain fittings such as electrical plugs. Other important by-products are cement, textiles, dyes and rubber products.

FIGURE 3.30 Brimham rocks, North Yorkshire

FIGURE 3.31 A tor, Darmoor

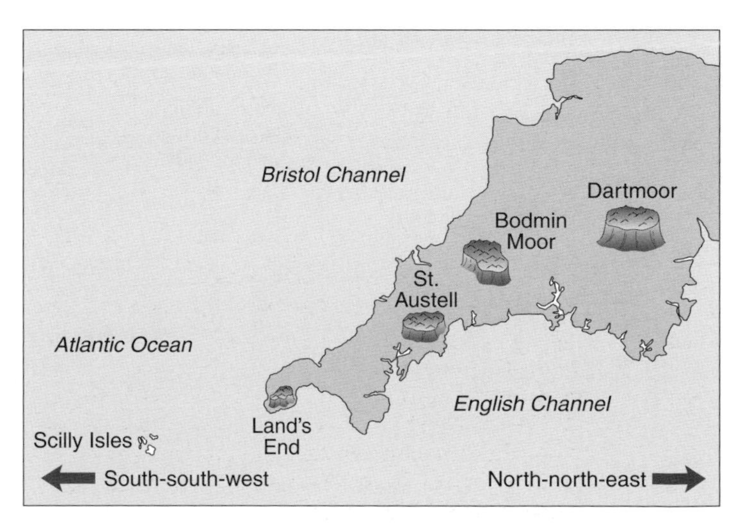

FIGURE 3.32 Batholith system, south-west England

FIGURE 3.33 The effects of china clay quarrying on the landscape of south Cornwall

ACTIVITIES

1 a) Summarise the importance of the china clay industry to south Cornwall.
b) Suggest various ways in which your daily, early-morning routine might involve a wide range of china clay by-products.
2 a) With the help of Figure 3.30, describe the appearance of the millstone grit tors at Brimham Rocks.
b) Use Figure 3.31 to compare the appearance of a typical granite tor with the formations at Brimham Rocks.
c) Account for any differences in appearance which you have outlined in part b) of this activity.
3 In what ways do china clay quarrying processes appear to be:
a) wasteful?
b) environmentally damaging?
c) less environmentally damaging than the more traditional methods used to quarry other industrial raw materials?

FIGURE 3.34 Structure of Carboniferous limestone

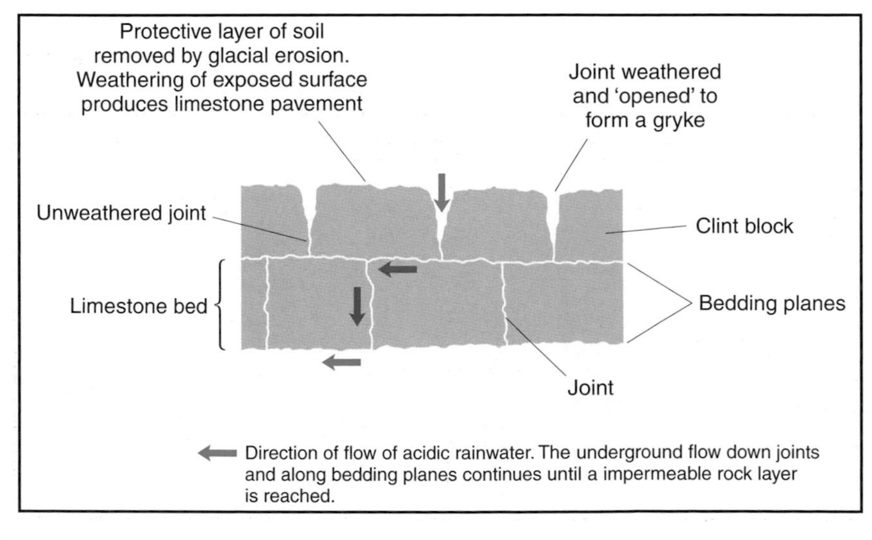

Study 3.6 Carboniferous limestone landscapes – Yorkshire Dales National Park

Limestones are a group of sedimentary rocks rich in calcium carbonate ($CaCO_3$). Limestones form beneath water – usually at the bottom of the sea. Many invertebrate, marine creatures take $CaCO_3$ from the seawater and use it to build their skeletons and shells. When these animals die, their remains fall to the bottom, the softer parts of their bodies decompose and layers of $CaCO_3$ fragments build up above the seabed. Over time, and given the correct conditions, these remains will then be compressed and compacted into solid rock.

The rocks of Britain include four major limestone types:

- Carboniferous limestone, which is hard, grey-coloured, well-jointed and contains many fossils.
- Chalk (Cretaceous limestone), which is a very pure form of limestone composed of coccoliths (minute calcite discs from the skeletons of algae). It is usually referred to as 'chalk' rather than as a particular type of limestone.
- Magnesian (Permian) limestone, which contains a higher proportion of magnesian carbonate than other limestones.
- Oolitic (Jurassic) limestone, which is composed of small, lime-coated spheres called ooliths.

The rest of this study examines the processes and features associated with just one of these types of limestone – Carboniferous limestone – within the context of the Pennine uplands of the Yorkshire Dales National Park. Carboniferous limestone forms especially distinctive landforms because of its permeability and structural characteristics (Figure 3.34). Being permeable means that water is able to pass through the rock – but only via the planes of weakness within it, notably the near-horizontal bedding planes between the rock layers, and joints which tend to be aligned vertically within the layers. There is much debate about the formation of these joints: some suggest that they occurred as the rock dried out on being uplifted out of the sea in which it formed; others favour the idea that they are fractures, caused by Earth movements at the end of the Carboniferous period. Limestone is a rigid rock, which may have cracked, instead of being folded, as

$H_2O + CO_2 + CaCO_3 = Ca(HCO_3)_2$
where:
H_2O is rainwater.
CO_2 is carbon dioxide gas in the air. This dissolves in the rainwater and turns it into weak carbonic acid.
$CaCO_3$ is calcium carbonate (limestone). This is insoluble in water.
$Ca(HCO_3)_2$ is calcium hydrogen carbonate, formed when the carbonic acid acts on the limestone and makes it soluble.

FIGURE 3.35 The formation of calcium hydrogen carbonate

a result of changing pressures within the Earth's crust.

Although initially pure, rainwater tends to absorb carbon dioxide during its descent to the Earth's surface, producing a very weak form of carbonic acid. Limestone does not react with pure water, but it can lead to a chemical reaction with even quite dilute strengths of carbonic acid (Figure 3.35). The result of this chemical reaction is to produce calcium hydrogen carbonate which, unlike the original limestone, is soluble and so can be dissolved and transported by rainwater.

The features most at risk from the acidic solution process are the lines of weakness mentioned earlier – the bedding planes and the joints. Solution of the bedding planes can lead to the collapse of large sections of bedded limestone, whilst the widening and deepening of joints produces **grikes** (also known as **grykes**). If the limestone is below layers of soil and vegetation, the additional acidity produced by vegetation decay and the waste produced by grazing animals will accelerate the solution process. One common effect of the solution process is the formation of surface depressions, called shakeholes. Erosion by a **glacier** or by meltwater may have completely removed any surface material and exposed the uppermost limestone layer to direct solution activity by rainfall. The result is a limestone pavement formation consisting of more resistant **clint blocks** separated by ever-deepening and widening grikes. Wind-blown soil is often deposited in grike bottoms, where sheltered, micro-climate locations allow ferns and hardy grasses to survive. The upper surfaces of the clint blocks are particularly vulnerable to solution weathering, with the consequences shown in the close-up pavement view in Figure 3.36.

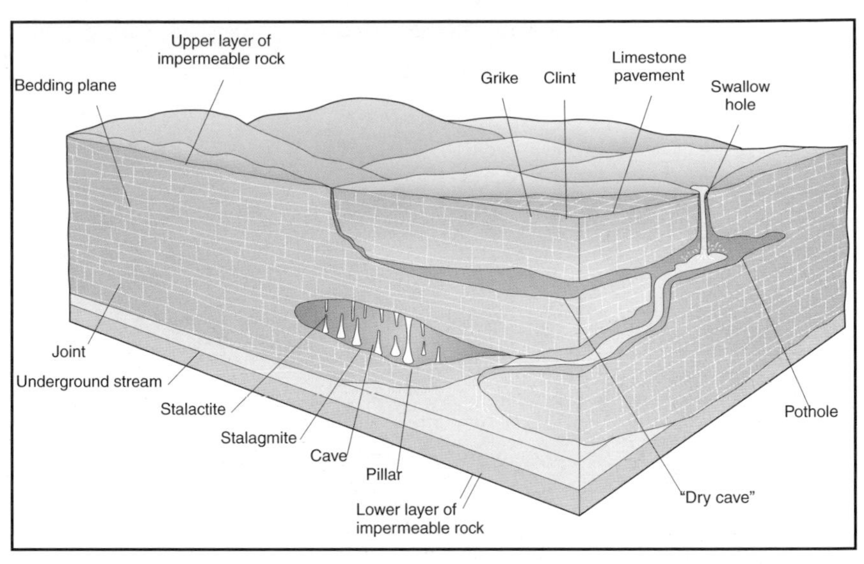

FIGURE 3.37 Drainage in carboniferous limestone scenery

FIGURE 3.38 The entrance to White Scar Cave, Yorkshire

FIGURE 3.36 Limestone pavement

FIGURE 3.39 Stalactite, stalagmite and pillar (column) formation

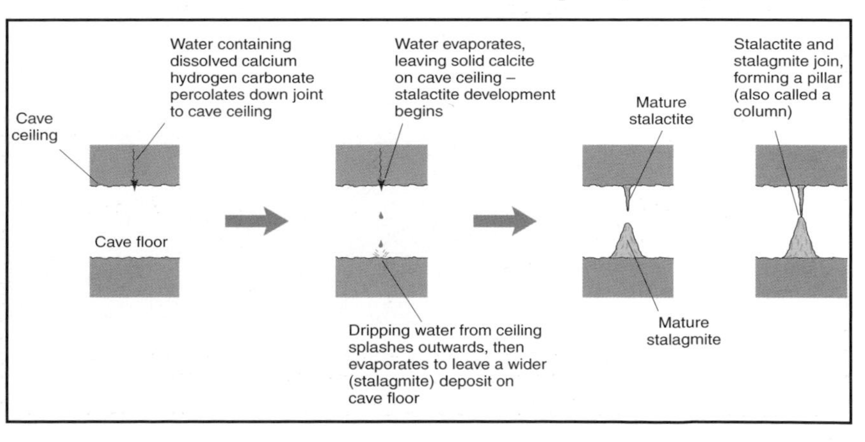

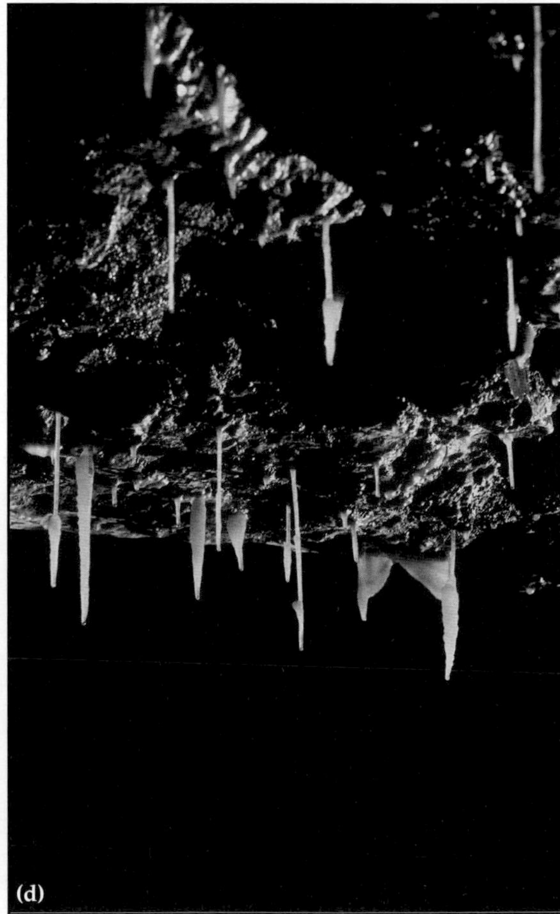

FIGURE 3.40 Internal features of White Scar Cave (a) Judge's Head (b) Arum Lily (c) Flow stone (d) Carrot stalactites

The mildly acidic water travels underground via the bedding planes and joints, until reaching a layer of impermeable rock such as slate. If there is a sufficient volume of water on this downwards passage, streams will be formed and these, in turn, will erode and dissolve the surrounding limestone to produce caves and cavern systems; these systems are invariably located immediately above the limestone/impermeable rock interface (Figure 3.37). White Scar Cave in Yorkshire was formed in this way – at the unconformity where Carboniferous limestone lies above the much older slate beds below it (Figure 3.38).

The underground water may become so saturated with dissolved calcium hydrogen carbonate that evaporation is able to take place within the caves. When this happens, the sequence shown by the formula in Figure 3.35 is reversed, which means that the dissolved calcium hydrogen carbonate becomes solid calcium carbonate, usually referred to as calcite. This re-solidified rock may take many forms. If the evaporation takes place on dripping water from a cave ceiling, the drips will form thin **stalactites** and much broader **stalagmites** below them. Figure 3.39 explains this process in greater detail, and shows how both of these features may eventually link to form **pillars** (also called **columns**). Much of the saturated water will flow down the sides of the cave instead of dripping from its roof – evaporating as it does so to form curiously-shaped flowstone features, like the examples from White Scar Cave, on the north-western flank of Ingleborough Hill (Figure 3.40).

Some underground streams are not the result of water percolating below ground, but are actually surface streams which have flowed from an area of impermeable rock onto an adjacent area of permeable limestone. The point where the stream disappears underground is called a **swallow hole**, also known as a **sink hole** (Figure 3.41) and it is often quite easy to track the alignment of such a rock change at the surface by following its line of swallow holes. Similarly, a change of rock from permeable limestone to an impermeable bedrock will force any underground streams to re-surface, when they become known as **resurgent streams**.

The last Ice Age produced a number of striking effects in Carboniferous limestone districts, apart from the scouring of upper rock layers which led to the pavement formations described earlier. Glaciers tended to enlarge existing river valleys, exposing limestone ridges on either side. The view down the Greta Valley shown in Figure 3.42 is a classic example of this type of feature. It exhibits an almost perfect symmetrical valley cross-section profile and its valley sides are lined with scree slopes composed of rock fragments – the result of freeze-thaw action on the limestone scars which have been fully exposed since the retreat of the ice some 10 000 years ago. Ice Age conditions froze underground as well as surface water, one of the results of this underground freezing being the creation of the vast Battlefield Chamber (Figure 3.43) within the White Scar Cave complex.

FIGURE 3.41 A swallow hole

Carboniferous limestone outcrops at the surface in much of central Scotland and northern England (Figure 3.44). Probably the best-known example of such outcrops within Britain are around the Yorkshire Dales National Park village of Malham. The landscape of the Malham area is the direct result of an unusual combination of geological history and a long period of more recent denudation.

About 345 million years ago, Britain was much closer to the Equator. Most of what is now northern Britain was then on the edge of a large continent, whilst the whole of southern Britain was covered by sea. The sea level was gradually rising and so submerging the continental edge. The advancing, shallow sea was rich in marine life and thick layers of calcium carbonate accumulated on the sea floor. Coral reefs similar to those on the Great Barrier Reef described in Study 2.8 developed and these have since become compacted by later deposits of sediment to produce 'reef limestone'.

Another important factor in the geological development of the northern Pennines was the existence of deep faults, which allowed extensive parts of the area to move vertically relative to each other. Subsequent Earth movements caused the land to the north of each of these faults to be uplifted. Crustal movements uplifted the entire area above sea level and exposed it to denudation which has since removed the more recent, upper layers of rock to lay bare the Carboniferous limestone beds below them.

The most impressive limestone (or karst) landscape feature in northern England is Malham Cove (Figure 3.45), a 70 m high sheer cliff of limestone in a curving, concave alignment. Dry valleys are less spectacular but a much more common feature of limestone landscapes; they are the result of a combination of the following factors:

- ice erosion
- meltwater erosion at the end of the last Ice Age

FIGURE 3.42 The symmetry of the Greta Valley cross section

FIGURE 3.43 The Battlefield Chamber, White Scar Cave

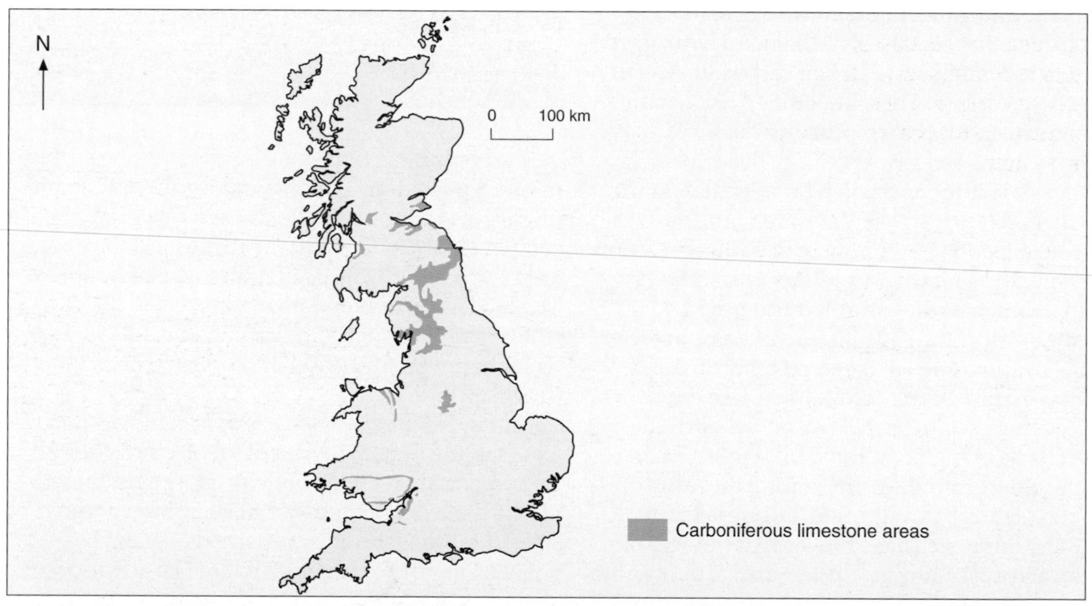

FIGURE 3.44 Carboniferous limestone outcrops in Britain

FIGURE 3.45 Malham Cove

- drier climatic conditions
- lowering of the water table beneath the valley floor.

Limestone pavements are a common surface feature of the Malham area, a particularly impressive example lying on the upper surface of the cove itself. The Goredale Scar gorge (Figure 3.46) is a meltwater-formed feature which contains a stream which could not possibly have eroded such a large valley. As in the case of the Greta Valley, Goredale Scar is lined with long, horizontal scars covered at intervals by large fans of scree fragments. To the north of the village lies Malham Tarn (Figure 3.47) – an unlikely feature in a permeable limestone region. The existence of this surface lake is due to a small outcrop of impermeable slate – another consequence of Earth movements along the North Craven Fault. Underground systems do exist in the Malham area, but access to them is restricted to experienced potholers; visitors wishing to experience karst cave scenery must go to locations further west within the Pennine system.

FIGURE 3.46 Goredale Scar

ACTIVITIES

1 a) Describe briefly how and when specific types of limestone rock were formed.
b) Describe the distribution of Carboniferous limestone deposits within Britain. Use an atlas to name the most important deposit locations.
c) Explain the geological significance of the chemical reactions by which calcium carbonate and calcium hydrogen carbonate are interchanged.
2 Describe the significance of 'permeability', with respect to Carboniferous limestone rock.
3 Investigate the chief differences between the 'erosion' and 'denudation' processes.
4 With reference to named examples, explain how both surface and underground Carboniferous limestone features are formed.

FIGURE 3.47 Malham Tarn

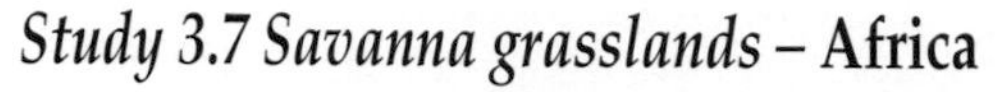

Study 3.7 Savanna grasslands – Africa

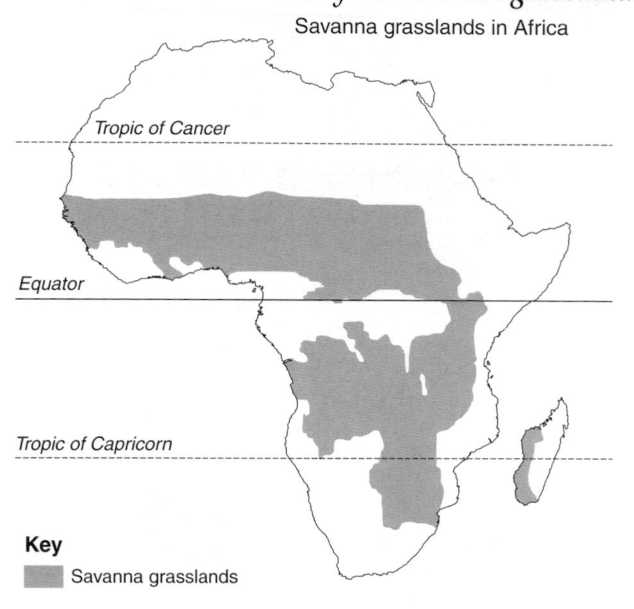

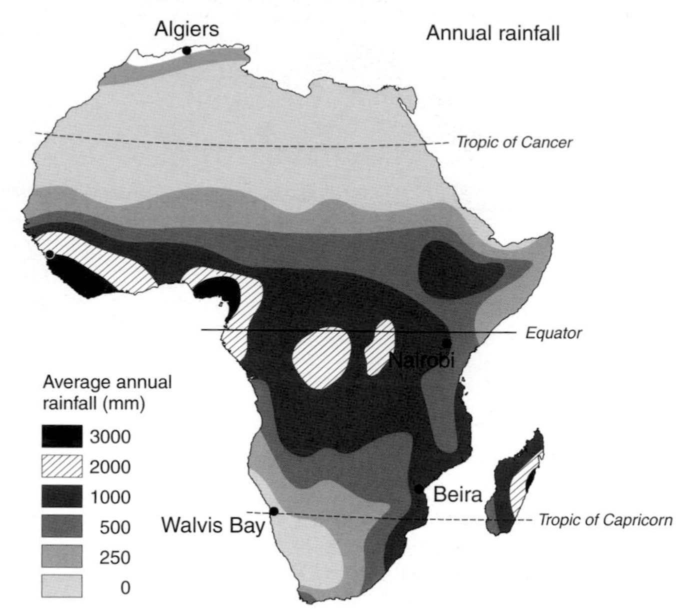

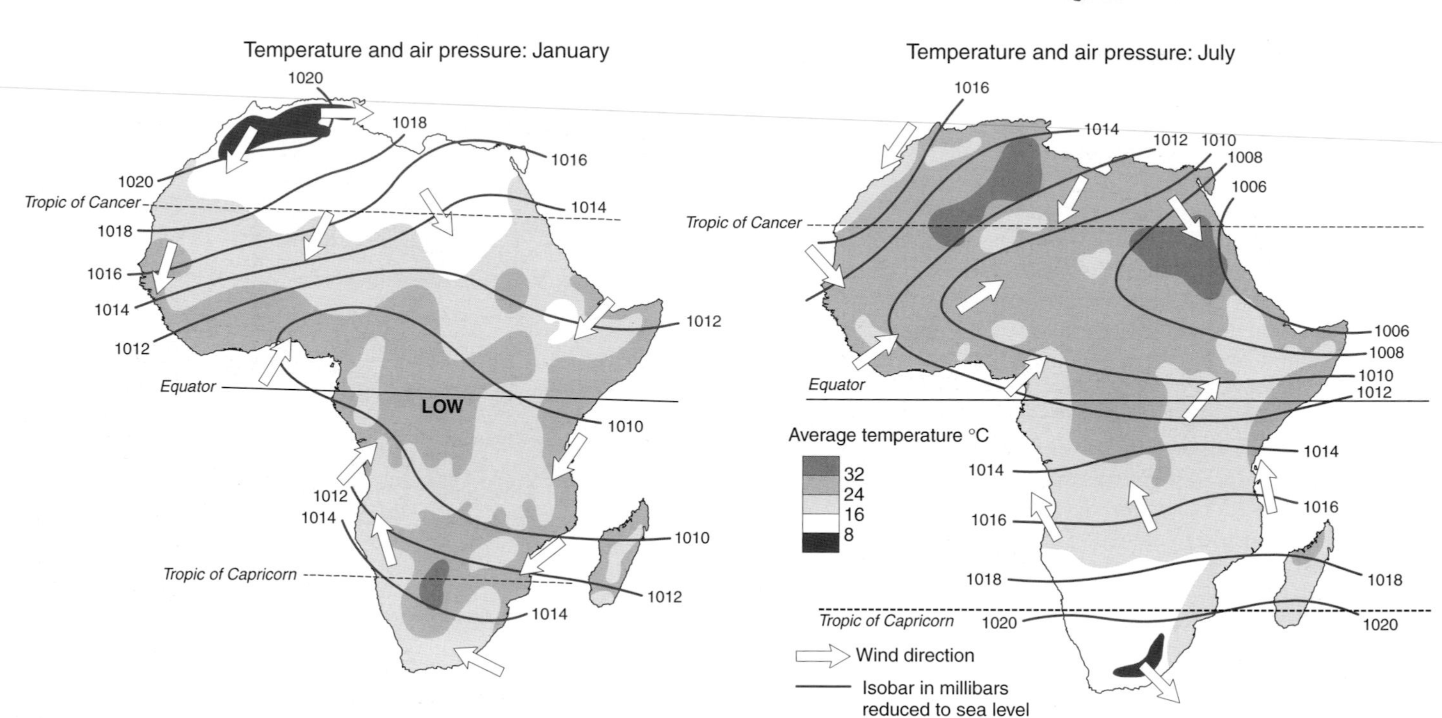

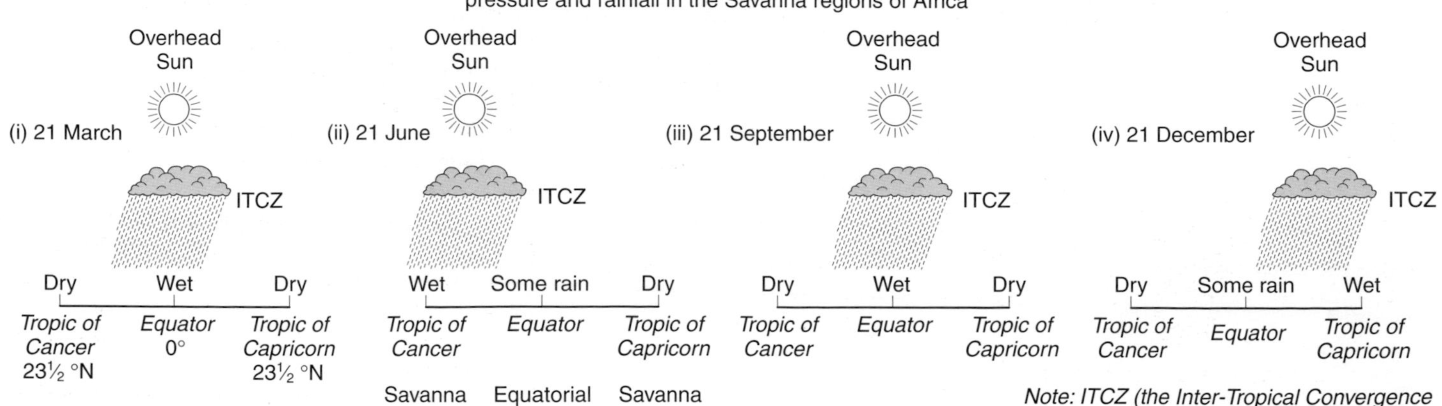

FIGURE 3.48 The African savanna biome

Note: ITCZ (the Inter-Tropical Convergence Zone) is where the opposing north-east and south-east trade winds meet

Desertification within Africa

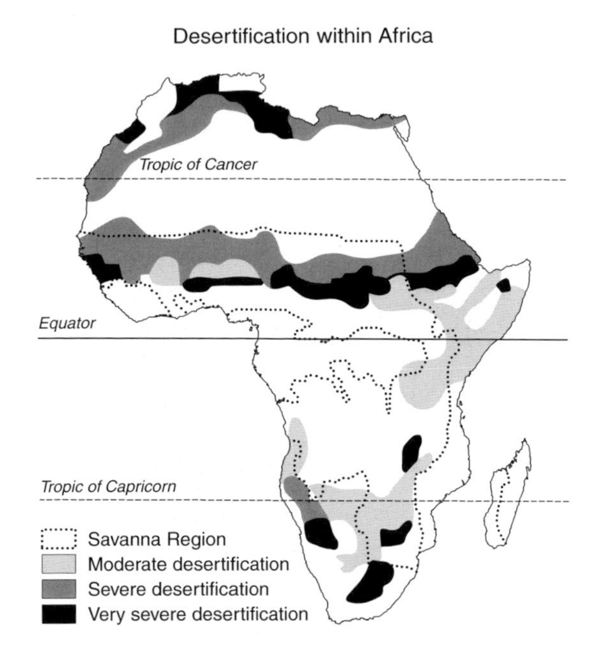

Soil erosion within Africa

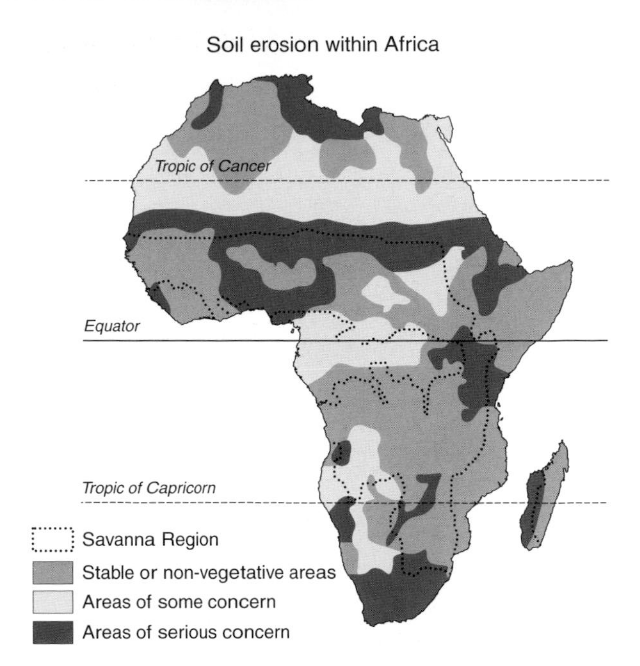

The effect on Africa's Natural Systems of exponential population growth

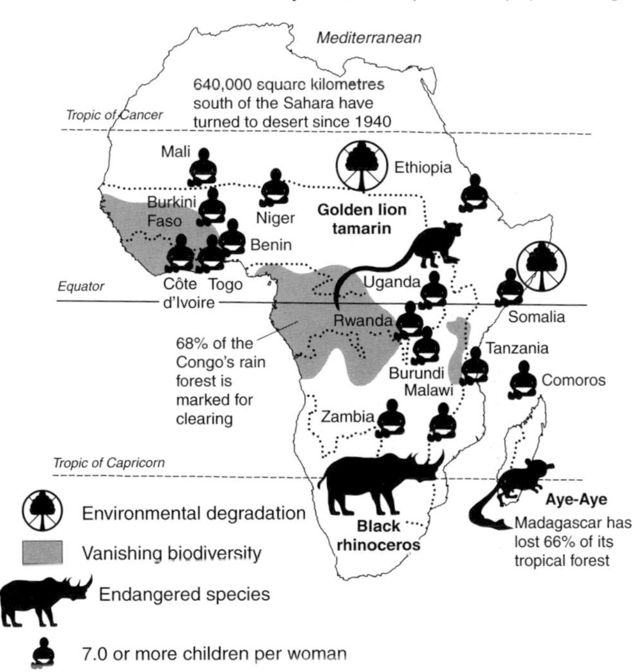

Variations in the elephant population of Africa

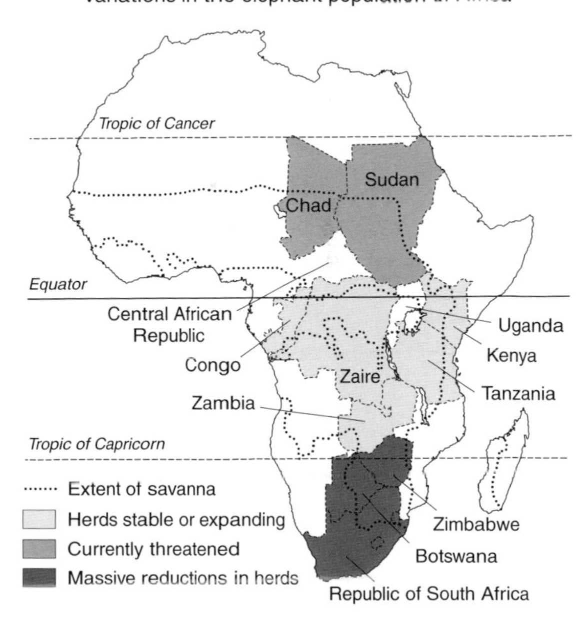

Soil moisture balance for African savanna south of the Equator

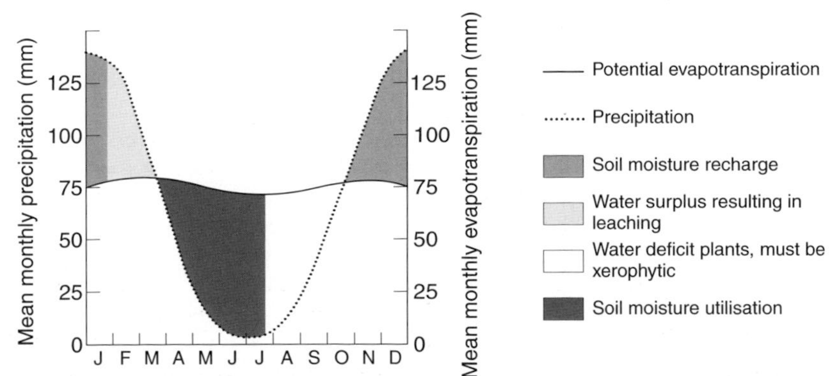

ACTIVITIES

Using the data in Figure 3.48 and the section on grassland biomes (pages 87–91) answer the following questions:

1 Describe and give detailed explanations for the distinctive characteristics of the African savanna biome.

2 'Human interventions in the savanna biome of Africa, coupled with increasingly unreliable rainfall, threaten both the natural ecosystems and continuing human survival in the region.' Discuss this statement with reference to specific named examples.

Unit 4
Natural environments 2

Ice formation and glacier characteristics

'Ice' is formed from snow, but only after passing through the sequences illustrated in Figure 4.1. On reaching the final stage in the sequence, ice tends to congregate in large masses, the nature of which depends entirely on the local relief and climate characteristics. These masses range in size but usually take the form of **ice sheets** or glaciers (Figure 4.2).

FIGURE 4.1 Ice formation

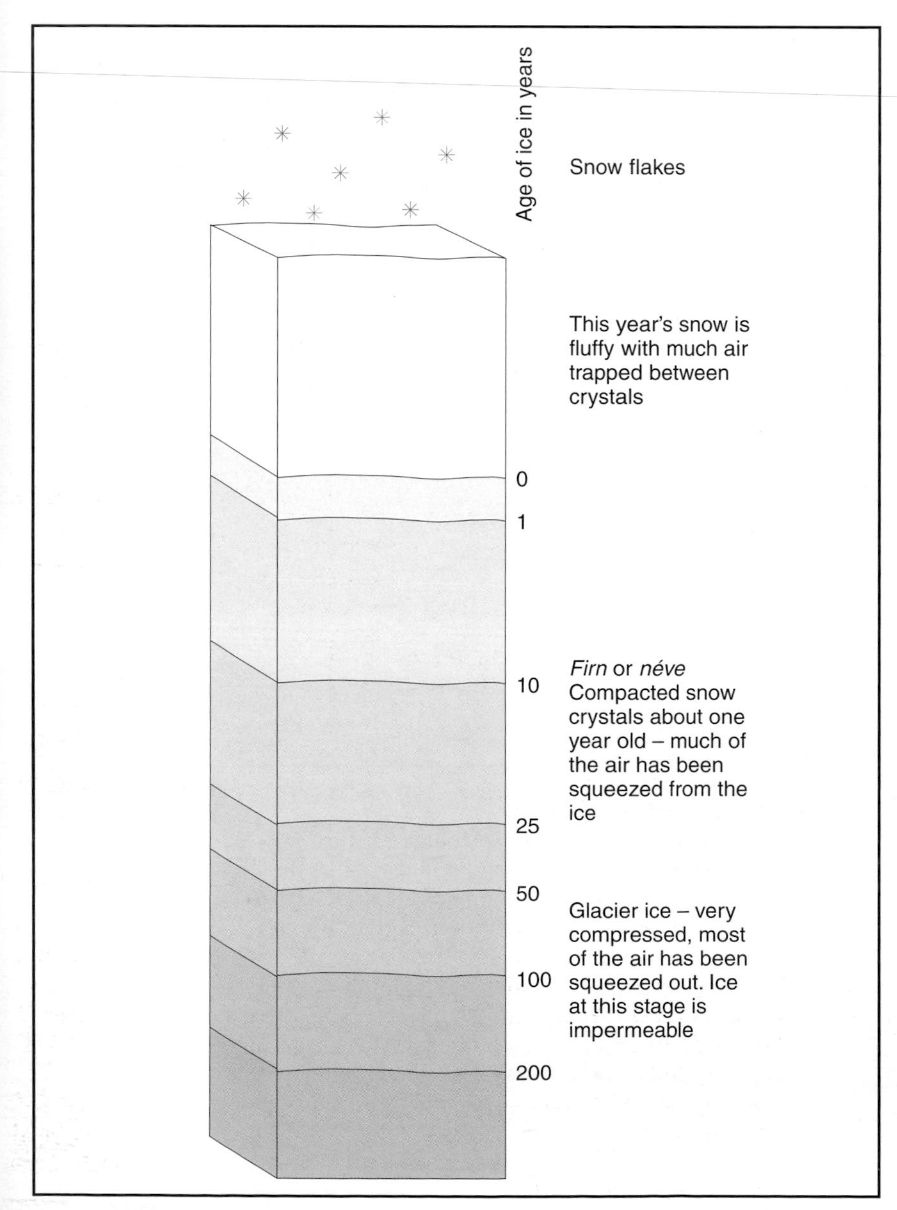

Major glaciers tend to be one of two types, although it is possible for a glacier system to exhibit the behaviour and features of both types if it passes from a zone which remains frozen throughout the year (due to its air temperature being permanently below freezing point) to another where higher temperatures lead to a seasonal pattern of surface thaw followed by re-freezing. The main contrasts between the two types of glacier are displayed in Figure 4.3.

All glaciers are climate-sensitive, but valley glaciers are vulnerable to even modest fluctuations in temperature. The 'Little Ice Age', approximately 1000 years ago (Figure 4.4) was so intense that it froze swift-flowing tidal rivers such as the Thames so deeply that people could walk across them in complete safety. The most recent phase shown on Figure 4.4 appears to be one of consistently rising

- **Cirque glaciers** are up to hundreds of metres in size. They develop from snow patches in hollows into armchair-shaped basins.
- **Ice caps** are small ice sheets, less than 50 000 km^2, which are sufficiently deep to bury the landscape.
- **Ice sheets** are more than 50 000 km^2 and are also capable of burying the landscape. The only examples are the Greenland and Antarctic ice sheets.
- **Ice shelves** are floating extensions to the Antarctic ice sheet; their thickness is greatest close to the land, where it may exceed 1000 m.
- **Niche glaciers** are usually tens of metres in size, and are found on steep slopes. They originated as snow patches and remain as small wedges of shallow ice.
- **Piedmont glaciers** occur where glaciers advance out of a mountainous region into a wider, lowland region.
- **Valley glaciers** are formed when ice moves out of a cirque into a pre-existing valley.

FIGURE 4.2 Types of ice mass

	Polar (cold) glacier	Temperate (warm) glacier
Location	Very cold regions in which no ice melt is possible (e.g. Antarctica); the relief is usually relatively gentle	Upland localities where summer temperatures are sufficiently high to allow some ice melt to occur (e.g. Alps); the relief is much steeper and includes glaciated valleys
Temperature profile	Winter; –30 –20 –10 0 +10 °C; Surface of glacier; Summer; PMP; Increasing depth; Base of glacier. Temperature at base of cold glacier is well below PMP (pressure melting point). Little or no meltwater beneath glacier prevents the glacier moving freely	Winter; –30 –20 –10 0 +10 °C; Surface of glacier; Summer; PMP; Increasing depth; Base of glacier. Temperature at base of temperate glacier is about the same as PMP. Meltwater beneath glacier can either be permanent or seasonal allowing the glacier to move freely (less friction)
Velocity profile	Ice surface; Internal flow; Basal flow; Glacier depth (m); Speed of flow; Base of glacier	Ice surface; Internal flow; Basal flow; Glacier depth (m); Speed of flow; Base of glacier (pressure melting point)

FIGURE 4.3 Comparison between the two main types of glacier

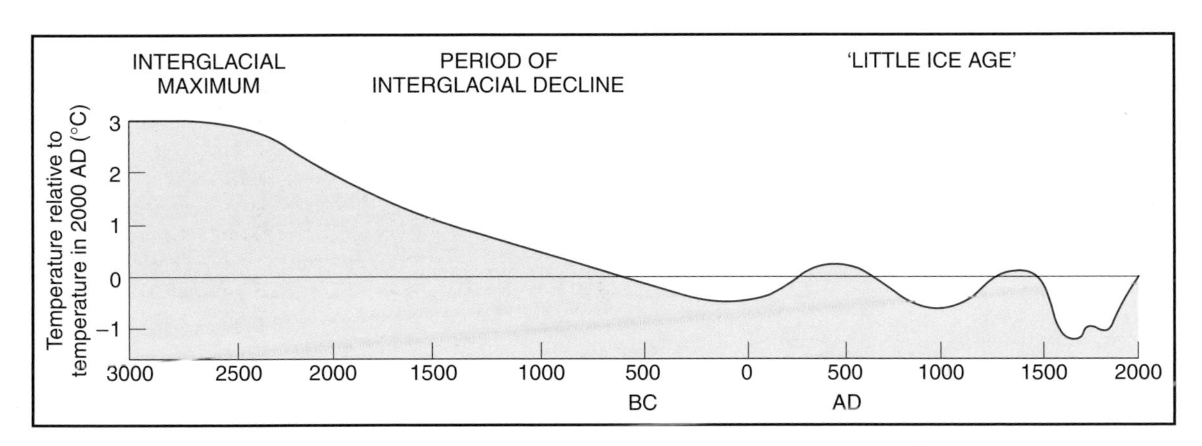

FIGURE 4.4 The Little Ice Age

global temperatures and observers now routinely monitor the retreat of glacier snouts throughout all the major high altitude and high latitude regions. The following data illustrate the extent of the trend:

- Mount Kenya lost 92 per cent of its ice mass during the twentieth century.
- Mount Kilimanjaro's glaciers have shrunk by 73 per cent during the twentieth century.
- Glaciers in both the European Alps and the Caucasus Mountains have decreased in volume by 50 per cent during the twentieth century.
- Glaciers in the Tien Shan Mountains have decreased in volume by 22 per cent within the last 40 years.
- Glaciers on South Island, New Zealand have decreased by 26 per cent since 1890.
- The number of glaciers in Spain dropped from 27 to 13 between 1980 and 2000.

Glaciers are highly complex natural features and every glacier is unique. However, there are a number of basic principles which determine glacier formation and the impact which they have on the valleys containing them. Figures 4.5 and 4.6 illustrate the most important processes involved and the erosion and deposition features which result from them.

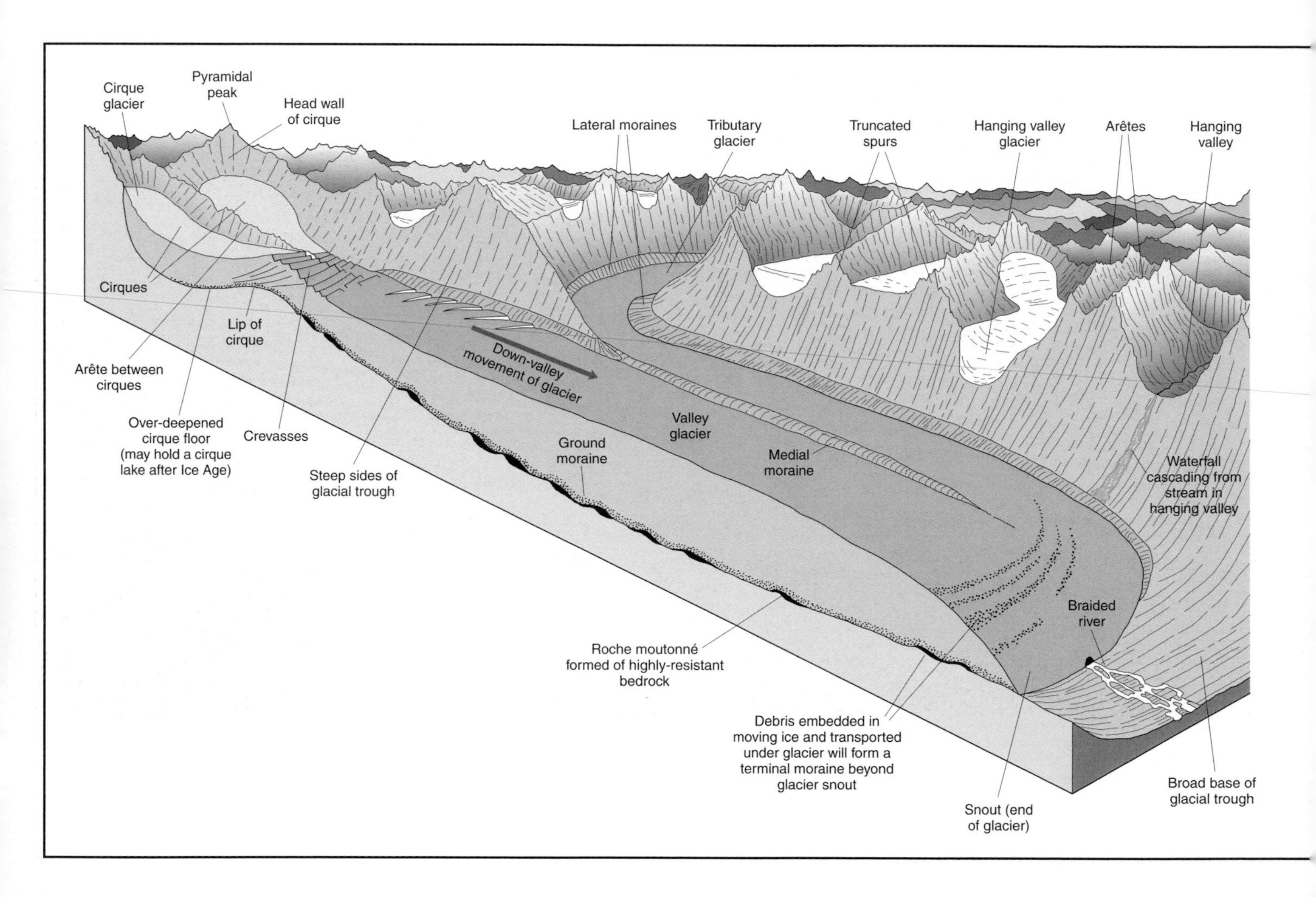
Cirque glacier
Pyramidal peak
Head wall of cirque
Lateral moraines
Tributary glacier
Truncated spurs
Hanging valley glacier
Arêtes
Hanging valley
Cirques
Lip of cirque
Arête between cirques
Down-valley movement of glacier
Over-deepened cirque floor (may hold a cirque lake after Ice Age)
Crevasses
Steep sides of glacial trough
Ground moraine
Valley glacier
Medial moraine
Waterfall cascading from stream in hanging valley
Braided river
Roche moutonné formed of highly-resistant bedrock
Debris embedded in moving ice and transported under glacier will form a terminal moraine beyond glacier snout
Snout (end of glacier)
Broad base of glacial trough

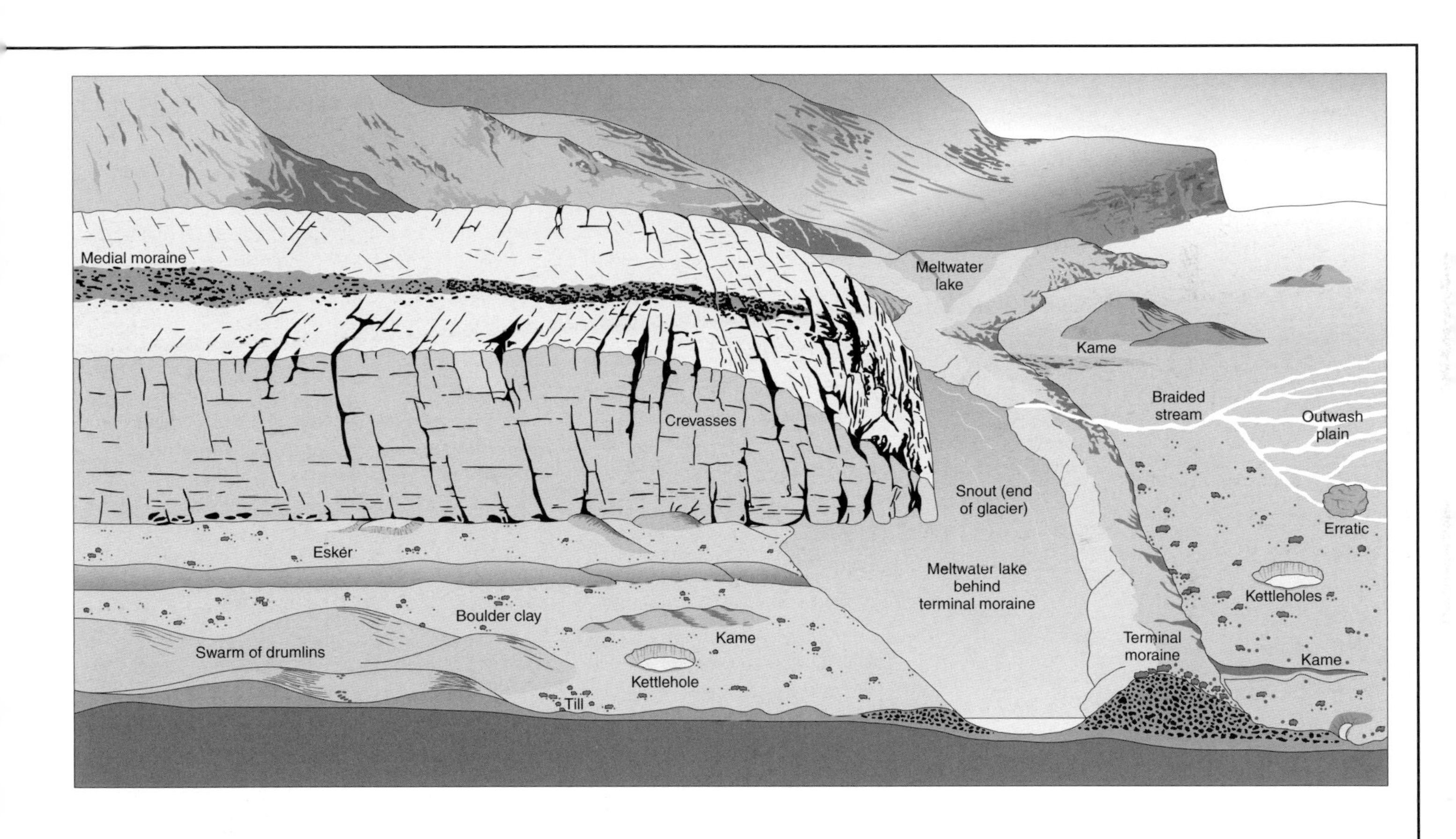

FIGURE 4.5 Glacial landforms

FIGURE 4.6 Erosion and deposition features of glaciation

Glacial erosion feature	Description of feature
Arête	Narrow, knife-edged ridge between two adjacent cirques.
Cirque	Large amphitheatre-shaped depression on a mountain side; has a steep back wall and (often) a cirque lake.
Glacial trough	Steep sided, straight and flat-bottomed U-shaped valley.
Hanging valley	Tributary valley perched high above the level of the floor of the main glacial trough; formed by a tributary glacier whose erosive power was significantly less than that of the main glacier; streams in hanging valleys often descend to the floor of the main glacial trough in the form of waterfalls.
Pyramid peak	Pointed peak below which are several outward-radiating arêtes.
Ribbon lake	Long, narrow lake occupying an over-deepened valley bottom.
Roche moutonnée	A rock mass, the upstream side of which is smooth and striated; the downstream side is much steeper.
Striations	Parallel scratch marks caused by glacial abrasion of bare rock surfaces.
Truncated spurs	Interlocking spurs whose valley ends have been removed by glacial abrasion.
Drumlin	Smooth, elongated mounds deposited by ice; their long axis is parallel to the direction of the ice flow.
Erratic	Rock transported by ice and then deposited in an area of different geology to that of its source.
Esker	Fluvioglacial feature consisting of a sinuous ridge of silt, sand and gravel deposited by meltwater flowing in sub-glacial channels or tunnels within the ice mass.
Kame	Fluvioglacial delta-like deposits of sand and gravel formed by melt water along the front of a stationary or slowly-receding glacier or ice sheet.
Kettlehole	Depression in which a small lake often occurs. Formed as a result of a block of ice, which has become detached from the main glacier or ice sheet, being covered by deposited debris before melting and leaving a hollow in the ground surface.
Lateral moraine	Deposited material found along the surface edge of a valley glacier; some of its material is frost-shattered fragments.
Medial moraine	Formed in the centre of the surface of a main glacier – the result of two lateral moraines joining where two tributary glaciers meet.
Outwash plain	Lowland area over which layers of debris have been deposited by meltwater streams flowing from a glacier or ice sheet; the larger particles tend to be deposited nearest to the meeting ice mass.
Terminal moraine	Material deposited at the front of a glacier where its snout remained stationery for a long period of time; terminal moraines often act as natural dams across valleys and help to create ribbon lakes.
Till	Layer of unsupported material deposited directly below a glacier; the material size can range from fine clay particles to large boulders.

ACTIVITIES

1 Describe any links which exist between each of the pairs of processes and features listed below. When doing this, make it clear that you understand the precise meaning of each term.

- Abrasion and plucking.
- Ablation and accretion.
- Medial moraine and terminal moraine.
- Snout and meltwater.
- Snow and firn.

2 a) With the help of an atlas, locate each of the glacier-depletion locations quoted in this section.
b) Consider whether or not these locations appear to indicate any global trends.
c) Suggest possible reasons for any trends you have just identified.

3 a) Divide into groups, each of which will be required to become thoroughly familiar with one particular aspect of the information contained in this section (e.g. ice formation).
b) Each group will then introduce its allocated topic to the other members of the whole class.

Study 4.1 Hazards in polar and high altitude regions – Antarctica and the Alps

The higher altitudes of most mountain ranges and both polar regions are subject to similar kinds of hazards. Sea level temperatures in polar regions are consistently lower than those in other parts of the world. Additionally, air temperatures decrease with increasing altitude – as any experienced climber will quickly verify. Air density is usually highest close to sea level, because that is where the air pressure is greatest, due to the increased density of air molecules. As every molecule of air contains the same quantity of heat energy, the presence of more molecules in the air will inevitably lead to warmer air at lower altitudes. The decrease in temperature with increasing altitude is called the **adiabatic lapse rate** and can take place at two different rates. If the air is particularly dry, the rate of temperature decrease will be 1 °C per 100 m increase in altitude. Moisture-laden air cools much more slowly – at a wet lapse rate of 0.4 °C per 100 m. The reverse processes are true for descending air, which will become warmer at an appropriate rate dependant on its moisture content.

Wind-chill is the term used to assess the loss of heat from the human body due to the combined effects of low temperatures and air movement. Inanimate objects such as bricks can only chill-down to the temperature of their surrounding air. Humans exposed to very low temperatures can wind-chill so severely that frostbite permanently damages their extremities and becomes life-threatening (Figure 4.7).

Polar and high altitude regions also experience low precipitation rates. This tendency is partly due to the significantly reduced moisture-holding capacity of colder air – which explains the reduced evaporation rates which apply to maritime regions dominated by cold sea currents. Polar regions are also subject to the constant descent of very cold, dry air masses which have been unable to attract moisture whilst flowing through the upper levels of the atmosphere.

All of the above climatic phenomena represent long-term hazards for the inhabitants of polar and high altitude regions. **Avalanches** represent a very different and smaller scale hazard which appears to have little connection with climatic (as opposed to local weather) characteristics. They are rarely predictable and are frequently the result of human activity. For example, a survey conducted between 1989 and 1995 by the Swiss Federal Institute for Snow and Avalanche Research concluded that 203 of the 236 avalanches reported during that period were triggered by sporting activities on Alpine mountain slopes. In a typical year, 26 people are killed by avalanches in Switzerland alone. The most severe recorded single avalanche event took place at Bloss, in the Swiss Alps, when 111 of the village's 376 inhabitants died. The worst season-long event was in early 1951, when a combination of heavy snowfalls and strong winds deposited a layer of damp, unstable snow on the existing snowfield; 130 avalanches took place, resulting in 280 deaths throughout the Austrian, French, Italian and Swiss Alps. Avalanches are a particularly destructive hazard, as wet, compact snow is very heavy and major avalanches may consist of 50 million tonnes of snow. Avalanches in Switzerland have been observed to descend at 350 km/hr.

The following information describes the nature of avalanches:

- Key factors in the occurrence of avalanches are

Wind speed (km/h)	Air temperature (°C) 45	40	35	30	25	20	15	10	5	0	−5	−10	−15	−20	−25	−30	−35	−40
Calm	45	40	35	30	25	20	15	10	5	0	−5	−10	−15	−20	−25	−30	−35	−40
8	43	37	32	27	22	16	11	6	0	−5	−10	−15	−21	−26	−31	−36	−42	−47
10	34	28	22	16	10	3	−3	−9	−15	−22	−27	−34	−40	−46	−52	−58	−64	−71
24	29	23	16	9	2	−5	−11	−18	−25	−31	−38	−45	−51	−58	−65	−72	−78	−85
32	26	19	12	4	−3	−10	−17	−24	−31	−39	−46	−53	−60	−67	−74	−81	−88	−95
40	23	16	8	1	−7	−15	−22	−29	−36	−44	−51	−59	−66	−74	−81	−88	−96	−103
48	21	13	6	−2	−10	−18	−25	−33	−41	−49	−56	−64	−71	−79	−86	−93	−101	−105
56	20	12	4	−4	−12	−20	−27	−35	−43	−52	−58	−67	−74	−82	−89	−97	−105	−113
64	19	11	3	−5	−13	−21	−29	−37	−45	−53	−60	−69	−76	−84	−92	−100	−107	−115
72	18	10	2	−6	−14	−22	−30	−38	−45	−54	−62	−70	−78	−85	−93	−102	−109	−117

Unpleasant

Frostbite likely. Outdoor activity dangerous

Exposed flesh will freeze within half a minute for the average person.

FIGURE 4.7 Wind-chill prediction

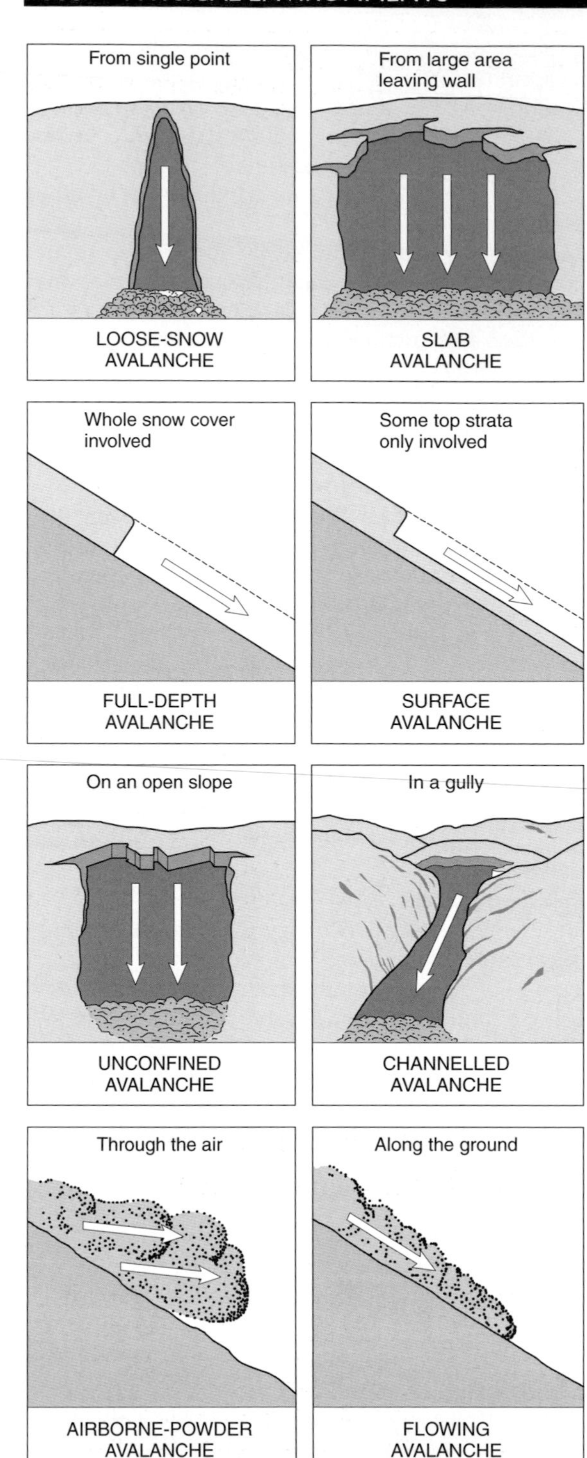

FIGURE 4.8 Types of avalanche event

slope gradient, height and depth of fresh snow, stratification characteristics of the snow layers, wind speed, air temperature and human activity.
- A combination of fresh, dry snow and strong winds creates the highest avalanche risk.
- 17° slopes are the most gentle gradients on which avalanches have been observed; it is rare for avalanches to be triggered on slopes with gradients below 30°; the most dangerous slopes are in the gradient range 30–45°.

ACTIVITIES

1 a) Explain the basic difference between the two adiabatic lapse rates.
b) Assuming that sea level temperature is 10 °C, use the dry and wet adiabatic lapse rates to calculate two separate summit temperatures for each of the following peaks:
- Ben Nevis (1343 m)
- Mont Blanc (4810 m)
- Vinson Massif (4897 m)
- Mount Everest (8848 m)

2 Calculate the wind chill, in °C, for each of the following situations:
a) Air temperature 10 °C; wind speed 30 km/hr.
b) Air temperature −15 °C; wind speed 65 km/hr.

3 List separately the various reasons for low air temperatures and low precipitation rates in polar and high altitude regions.

4 Prepare and then deliver a short talk entitled 'The nature of avalanche events'.

Study 4.2 The Arctic tundra biome – Alaska

Approximately 10 per cent of the world's total land area is Arctic tundra. As Figure 3.2 shows, the inhospitable region which supports the Arctic biome lies between the polar wastes to its north and the coniferous-forested taiga region immediately to the south. The chief characteristics of the tundra are:
- Very low temperatures and low precipitation (Figure 4.9).
- Strong winds and high wind-chill.
- A very short growing season (only 6–8 weeks), during which there is almost continuous daylight.
- Nutrient-deficient soils (Figure 4.10).
- Vegetation is restricted to a thin cover of low-growing grasses, lichens and mosses plus occasional dwarf shrubs. Leaves tend to be coated with a layer of wax, to reduce the loss of heat and moisture. Some lichens dehydrate seasonally so as

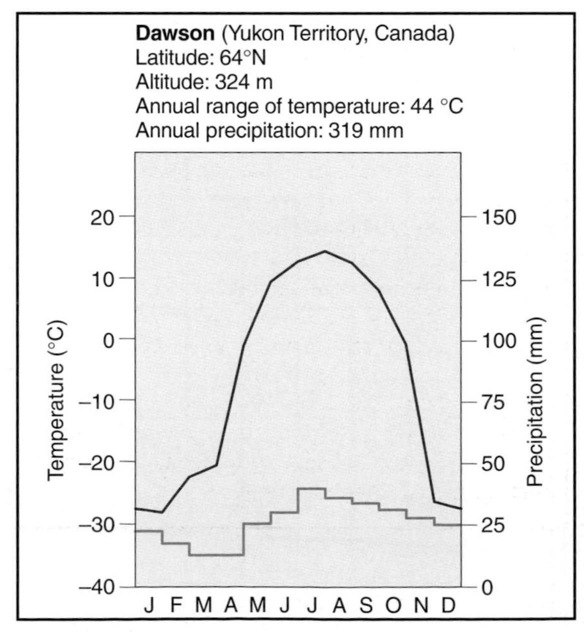

FIGURE 4.9 Tundra climate graph for Dawson, Canada

to avoid frost damage in winter. Many tundra plants store food in underground bulbs and tubers.

- Insect life is prolific in summer and blackflies and mosquitoes thrive around the seasonal, shallow pools. These insects form the principal source of food for the plentiful migratory waterfowl (Figure 4.11).
- In North America, caribou migrate each year to feed on fresh seasonal plant growth; reindeer do likewise in northern Eurasia. Predators (e.g. the Arctic fox, lynx, snowy owl and wolf) feed on the grazing mammals and much smaller species such as hares, lemmings and voles which are indigenous to this region due to their ability to burrow underground. Other forms of animal protection include thick fur, a compact body shape to minimise the area exposed to the elements and the evolution of white fur or feathers to evade detection.

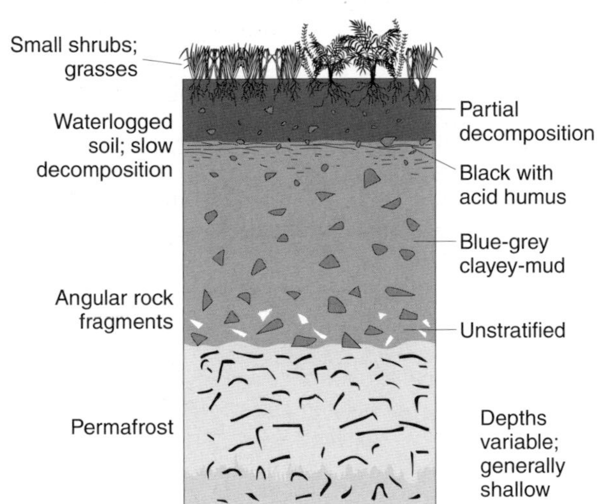

FIGURE 4.10 Tundra soil profile

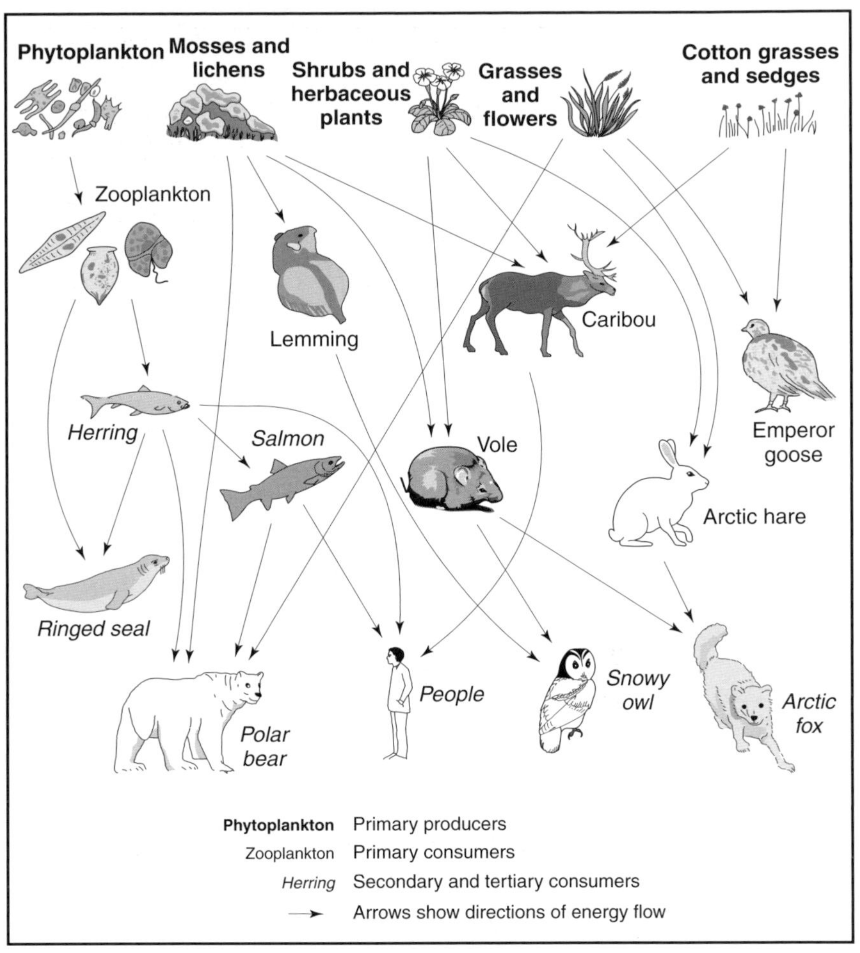

FIGURE 4.11 Tundra food web

ACTIVITIES

1 a) Use Figure 3.2 and an atlas to identify the extremes of latitude between which most of the Arctic biome occurs.
b) What variations in latitude extremes occur within and between the two continental areas?
c) What reasons for such variations might be expected to occur in:
- coastal areas?
- inland areas?

2 In what ways has the tundra's animal life adapted to:
- low temperatures and wind chill?
- low precipitation?
- other environmental factors?

3 According to Figure 4.10, in what ways are tundra soils different from those usually occurring in lowland Britain?

4 Investigate, and then write a report on, the *EXXON VALDEZ* oil taker spillage which took place in 1989 and subsequently devastated large areas of the Alaskan coastline.

Study 4.3 Nunatak ecology – Kluane National Park Reserve, Canada

A **nunatak** (the Inuit word meaning 'land attached') is an isolated mountain peak within a glacier or ice sheet (Figure 4.12). The nunataks in the Seward Glacier (1040 km^2 and 900 m deep) in Yukon Territory's St Elias Mountains are inhospitable places, exposed to hurricane-force winds and temperatures forced down by wind-chill. They emerged from the ice as recently as 12 000 years ago – too short a period for any natural evolution of wildlife to have taken place there. The bare rock faces of these nunataks are devoid of any form of life, but more sheltered parts do support a small range of plants, animals and birds. After arriving by various means, these life forms have either survived the following winter or died and so become sources of food for the few who did survive. Nunatak ecology shares some similarities with that of the tundra region as a whole, but is considerably more fragile.

Snow-buntings are one of the few birds to breed on the Yukon nunataks, which lie on an important seasonal migration route, and about 100 species of birds including sparrows, thrushes and warblers use nunataks as resting points. Many of these birds lose their way, become exhausted or can no longer fly because their wings have become iced-up. Once grounded, they cannot survive the onset of winter. It has been known for entire flocks of swans to crash into a nunatak rock face during night-time storms. Ravens are able to exist by the scavenging casualties' corpses.

It is the plant life which is the real foundation of the nunatak ecosystem. Lichens, mosses, sedges, heathers and isolated alpine flower meadows somehow manage to exist in shallow rock depressions having a southerly aspect. Poppies, daisies, dandelions, saxifrages and nitrogen-fixing oxytropis are other common plants, which exist in conjunction with insects such as moths, dragonflies and butterflies.

Pikas are the only year-round mammal residents. They obtain shelter amongst the boulders and jagged, frost-shattered rocks. Pikas are, by nature, vegetarians, but are often forced to supplement their meagre nunatak diet with dead birds – especially their brains, which provide easily digestible fat and protein. Pikas nunatak-hop their way across glaciers until they find a suitable habitat having little competition from other pikas for the scarce food available.

ACTIVITIES

1 Design a food web incorporating all the forms of wildlife on nunataks in north-west Canada.

Study 4.4 Changing economies of indigenous peoples – The Inuit of northern Canada

About 4000 years ago, groups of hunters moved across the Bering Strait to what is now Alaska, northern Canada and Greenland. About 1200 years ago, the Inuit people of Canada descended from one of these groups, the Thule. The Inuit who chose to live on the western Arctic coast of Alaska hunted whales, which were very numerous there in pre-commercial hunting times. The smaller communities which had migrated further east relied on seals, walrus, caribou and musk ox. Both groups were able to survive in such hostile environments because they acquired great confidence and patience as individuals – as well as the ability to co-operate very closely together within their hunting groups.

The Arctic had considerable military significance during the Cold War which existed between the then USSR and the NATO (North Atlantic Treaty Organisation) defence alliance. The Arctic was of supreme importance because it provided a buffer zone in which either side's inter-continental missile attacks could be detected.

A century ago, the Inuit lived a stone-age existence, hunting seals and caribou with weapons made mainly of bone and rock. In the 1950s and 60s, many of them left their hunting camps and semi-nomadic lives to live in permanent settlements with schools and clinics funded by the Canadian Government. Lack of employment has since become a major problem, with up to 50 per cent of the adult population without productive work in some areas. The result has been alcoholism, drug abuse, a suicide rate six times higher than the national average (particularly amongst younger people) as well as long-term boredom; it is perhaps not surprising that 56 per cent of the current Inuit population are under 25 years of age.

A particularly striking example of social disintegration was reported in the early 1990s. It concerned a community on the east coast of Greenland comprising 2700 Inuit and 100 Danes called Tasiilap. The Greenland 1960 Report proposed the creation of a single urban industrialised society out of previously very dispersed seal-hunting villages. The reorganised community then suffered the world's worst alcohol problem and suicide rate (seventy times higher than that in Britain). Tasiilap's other problems included a high HIV rate, wife-battering, indebtedness and declining educational standards.

FIGURE 4.12 Nunataks in the Seward Glacier, Yukon Territory

On 1 April 1999, the Inuit people celebrated a fresh start – with the creation of their own territory, called Nunavut (Figure 4.13). The main details of the agreement which led to this historic and exciting challenge for its indigenous people are listed below:

- It consists of 1994 million km^2 (one-fifth of the entire land area of Canada and over 80 times that of Britain). This land used to be part of the Canadian state of Northwest Territories and the Inuit now exercise a degree of control over mineral extraction throughout this huge area.
- 350 000 km^2 of the above land is owned by the Inuit, who have full mineral extraction rights within 10.5 per cent of the area. This is very significant, because the Inuit chose these wholly-owned areas knowing that they contain 80 per cent of Nunavat's current deposits of valuable minerals, including copper, zinc, gold and silver.
- Nunavut remains part of Canada, but its 24 000 inhabitants (85 per cent of whom are Inuit, the remaining 15 per cent being non-aboriginal) will have total control over such internal matters as education, health, social services, housing and the maintenance of law and order. Defence, finance and currency issues will continue to be the responsibility of the Canadian Federal (national) Government.
- As part of the agreement, the Inuit undertook never to make any further claims on land. To compensate them for this undertaking, the Inuit will receive the equivalent of £600 million over the next 14 years. This payment will be held in trust and gain interest which will be used for business start-up loans, student scholarship grants and income-support supplements to allow hunting families to continue their traditional way of life. This last point is very significant, because all Inuit feel genuinely happy only when living off their natural environment. In other words, the land remains at the heart of Inuit culture, in spite of the dramatic changes which have transformed the life-style of the majority of their people.
- The Nunavut Agreement gives Inuit a legal right to hunt wildlife on land and water throughout their territory. The main species to which this relates are: seals (part of the Inuit's staple diet), caribou, polar bears, muskox (now on the increase once more), beluga whales, foxes, hares and lemmings.
- Because Nunavut has a very small tax revenue base, it receives up to 95 per cent of its annual budget from central government funds.
- Nunavut has its own capital, Iqualuit.
- The language of government is the Inuit language, which is Inuktitut.

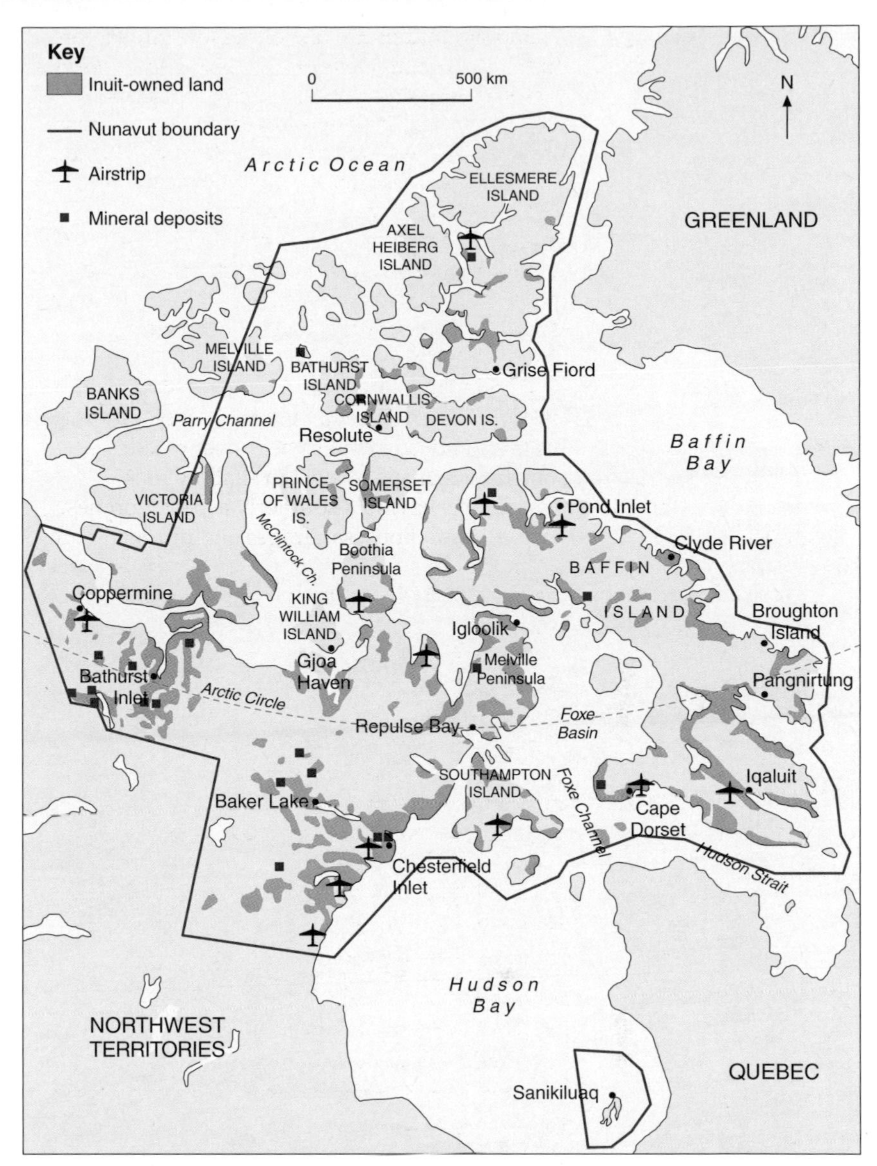

FIGURE 4.13 Nunavut

Tourism is now a major contributor to the Inuit economy, with 20 000 visitors injecting $35 million every year. Three new national parks have been created to conserve much of the natural environment and most settlements have created services such as transport facilities, accommodation, and clothes shops required by the ever-expanding group of eco-tourist visitors. Kayaking, mountain climbing, floe-edge tours and dog-team expeditions are proving to be increasingly popular attractions, although many of the younger Inuit had to learn the necessary traditional skills from scratch before qualifying as kayak instructors.

FIGURE 4.14 Antarctica

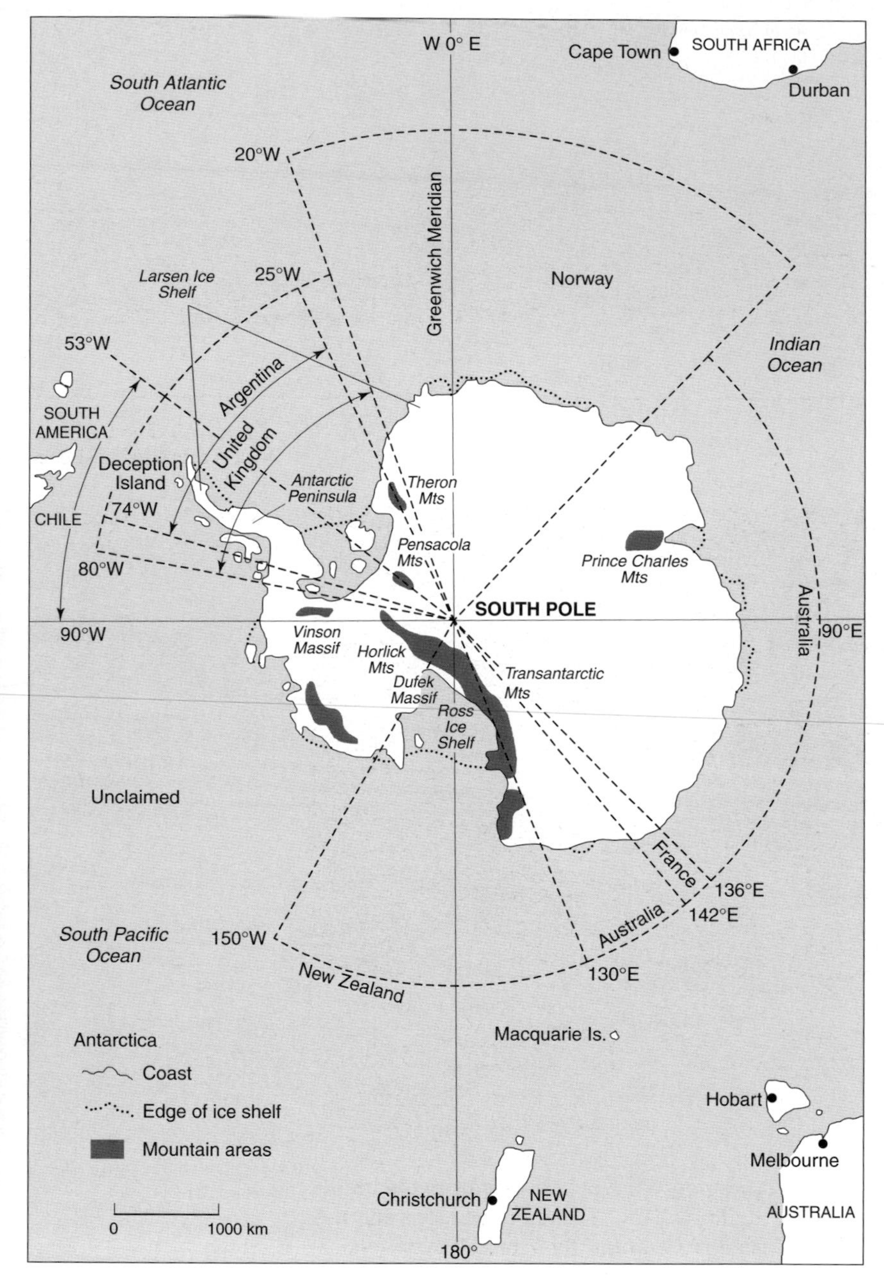

ACTIVITIES

1 Suggest reasons why Iqaluit was chosen as the site for the new capital of Nunavut in preference to:
a) Igloolik
b) Resolute.

2 In what type(s) of locations are most of the 'wholly-owned' Inuit lands situated?

3 Suggest reasons why the Federal Canadian Government was willing to make such major financial and other concessions to the Inuit as part of the 1999 Agreement.

4 In what ways might the Nunavut Agreement of 1999 be regarded as an effective example of planning for 'sustainable development'?

Study 4.5 The protection of cold wilderness regions – Antarctica

Antarctica is the world's greatest remaining 'wilderness' region, by which we mean that its current state is dictated much more by natural forces than any human influence. The following information gives you an impression of its chief natural characteristics which, collectively, have led to it being given unique protective status by the international community.

Antarctica is a true continent – unlike the Arctic, which is a relatively thin expanse of floating ice beneath which submarines can pass with relative ease. Antarctica has the highest average elevation of any continent, 2300 m, compared with only 900 m for Asia, whose average height includes Mount Everest. Antarctica's highest peak is the Vinson Massif (Figure 4.14) and two of its 70 volcanoes are still active; Deception Island last erupted in 1967, 1969 and 1970 and Mount Erebus, on Ross Island, has a permanent molten lava lake within a crater on its 3794 m summit.

Most of Antarctica is covered by an **ice cap**, which has accumulated very slowly over many thousands of years through the compacting of fallen snow. Its immense ice sheets, some of them 3 km deep, retain 70 per cent of the world's fresh water and 90 per cent of its ice. If these sheets were to melt completely, global sea levels would rise by approximately 73 m. There are two main continental ice sheets and they function in very different ways. The East Antarctic Ice Sheet is land-based and is relatively stable, in that it responds very slowly to any climatic changes. Much of the ice mass in the West Antarctic Ice Sheet lies as much as 2500 m below sea level (Figure 4.15); this marine-based ice sheet is much more climate-sensitive and would certainly be the first of the two to thaw in the event of rapid global warming. In fact, there is evidence that some of Antarctica's coastal ice shelves are now beginning to disintegrate, the most spectacular example of this being in 1995, when large sections of the Larsen Ice Shelf broke loose

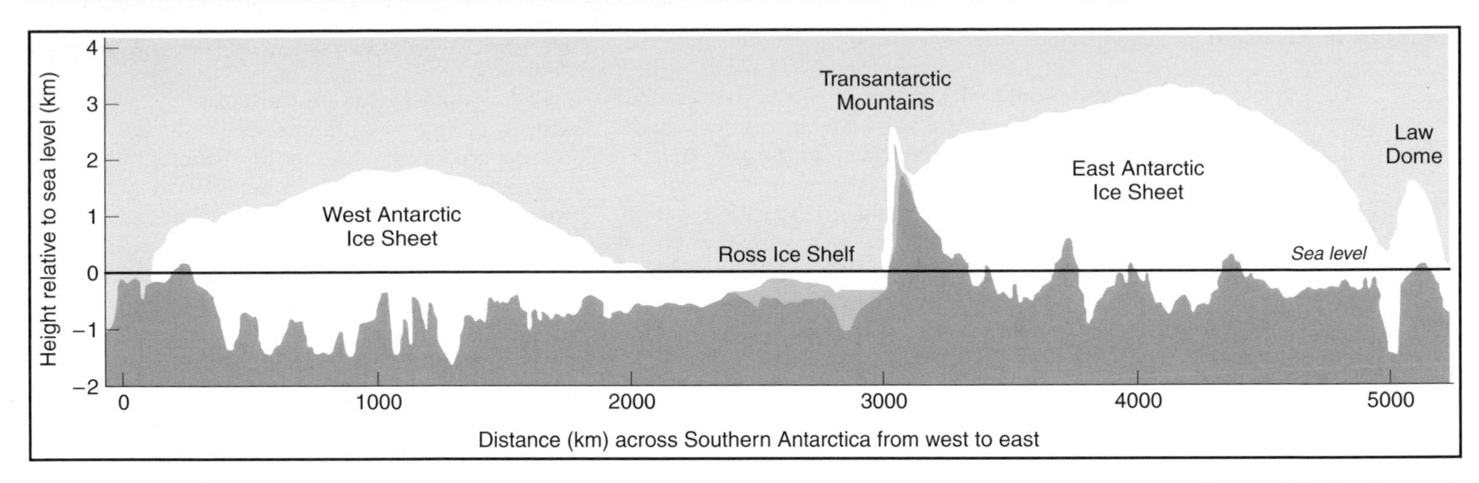

FIGURE 4.15 The East and West Antarctic Ice sheets

from the shore. East Antarctica is significantly colder than the western parts of the continent, partly due to its higher altitude and the intensely cold katabatic (gravity-induced) winds which blow down-slope, but also because its ice reflects 80 per cent of all incoming radiation (as opposed to only 5 per cent for the open sea).

Antarctica's continental **ice budget** has a number of component inputs and outputs. For example, it is estimated that the frozen water which falls as snowflakes at the South Pole takes 50 000 years to reach the coast. Some of this ice is lost through ablation as water vapour in the very dry air above it. Any remaining ice which does not remain part of a continuous sheet forms glaciers as it begins to approach the coastline. Antarctica's largest example of this type of feature is the 40 km-long Lambert Glacier, in the Prince Charles Mountains. At the coast, icebergs 'calve' off the ice sheet (Figure 4.16). One particularly large example broke free of its ice sheet in 1963; this 110 km × 75 km mass was moved great distances by ocean currents until finally melting seven years later. The Circumpolar Current around the Antarctic continent is generated by strong prevailing westerly winds which encircle the world in the higher southern latitudes.

In summer, the ice sheets and glaciers cover 98 per cent of the continental land mass, but their area doubles in winter due to the seasonal freezing of coastal waters. This additional iced area helps to keep the Antarctic cold by delaying the warming effect of the Sun. Also, when it does finally melt, the cold water from the disappearing floating sheets and icebergs seeps down to the ocean bottom, where it joins the global current circulations and helps to offset the global heat input of the tropical regions (where insolation is significantly greater throughout the year).

Antarctica's remarkable natural beauty has now prompted a modest but increasing tourist industry; over 10 000 tourists (50 per cent of them USA citizens) now visit the area during each Southern Hemisphere summer. The first tourists arrived in 1958 but the volume of trade increased quite sharply in the mid 1950s, when a number of low-priced, ice-strengthened ships became available following the disintegration of the former USSR.

FIGURE 4.16 Iceberg calving

Another factor is the popularity of high-quality televised TV natural history documentaries. Antarctica is also benefiting from the increased affluence of many countries, the ease and cheapness of travel and the trend towards ever-more exotic holiday locations. Most visitors travel by ship from ports in Argentina, New Zealand, Tasmania and the Falkland Islands; Quantas, the Australian national airline, operates continental over-flights. In 1991, seven tour operators formed the International Association of Antarctica Tour Operators (IAATO), with the aim of regulating their trade, aiding the exchange of up-to-date climatic and other relevant information and ensuring some consistency with regard to environmental and safety issues.

To date, there is little evidence to suggest that the current level of tourist activity is having any significant, detrimental effect on local wildlife populations such as penguin colonies. The typical visitor does appear to show a degree of sensitivity towards environmental issues and all visitors are fully briefed on good-visitor practise before their first onshore landing – particularly with regard to litter-dropping and the picking of mosses, lichens and flowers.

Waste disposal is difficult in all extremely cold environments, and is particularly so in Antarctica.

The absence of natural decomposers such as worms, due to the low temperatures, inhibits the natural degradation of rubbish; waste freezes solid and becomes instantly inert. Metals cannot rust because of the lack of moisture in the air. An example of the extraordinary powers of ice to preserve both bodies and materials made the news headlines in August 2000 when a retreating glacier in northern Iceland exposed the body of a wartime RAF officer and the remains of his bomber aircraft which had crashed into the glacier in May 1941. Even the rubbish left by Scott's British Antarctic Survey Expedition of 1910–13 and Shackleton's exploration team of 1914–17 is so intact that it has now been made the subject of a heritage preservation order. The British Antarctic Survey's current policy is that all waste from its own operations (except for sewage and food waste) must be stored until it can be transported from the continent for final disposal. Antarctica's own generation of pollutants is extremely low – due to the absence of industrialisation and, as yet, very limited communications networks. It is, however, not immune from pollution transported from other global regions by wind and water currents. It is these very low levels of pollution which make Antarctica such a valuable base for the study of global pollution trends. One major pollution-related concern is the presence of increasing ozone 'holes' in the atmosphere above Antarctica.

One obvious reason for the lack of industrialisation in Antarctica is the climate, which presents great difficulties for any kind of human and mechanised work. Remoteness is an additional problem, but experience in Alaska, Siberia, Sweden and the North Sea – as well as the Arizona and Sahara Deserts (for very different reasons) – has shown that the challenges presented by climate and remoteness can be overcome when the potential for profit and expansion is seen to justify the necessary investment and technological development.

Antarctica does possess mineral wealth and it is estimated that the Ross and Weddell Seas alone contain oil reserves comparable with those in the British sector of the North Sea. Deposits of copper, chromium, nickel and platinum are known to exist in the Dufek Massif, but there is still some doubt about whether they are sufficiently extensive to be commercially viable. Gold and chromium deposits are also believed to exist on mainland Antarctica and 3 m thick exposed coal seams have been discovered on the edge of the Transantarctic Mountains.

The apparent paradox of coal measures in a frozen environment is easily explained by examining the disintegration of Gondwanaland and the subsequent southerly track of the Antarctic land

FIGURE 4.17 The Southern Ocean biome

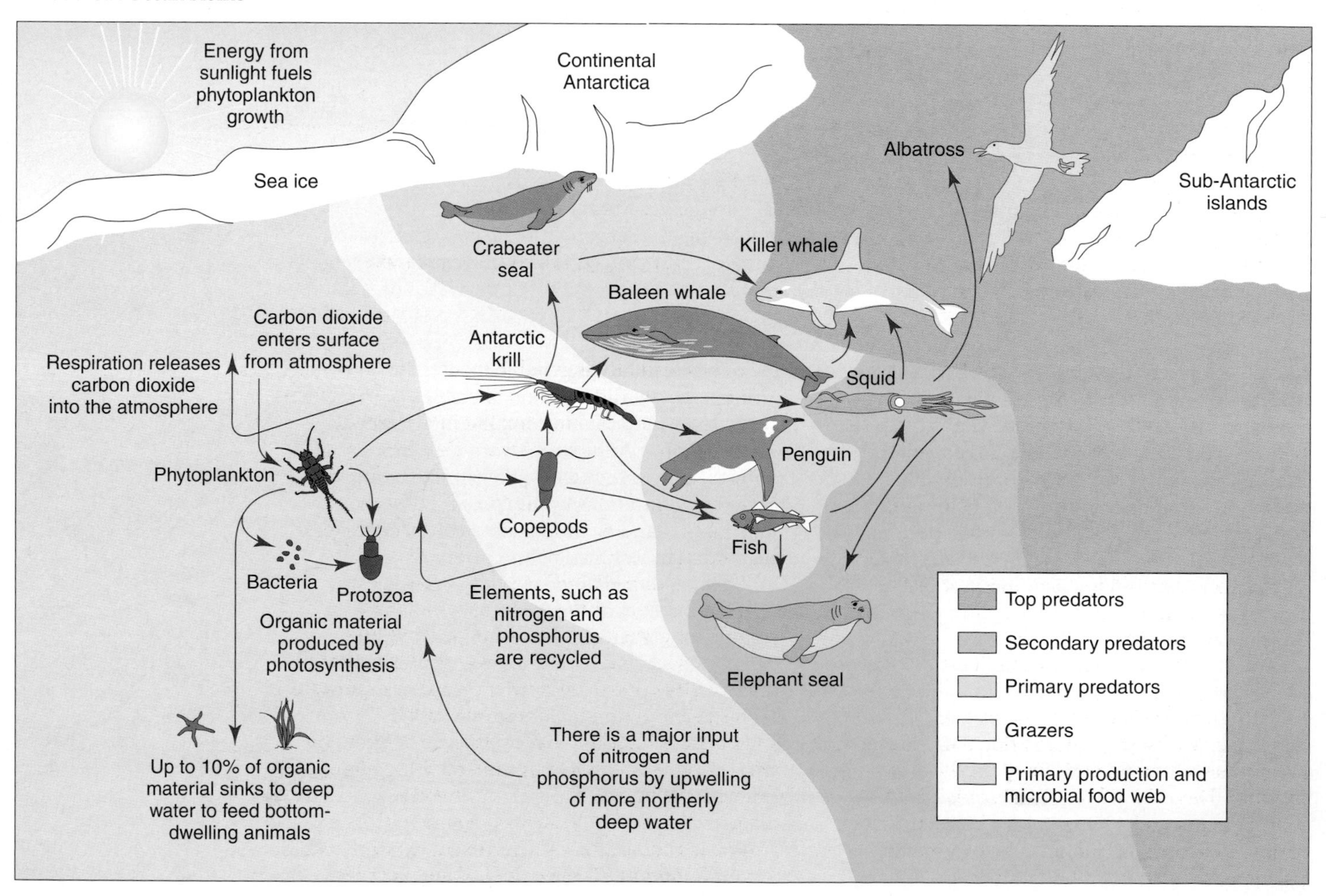

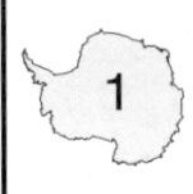
Military activities in Antarctica are prohibited although military personnel and equipment may be used for scientific research and other peaceful purposes.

Freedom of scientific investigation and cooperation in Antarctica shall continue.

Free exchange of information on scientific programmes and scientific data; scientists to be exchanged between expeditions and stations when practicable.

Existing territorial sovereignty claims are set aside. Territorial claims are not to be recognised, disputed or established by the Treaty. No new territorial claims can be made whilst the Treaty is in force.

Nuclear explosions and radioactive waste disposal are prohibited in Antarctica.

The Treaty applies to all land and ice shelves south of latitude 60°S, but not to the "high seas" within this area.

All Antarctic stations (and all ships and aircraft operating in Antarctica) have to make themselves available to inspection by designated observers from any Treaty nation.

Personnel working in Antarctica shall be under the jurisdiction only of their own country.

Treaty nations will meet regularly to consider ways of furthering the principles and objectives of the Treaty. Attendance at these meetings will be limited to countries showing substantial scientific research activity in Antarctica. Unanimous approval will be necessary at these meetings for any new measures to become effective.

All Treaty nations will try to ensure that no one engages in any activity in Antarctica contrary to the principles and purposes of the Treaty.

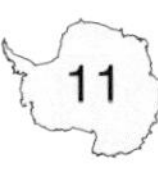
Any dispute between Treaty nations, if not settled by agreement in some other form, shall be determined by the International Court of Justice.

The Treaty may be modified at any time by unanimous agreement. After 30 years (e.g. after June 1991) any Consultative Party may call for a conference to review the operation of the Treaty. The Treaty may be modified at such a conference by a modified decision.

The Treaty must be ratified by any nation wishing to join. Any member of the United Nations may join, as well as any other country invited to do so by the Treaty nations. All notices of accession and ratification are deposited with the Government of the United States of America, which is designated as the Depositary Government.

The Treaty will be published in English, French, Russian and Spanish.

Since the signing of the Antarctic Treaty on 1st December, 1959, four further international agreements have been adopted. These are:

1. Agreed Measures for the Conservation of Antarctic Fauna and Flora (1964).
2. Convention for the Conservation of Antarctic Seals (CCAS) (1972).
3. Convention on the Conservation of Antarctic Marine Living Resources (CCAMLR) (1982).
4. Protocol on Environmental Protection to the Antarctic Treaty (1998).

A fifth agreement – the Convention on the Regulation of Antarctic Mineral Resources Activities (CRAMRA) – was agreed in 1988 but has never become mandatory.

FIGURE 4.18 Provisions of the Antarctic Treaty, 1959

mass to its present, polar location. The coal measures formed when its land mass was densely vegetated with sub-tropical forests and ferns. The past proximity of Antarctica to mineral-rich South Africa and South America within Gondwanaland strengthens the case for the polar land mass to contain a wide range of minerals.

Antarctica's land-based ecosystems are very primitive – certainly in comparison with the Southern Ocean biome which encircles it (Figure 4.17). Lichens, algae and mosses tend to grow only where water is available during the short growing season. There are only two species of flowing plants – grass species called Antarctic Hair Grass and Antarctic Pearlwort (both restricted to the Peninsula area of the mainland and the more sheltered islands). The animal life consists only of insects, worms, mites and protozoa (single-celled life forms). A species of midge is the continent's largest permanent invertebrate inhabitant.

It was virtually inevitable that the international community would devise some means of protecting the unique natural environment of Antarctica against short-term exploitation. On 1 December 1959 the Antarctic Treaty was signed and came into force on 23 June 1961. This treaty, whose 14 provisions are listed in Figure 4.18, was signed by 12 nations.

Some countries – particularly France, New Zealand and Australia – favoured declaring Antarctica a World Park, i.e. an international wilderness area in which only scientific research and strictly-controlled tourism would be allowed. Greenpeace has adopted a similar policy, but would prefer only 'high-quality' scientific activity to be permitted, as long as such research is conducted in the spirit of international co-operation. Greenpeace's other priorities include total wildlife ecological community protection, a complete ban on weapons and military activity of any kind and

Old threats to Antarctic linger as new curbs come into force

A continent was put off-limits to mining and oil exploration for the next 50 years when the Antarctic protocol came into force yesterday.

The protocol – now ratified by Japan, the last of the 26 states claiming an interest in the continent to do so – protects the Antarctic and the sea bed south of latitude 60°.

All explorers, scientific expeditions and tourist ventures will now have to ask permission to enter the region and make an environmental damage assessment before doing so.

But the region is still threatened by pirate fishing vessels plundering Antarctic waters in defiance of other international agreements designed to protect wildlife and natural resources.

An Australian fishing boat, the Austral Leader, reported an attempted ramming at the weekend by a Panamanian-registered pirate ship it confronted off Heard Island. The Australian navy is now having to sail 1200 km to the remote island to police the area. In theory, the Southern Ocean, the roughest and most inhospitable in the world, is governed by the Convention on the Conservation of Antarctic Marine Living Resources (CCAMLR), part of the same Antarctic Treaty that gave birth to the protocol.

But pirate ships are flouting the rules and, although French paratroopers boarded three of them last year, there is no way of policing the whole ocean.

The pirates are after the Patagonian toothfish, a deep-water, slow-breeding oily fish which fetches high prices in Japan. Although it is found throughout the Southern Ocean off the continental shelf of sub-Antarctic islands, stocks are rapidly being exhausted.

Chilean waters were fished out in 1992, Argentine waters by 1995, and last year the toothfish was said to be "almost gone" from South Africa's Prince Edward sub-Antarctic islands.

The pirates are estimated to have taken £200 million worth of the fish from the waters. Australian ships, legally fishing for toothfish with a quota set by the CCAMLR, have reported pirates arriving from South Africa in the last few days. Euan Dunn, of The Royal Society for the Protection of Birds, part of an international network of organisations monitoring the toothfish trade because of the damage the fishing does to other forms of wildlife, says that whole colonies of albatross are in danger of being wiped out. They get caught on the long lines of hooks used to catch the toothfish. Albatross are attracted by the squid used as bait on the lines.

The Guardian, 15 January 1998

FIGURE 4.19 Threats to Antarctica

recognition of the value of what it regards as the world's last great and virtually undisturbed wilderness region. It totally supports the **precautionary principle**, which requires a total abstinence from any activity which has the potential to cause harm, unless it is subject to the strictest possible levels of monitoring and evaluation. Such restrictions should, in the opinion of Greenpeace, apply especially to the commercial exploitation of marine resources. However, the provisions of the 1959 treaty failed to protect offshore waters, leaving fish and whales to remain highly vulnerable (Figure 4.19). The Antarctic Treaty has since been considerably strengthened by a series of measures which are also included in Figure 4.19. One such measure, the so-called Environmental Protocol, which came into force in 1998, now protects Antarctica by one of the most stringent sets of environmental regulations operating anywhere in the world.

ACTIVITIES

1 Draw a simple flow-diagram to illustrate the components in Antarctica's ice budget.

2 List the chief minerals likely to exist in Antarctica and, following research of your own, add the chief industrial uses and by-products of each material.

3 Draft and then present an introductory talk which tour operators belonging to IAATO can give to parties of visitors before their first landing on Antarctica. Your talk must include a list of do's and don'ts and detailed reasons for the importance of each of these rules. The initial planning, research and actual presentation can all be undertaken on a whole-group basis.

4 Suggest reasons why Antarctica is much colder than the Arctic.

5 Explain why so many nations should feel obliged to be involved in research in Antarctica. You might consider some of the following countries (which have been selected at random) in your answer: Argentina, Australia, Belgium, Brazil, China, India, Japan, Norway, Poland Russia, Britain and the USA.

Study 4.6 Regional glaciation – Western Europe

The Pleistocene period consisted of a number of phases of very low temperatures interspersed with much milder inter-glacials (Figure 4.20). In common with other regions at similar latitudes, the land surface of northern Europe was greatly modified by a wide range of glacial erosion, transportation and deposition processes (Figure 4.21). Figure 4.22 shows the most southerly margins of the European ice sheets at the coldest climatic phases, when the sheets were poised between advance and retreat.

The constantly and most deeply glaciated highland regions such as Scandinavia now provide many classic examples of prolonged erosion. The fjord coastline of Norway is one of the most striking of these examples (see Study 4.8). Glacial erosion also had a profound effect on the lower-lying areas of northern Europe and Finland's present lakeland landscape is the result of differential erosion of alternating surfaces of harder and less resistant rocks (Figure 4.23).

The margins of the ice sheet were dominated by the deposition of transported solid material, as well as meltwater, which was capable of transporting sands and gravels considerable distances. Denmark now has a variety of landscapes and soils as a result of its location on the ice sheet's edge (Figure 4.24). A summary of the many positive and negative impacts of glaciation within northern and central Europe is given in Figure 4.25.

Abrasion The erosion of surface material by the action of rocks, boulders and other forms of loose debris as they are transported by moving ice.

Erosion A general term for the loosening and removal of surface material, whether by the ice itself or debris which forms part of a moving ice mass.

Deposition The dropping of transported material when an ice mass melts.

Differential erosion The varying rates of erosion which take place as an ice mass moves over surface areas having different degrees of hardness.

Freeze-thaw action (also known as **frost-shattering**) A form of mechanical weathering by which water retained in rock crevices expands in sub-zero overnight temperatures to such an extent that it fractures part of the rock mass around it.

Plucking A form of glacial erosion which occurs when glacier ice freezes around a rock on a valley side (or bottom) and forces it loose as the ice continues to move towards lower land.

Transportation The movement of eroded or weathered material by a glacier or ice sheet.

FIGURE 4.21 Glacial erosion, transportation and deposition

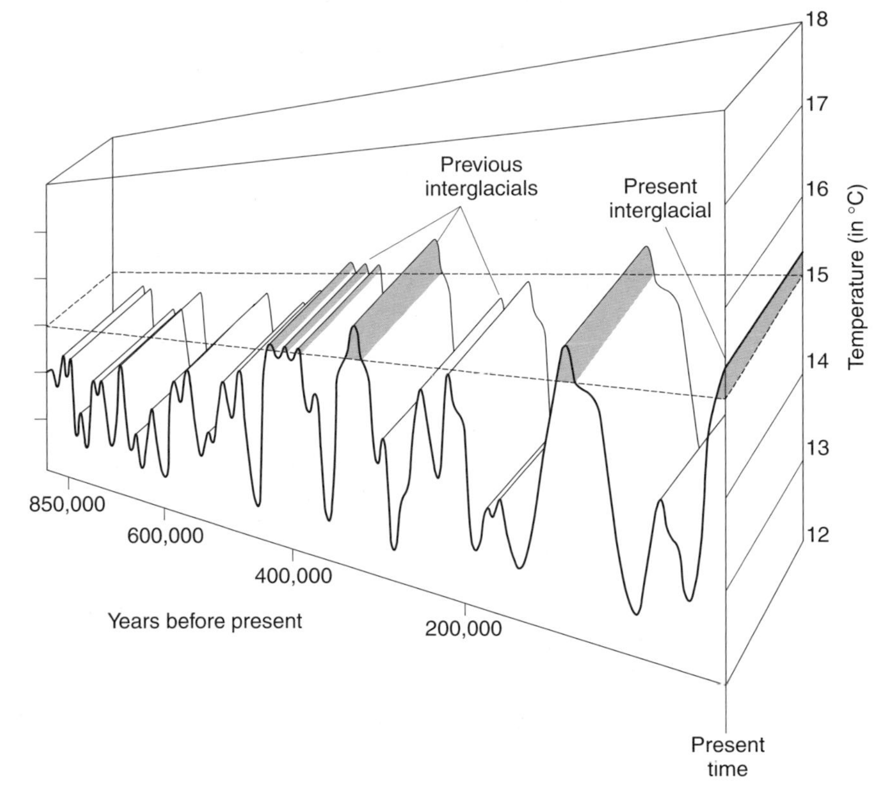

FIGURE 4.20 Glacials and interglacials in the Pleistocene

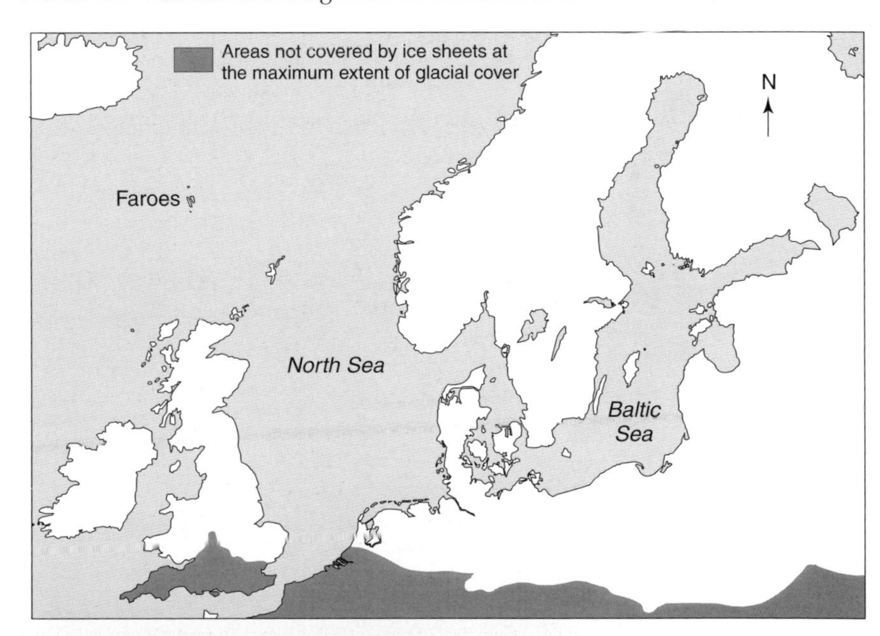

FIGURE 4.22 Glacial limits in northern Europe

FIGURE 4.23 Glacial landscape in Finland

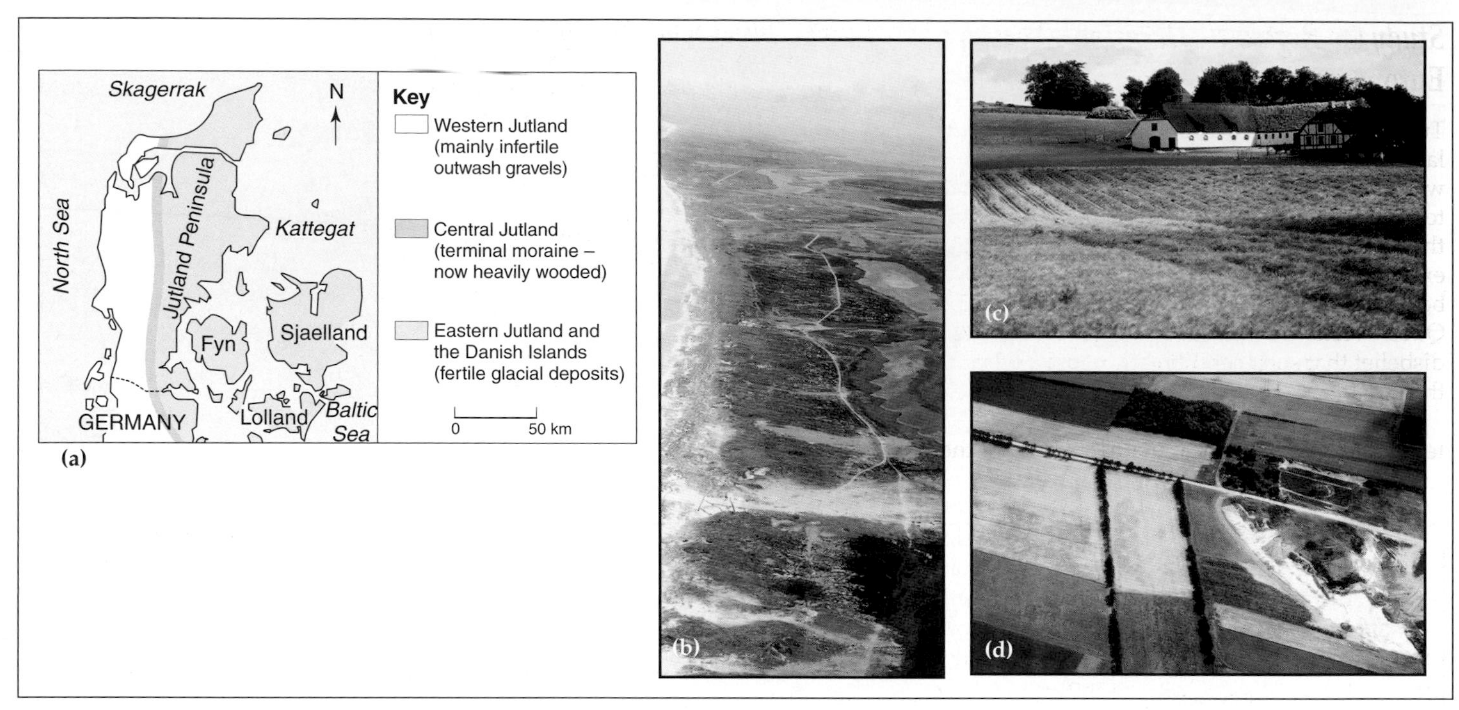

FIGURE 4.24 Glacial regions in Denmark

FIGURE 4.25 Advantages and disadvantages of glaciation

Glacial feature	Potential advantages	Potential disadvantages
Boulder clay soils	High fertility – often produce excellent arable land	Poor surface drainage often results in bogs and marshes
Glacial breeches and lake overflow channels	Provide natural routeways through upland areas	–
Lakes	Provide natural routeways through upland areas; tourist attractions; old lake beds provide rich alluvium-type soils; corrie lakes have HEP potential	Flood valley bottoms reducing land available for settlement, transport routes and mixed farming
Outwash plains	Sand and gravel deposits may be quarried to provide building materials	Sands and gravels are infertile – can only support heath land
Valleys	Provide natural routeways through upland areas; waterfalls on valley sides have HEP potential; popular tourism and winter-sports locations; 'alp' benches on upper valley sides provide summer grazing using **transhumance**	–

ACTIVITIES

1 With the help of an atlas, quote a series of named places which identify the borderline of the Scandinavian Ice Sheet's most southerly extent.
2 Extend the table in Figure 4.25 to include named examples taken only from northern and central Europe of the various advantages and disadvantages which are listed.
3 Produce a table which summarises the range of glaciated landscapes contained in this study. Your table will need to include landscape types, locations, characteristics and formation processes.

Study 4.7 High altitude zones – The European Alps

Tundra conditions occur not only in the higher latitudes, but also within much warmer regions, where altitude is the key factor in determining local temperatures. A well-known historical example of this occurred in the nineteenth century, when explorers reported that snow-capped peaks had been observed on mountains in central Africa. Queen Victoria responded by expressing total disbelief that such conditions could occur so close to the Equator.

Subsequent research into the effect of altitude on temperature has led to an understanding of the lapse rate phenomenon – by which local environments in highland localities may follow a very similar biome succession to that which takes place, at sea level, between equatorial and polar regions. Figure 4.26 traces this biome progression at a location in the south-west region of North America.

The Alpine region of central Europe is a prime example of the impact of lapse rates on high altitude areas. Glaciers are a permanent Alpine feature, although seasonal variations in temperature do determine the volume of ice in each glacier. The Aletsch Glacier system (Figure 4.27) includes the Alps' longest glacier; its surface area is 171 km^2 and its dimensions are 24 km (length) $\times$ 1.6 km.

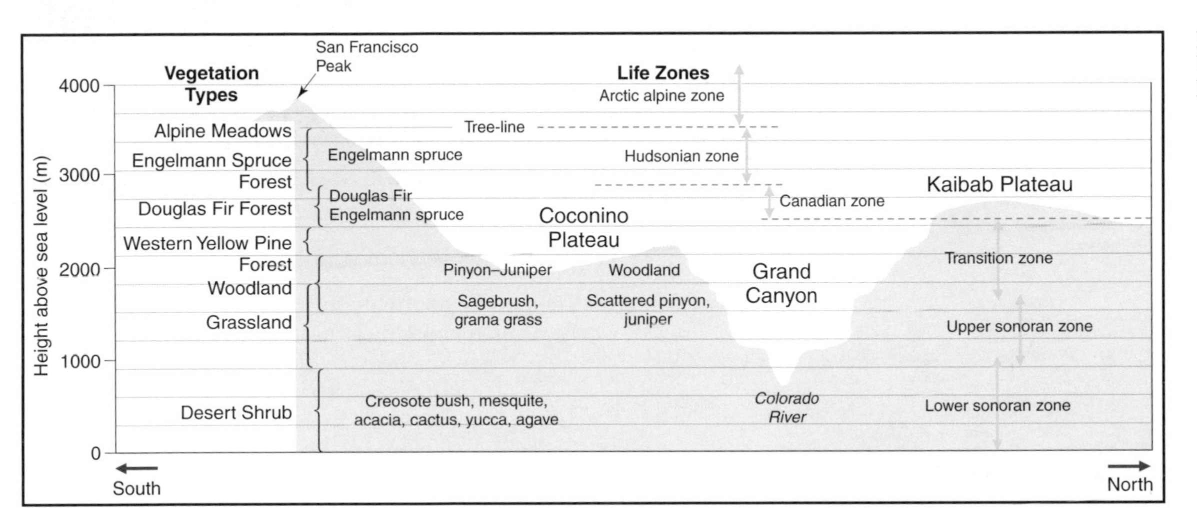

FIGURE 4.26 Biome/altitude links in North America

FIGURE 4.27 The Aletsh Glacier

FIGURE 4.28 A hydro-electric power station

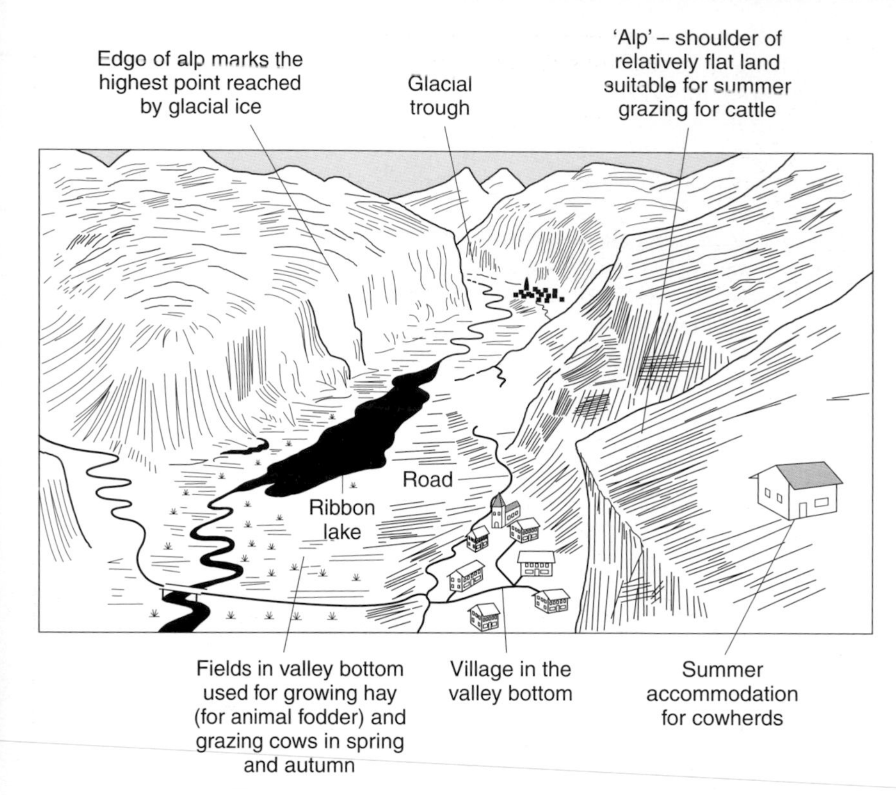

FIGURE 4.29 Transhumance in the Alps

FIGURE 4.30 The composition of the Earth

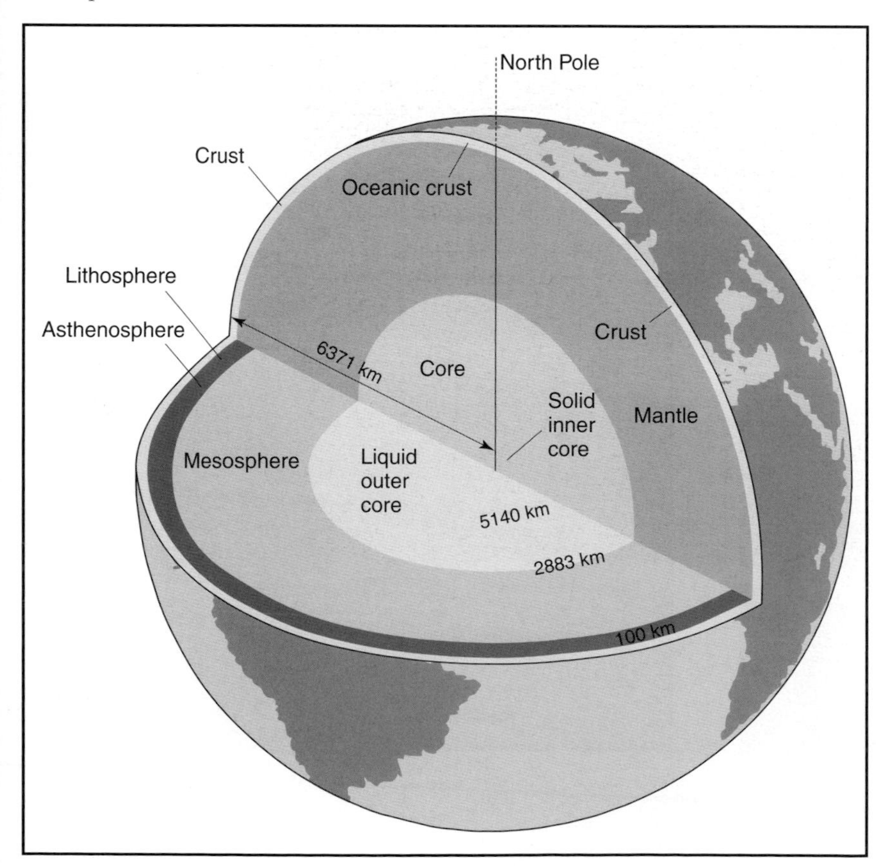

Alpine conditions have inevitably dominated the economy of Europe's highest altitude region:

- High rates of orographic rainfall and glacier meltwater have provided ideal conditions for hydro-electric power generation (Figure 4.28).
- Power-intensive industries have been stimulated by HEP, e.g. aluminium-smelting.
- Tourism is a major, if seasonal, contributor to the local economy.
- The local farming community has traditionally employed transhumance techniques to utilise the higher grazing pastures which are seasonally-available to them (Figure 4.29).

ACTIVITIES

1 Explain why Austria and Switzerland are two of the world's more wealthy countries – in spite of their high altitude and many other environmental difficulties.

Study 4.8 Isostatic change – Norway and north-western Scotland

The Earth's relatively thin crust (its outer layer) floats on the mantle – the much deeper, molten layer of magma below it (Figure 4.30). Floating in this way gives the crust a degree of flexibility which allows two types of movement to take place:

1 Lateral movements, by which continental land masses are able to drift slowly across the mantle's surface. The fragmentation of Gondwanaland was a direct result of prolonged continental drift. Another important impact is the occurrence of earthquakes and volcanoes, whose global distribution pattern follows the crust's lines of weakness along plate margins; this topic is discussed more fully in Unit 5.

2 Vertical movements, i.e. adjustments for the crustal weight bearing down on any particular point on the mantle's surface; the heavier the weight, the greater the sinkage into the less dense rock below it (Figure 4.31). Sinkage adjustments will take place whenever the weight burden changes sufficiently, but they usually occur very slowly and it takes a long period of time before full equilibrium is restored.

Major vertical movements have been caused by ice melting after the last Ice Age. The huge additional weight of ice during the Ice Age was sufficient to depress very large areas of the Earth's crust in the higher latitudes, as highly compressed ice is one-third the equivalent weight of solid rock and some ice sheets were several hundred metres thick. The melting of the ice sheets initiated a long term isostatic readjustment which continues to the present time (Figure 4.32).

Type of characteristic	Continental crust characteristics	Oceanic crust characteristics
Thickness	35–70 km	6–10 km
Age of rocks	1500 million years and over	Under 200 million years
Density of rocks	2.6	3.0
Appearance of rocks	Light-coloured	Dark-coloured
Rock types	Many types/ granite most common	Few types/ chiefly basalt

FIGURE 4.31 Continental/oceanic crust comparison

Observable changes in sea level relative to the local shoreline (as opposed to short term, predictable tidal height fluctuations) are the product of independent vertical movements of both land and water. Post-glacial land isostatic adjustments may, therefore, be disguised to some extent by changes in the sea level attributable to global warming and its effects on the world's remaining ice sheets and glaciers. Net rises in sea level have submerged river valleys creating rias (Figure 4.33) and **fjords** (Figure 4.34). One of the most distinctive features of net rises in land level are **raised beaches** (Figure 4.35). Close observation of such raised beaches often reveals evidence of relict features of active shorelines such as caves.

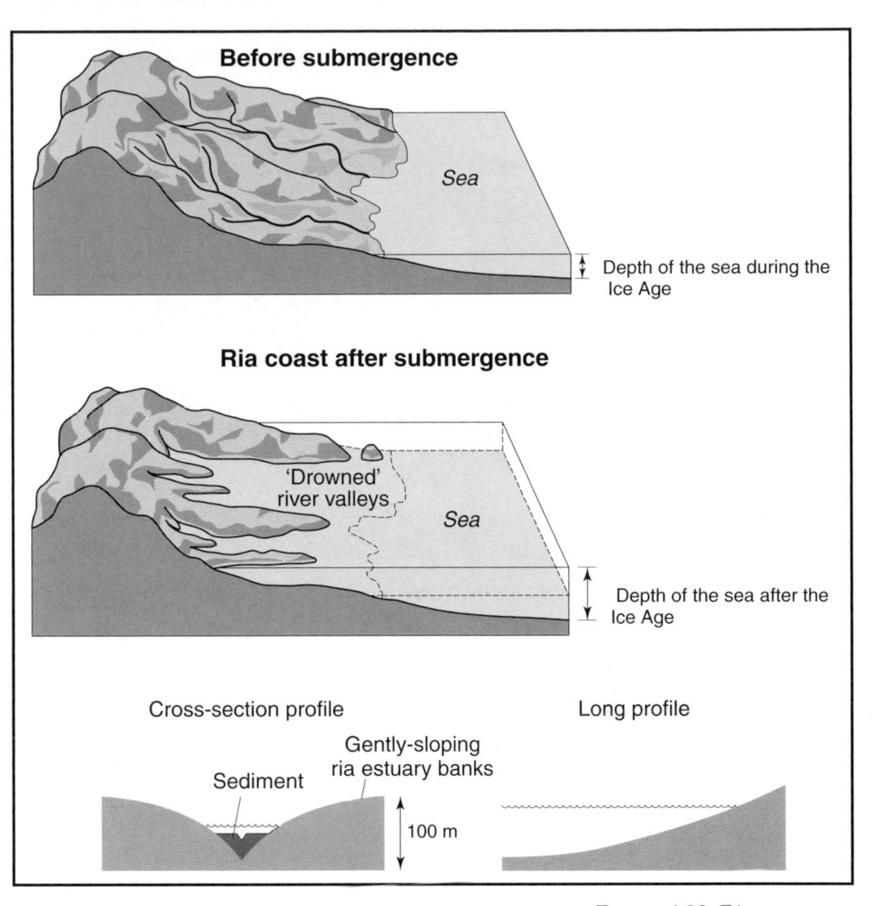

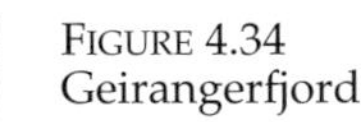
FIGURE 4.33 Ria formation

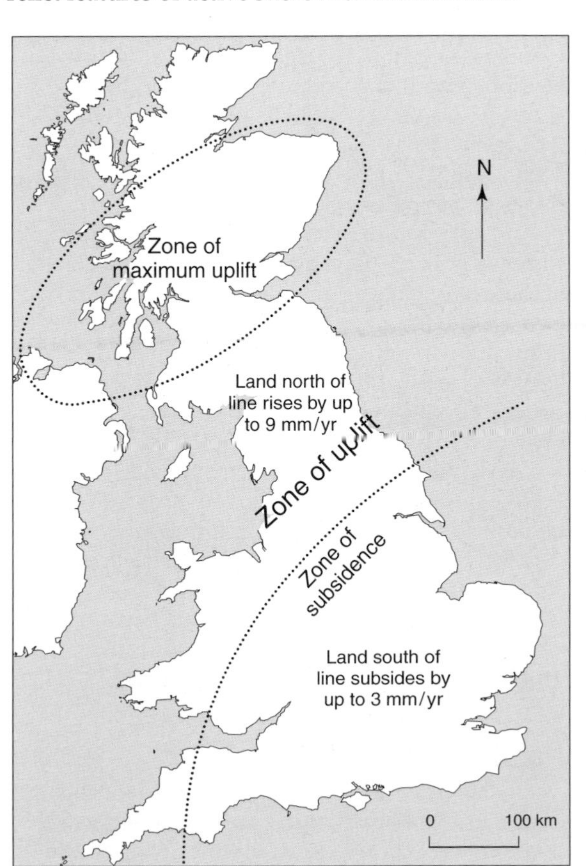

FIGURE 4.32 Isotatic adjustment in the British Isles

FIGURE 4.34 Geirangerfjord

FIGURE 4.35 Raised beach

ACTIVITIES

1 Summarize the nature of isostatic change and its chief causes.
2 Compare the appearances of ria and fjord-type coastal features
3 a) Describe the appearance of typical raised beaches.
b) List the types of relict coastal features which you might expect to see on raised beaches
c) Suggest ways in which raised beaches often prove to be valuable assets to island communities.

FIGURE 4.36 The Snowdon Horseshoe

Study 4.9 Relict glaciation features – Snowdon and the Llanberis Pass

About 500 million years ago, the North Wales region which is now called Snowdonia formed part of a mountain chain extending from Anglesey to north-west Scotland. This region was subsequently flooded by sea water. During the period when it was well below sea level, a variety of marine deposits were laid down and these provided the source material for the types of rocks which have dominated the Snowdonia region ever since. Increasing volcanic activity later metamorphosed these sedimentary rocks – a process which greatly increased their resistance to future erosion. Later still, the Silurian and Devonian periods witnessed the creation of the fold mountains of the Caledonian Orogeny and, at about the same time, shales were being metamorphosed into the slates for which Snowdonia was to become world-famous as a provider of slate roofing material.

Approximately 2 million years ago, major ice sheets began to cover most of Britain and extend across the present North Sea area into Scandinavia. Their moving ice masses created many classic types of glaciation features in Snowdonia – as well as the much larger area of north-west Scotland. Snowdon (Yr Wyddfa) at 1085 m above sea level, is the highest mountain in England and Wales, and now exhibits many fine examples of the cirque/arête type of high-altitude landscape. (In Wales, cirques are known as cwms).

The present day appearance of Snowdon and its mountain neighbour Garnedd Ugain indicates that neither summit was completely covered by ice sheets. However, Snowdon's uppermost area was the source of six major glaciers, whose erosive action resulted in the development of five long ridges which radiate outwards over an 80 km^2 area. Today, no fewer than ten cirques encroach on the summits of Snowdon and Gardnedd Ugain. Seven of these cirques have an easterly aspect, whilst two face north-west and a third faces south-east. The positions of all ten cirques were determined by deep fractures within the crust. Three of Snowdon's cirques have become popularly known as the Snowdon Horseshoe (Figure 4.36).

Glaslyn was the source of a cirque glacier which spilled over its lip and then downwards into Cwm Llydaw, to which its own ice accumulation was added. The combined Llydaw-Glaslyn Glacier then eroded an undulating basin for almost 2 km along a series of fractures orientated along an east-north-east axis. Outcrops of abraded rock indicate that ice over 400 m thick occupied the Horseshoe in the Late Devensian period. Most of this ice was later diverted northwards into the Llanberis Pass (Figure 4.38).

On Snowdon's north-eastern flank, the deep trough of Llanberis Pass has an almost perfect U-shape cross-section – the result of prolonged and intensive glacial activity. The trough's contour patterns (Figure 4.38) reveal very clearly the

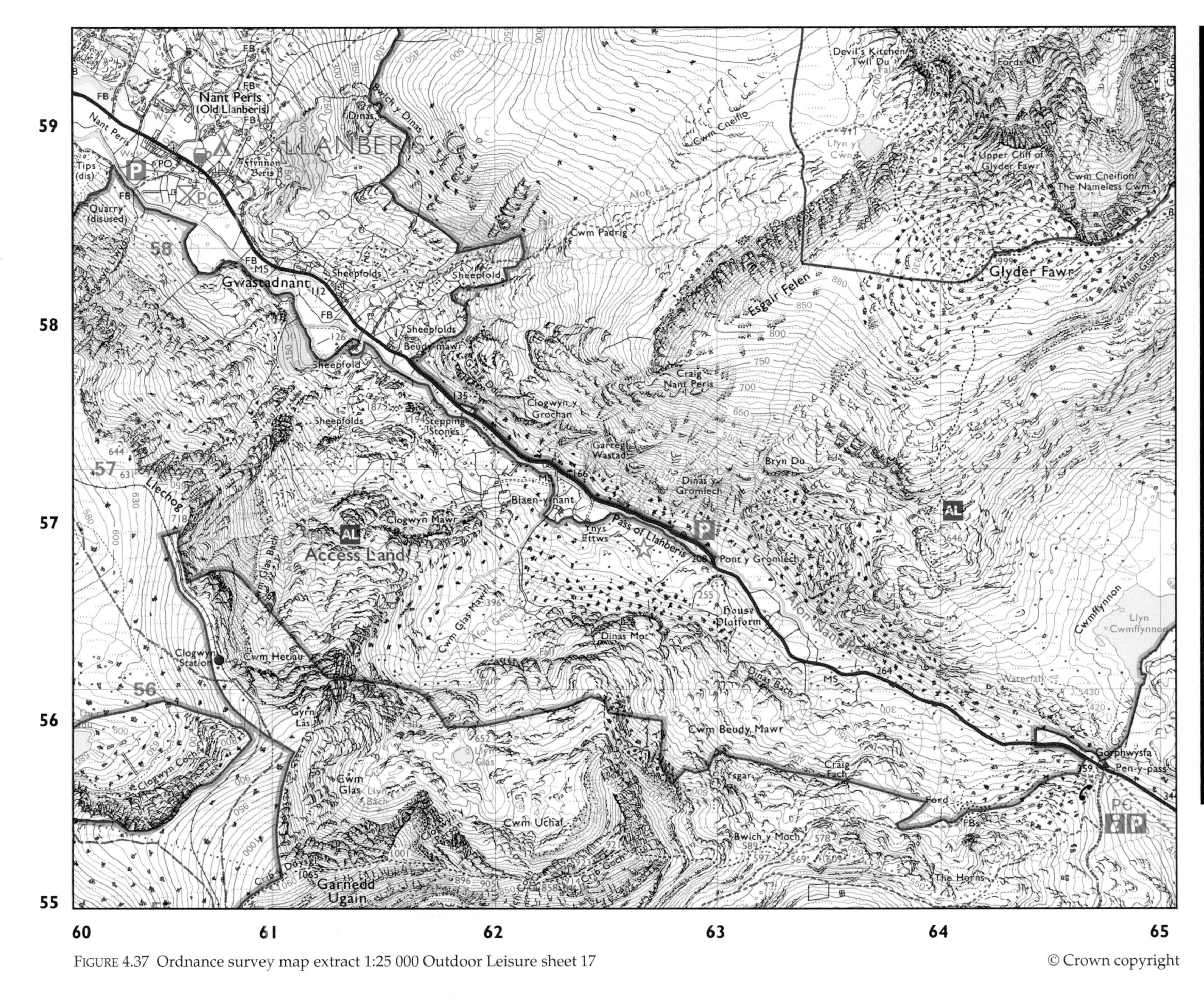

FIGURE 4.37 Ordnance survey map extract 1:25 000 Outdoor Leisure sheet 17

FIGURE 4.38 Llanberis Pass

FIGURE 4.39 A roche moutonée in Llanberis Pass

FIGURE 4.40 Perched blocks in Llanberis pass

steepness of its sides and how level the valley bottom is at any point between Pen-y-Pass and Llyn (Lake) Peris. Figure 4.38 enables a visual comparison to be made with the map evidence in the Ordnance Survey map extract.

Llanberis Pass provides one of Britain's finest examples of glacial trough formation. It is particularly well-known amongst glaciomorphologists for being one of the locations which led an eminent Victorian geologist, Sir Andrew Ramsay, to become the first person to suggest that large rock basins on mountain sides were the product of glacial action. Llanberis Pass includes two quite distinct sections:

FIGURE 4.41 Dinorwig slate quarry and HEP station

- A narrow 6 km-long trough from Pen-y-Pass (359 m) to the eastern tip of Llyn Peris (100 m)
- An 8 km-long rock basin containing Llyn Peris and Llyn Padarn.

Both sections lie on major NW–SE fracture lines, which provided an ideal route for the ice on its descent from Snowdon to the Irish Sea. **Striation** patterns on the bare rock surfaces towering 800 m above the trough floor provide evidence of intense glacial scour – as well as the impact of two very different directions of ice flow. The older and much deeper striations cut across Snowdon's tributary cirques; in doing so, they follow the NW–SE axis of Llanberis Pass. The more recent striations indicate the relatively minor flow of cirque glacier ice from Snowdon's own flanks.

Scree slopes cover extensive areas of the valley sides within the trough section of the pass, and till deposits form a thin veneer above its bedrock floor. **Roche moutonnées** are evident above Pont-y-Gromlech and Ynys Ettws (Figure 4.39). Perched blocks – some similar in size to that of a small house! – are a major feature of the valley and are located where they could not possibly have fallen from the adjacent cliffs, leaving transportation by ice as the only realistic explanation.

The Llanberis Pass outlet glacier gained in erosive power as it accelerated, to such an extent that it was able to over-deepen the lower section of the pass, which is now occupied by Llyn Peris (3.2 km long; maximum depth 35 m) and Llyn Padarn (3.2 km; 29 m). Both lakes contain substantial sediment deposits and are now separated due to the deposition of a post-glacial fan delta of the Afon (River) Arddu at Pont-y-Bala. The extensive alluvial plain near Cwm-y-Glo to the west of Llyn Padarn owes its existence to similar in-filling and reverts briefly to a shallow lake after prolonged rainfall.

The northern slopes of the lake section of the pass became world-famous in the nineteenth century, for the production of high-quality slates which were then in great demand as roofing material. The most favoured quarrying technique at that time was to work tiers of terraces 20–25 m high (and a similar distance wide) up the valley sides (Figure 4.41). At its peak in the 1870s, Dinorwig Quarry employed 3000 men on terrace work alone and Llanberis village on the opposite side of Llyn Peris quadrupled its population between 1831 and 1891. The quarry closed in 1969, when its total workforce had already been reduced to only 300, but the visible scars of its open cast workings are unlikely to be healed in the foreseeable future. Llanberis is the starting point of the Snowdon Mountain Railway (completed in 1896), whose track to the summit follows one of the smoother and gentler arêtes to have survived the glacial period.

ACTIVITIES

1 Draw an annotated sketch of the scene in Figure 4.36, showing both the types and names of all its main physical features.

2 a) Using detailed map and photographic evidence, contrast the natural physical characteristics of the 'rock-basin' and 'trough' sections of Llanberis Pass.

b) Provide reasoned explanations for the major differences which you included in your previous answer.

3 Quoting evidence from the visual resources contained in this study and from any other resources to which you have access, suggest how the Snowdonia National Park shown in Figure 4.37 might have been/currently is being used for each of the following purposes:

- agriculture
- transport links
- mineral extraction
- leisure and tourism
- electricity generation.

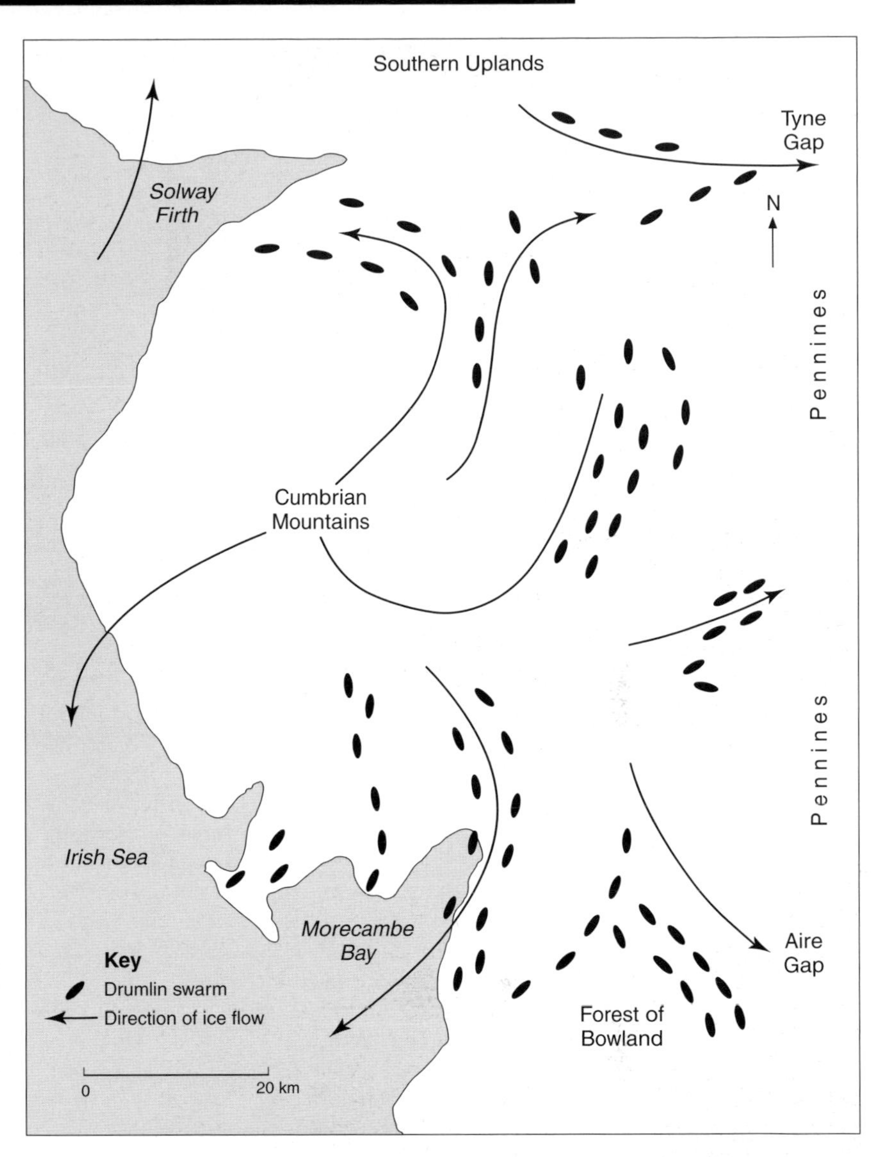

FIGURE 4.42 Drumlin swarms in the Lake District and north Lancashire

FIGURE 4.43 A drumlin swarm

Study 4.10 Drumlins – Cumbria and north Lancashire

Drumlins are one of the most common and most easily recognisable features of glacial deposition in low-lying areas; favoured locations range from the bottom of glaciated highland valleys to broad coastal plains. Drumlins usually occur in groups which, from a distance, tend to have a rounded (and somewhat chaotic appearance) often described as 'basket of eggs' topography. Drumlins provide excellent grazing land and their crowns are popular places on which to site farm buildings or hamlet-size settlements. The lower land between the drumlins, being clay, is often too waterlogged for effective arable use.

The process by which drumlins were formed is still the subject of much discussion. One possible explanation is the deposition of glacial till around a large rock in the path of a moving body of ice; another is that they were formed from excess material deposited by a glacier which could no longer transport its existing load of eroded debris. It seems likely that, whichever process created them initially, final shaping may well have been carried out by the advance or retreat of later glaciers. The fact that the long axes of drumlins within a swarm always follow a similar direction indicates a shared cause, whether this be moving ice alone or a combination of ice and meltwater flows. Figure 4.42

FIGURE 4.44 Exposed boulder clay

locates the main drumlin swarms on the edges of the Lake District and indicates their clear link with the routes taken by the region's major ice flows. Figure 4.43 shows a typical drumlin swarm. The close-up picture in Figure 4.44 gives some detail of the boulder clay type of glacial till of which most drumlins are formed.

It is relatively easy to undertake some basic analysis of drumlin swarms. Figure 4.45 shows the terms used to describe the key dimensions, from which the elongation ratio may be obtained, by dividing the maximum length by the maximum width. As most drumlins exceed 25 m in height and their long axis may exceed 1 km, they are sufficiently large for their contour patterns to show up clearly on Ordnance Survey maps, particularly those printed to at least a scale of 1:25 000. These often provide spot height data for drumlin crowns in addition to contour intervals of only 10 m.

ACTIVITIES

1 Describe the apparent relationship between drumlin swarm location and ice flow routes as shown in Figure 4.42.

2 Discuss the economic potential (and drawbacks) of drumlin formations.

3 This activity takes the form of a detailed investigation, based on a drumlin swarm which is either located in your local area or included on large scale Ordnance Survey maps which you are able to use. Possible lines of research, all of which must be fully analysed in writing, are:

- Making comparisons of the heights of the crowns of the drumlins (e.g. height range and mean height).
- Estimating their long axis-orientations, which may be displayed on a compass rose diagram.
- Measuring and tabulating their two axes. This tabled information could form the basis of a Spearman's Rank correlation coefficient analysis exercise; alternatively, it could be used to provide a series of elongation ratios which you can process and display using appropriate methods of your own choice. Figure 4.46 provides a worked correlation coefficient example, then suggests suitable information pairings for you to use with the data collected.
- Investigating the various land uses to which your drumlin swarm has been put. If you have chosen a local example, it should be possible for you to undertake some interview work with land owners and tenants in that area. Your questions should be worded so as to obtain details of both past and present land use, as well as the reasons for any changes in land use which have taken place.
- Investigating the likely reasons for the existence of a drumlin swarm in your chosen location. Doing this could involve some in-depth research of the physical characteristics of the surrounding regional environment.

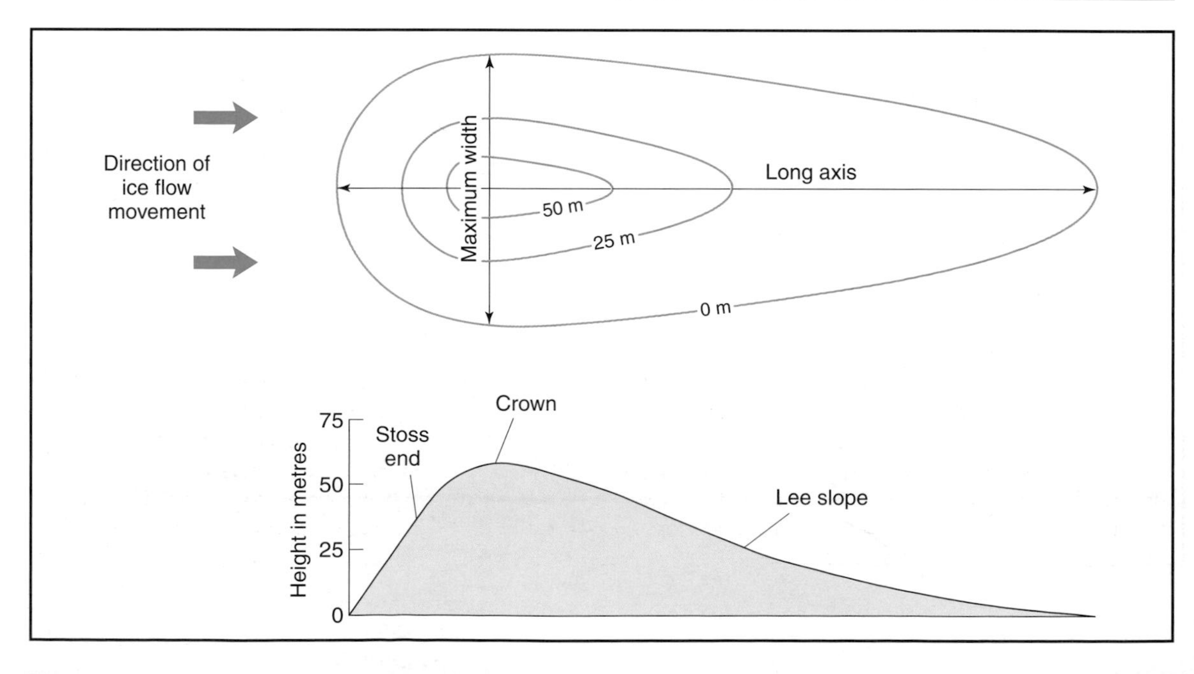

FIGURE 4.45 Drumlin features and dimensions for use with Activity 3

FIGURE 4.46 Spearman's Rank correlation coefficient worked example for use with Activity 3

Example of the use of Spearman's Rank correlation coefficient, based on data of a ficticious drumlin swarm.

- Obtain data and display it in the form of a table:

Drumlin reference number	Drumlin length (m)	Drumlin width (m)	Drumlin height (m)
1	605	337	40
2	483	418	38
3	570	370	41
4	668	406	46
5	710	420	37
6	570	360	40
7	467	290	32
8	450	304	35

- Decide which pairs of datasets you wish to compare
- Make a copy of the table below for each comparison you wish to make; amend the headings of the second–fourth columns of each table as appropriate; follow the rest of the operations in the sequence for each comparison

Drumlin number	Drumlin length in m.	Drumlin length rank order	Drumlin height in m.	Drumlin height rank order	Difference between the two rank orders (d)	d^2
1	605	3	40	3=	0	0
2	483	6	38	5	1	1
3	570	4=	41	2	2	4
4	668	2	46	1	1	1
5	710	1	37	6	5	25
6	570	4=	40	3=	1	1
7	467	7	32	8	1	1
8	450	8	35	7	1	1

Total of all squared differences (Σd^2) 34

- Insert data in the second and fourth columns
- Rank both sets of data in descending order to complete the third and fifth columns
- Enter in the sixth column the difference between the two rank orders for each drumlin (d)
- Square each difference (d^2) and write it in the seventh column
- Find the total of all these squared differences – this is known as Σd^2
- Use this standard formula to calculate the SRCC (R) required:

$$R = 1 - \left(\frac{6 \times \Sigma d^2}{n^3 - n}\right)$$

where n is the number of lines in the rank order table

- Enter the data in the formula to obtain the SRCC:

$$R = 1 - \left(\frac{6 \times 34}{512 - 8}\right)$$

$$= 1 - \left(\frac{204}{504}\right)$$

$$= 1 - 0.4$$

$$= 0.6$$

- Assess the SRCC obtained, using these guidelines:

  O = No correlation (i.e. the two sets of compared data appear to exhibit a totally random pattern of 'linking' with each other)

  Up to ± 0.5 = Weak correlation

  ± 0.5 to ± 0.7 = Significant correlation

  ± 0.7 to ± 0.9 = Strong correlation

  ± 1.0 = Perfect correlation (i.e. both rank orders in the seven-column table are either the same or are in perfect reverse order)

The same technique can be repeated for each further pair of data to be investigated (e.g. drumlin length/width and drumlin width/height). On completion, it is useful to display all the SRCC calculations in descending order in a further table; such a table should provide a sound basis from which to start written observations.

Unit 5
PLATE TECTONICS, VOLCANOES AND EARTHQUAKES

The evidence for continental movement

During the early seventeenth century, the respected philosopher and scholar Francis Bacon observed from map evidence that the eastern coastline of South America appeared to 'fit' into the western coastline of Africa. During the latter part of the seventeenth century and throughout the eighteenth and nineteenth centuries, other leading philosophers and scientists also pointed out observable similarities in the shape of the coastlines of other adjacent continents. However, it was not until the early twentieth century that scientists and explorers had amassed enough primary evidence for a theory of 'continental fit' to be put forward. In the event, two totally independent scientists advanced almost identical theories at approximately the same time; both supported by remarkably similar evidence. The work of Albert Wegener is associated with the embryonic ideas about continental movement, whilst the work of his American counterpart, F.B. Taylor, is now less widely referred to.

FIGURE 5.1 Faunal evidence of continental drift

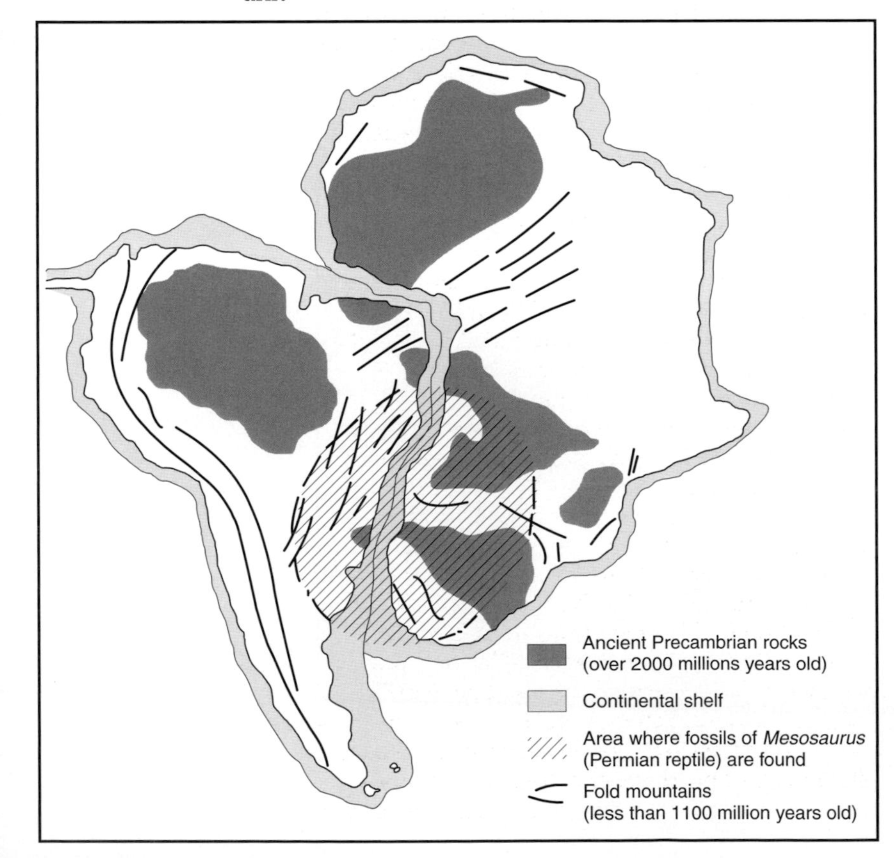

In 1912, Wegener (a professor of meteorology and geophysics in Germany), proposed a revolutionary theory that, at some unknown time in the distant past, all the continents had been joined as a super-continent which he called **Pangaea**. Wegener's evidence suggested that Pangaea had existed about 250 million years ago, during the Carboniferous Period. Wegener suggested that the initial 'split' of Pangaea had created two super-continents which he called **Laurasia** (or the Northern Continent) and **Gondwanaland** (the Southern Continent). These two land masses later divided, eventually producing the pattern of continents which we recognise today.

Wegener's 'evidence' was drawn from a number of scientific disciplines:

1 Palaeoecology and palaeobiology (which study respectively the habitats of fossilised animals and plants and the distribution of flora and fauna throughout geological history). Both fossilised flora and fauna distributions indicate that similar species co-existed in similar habitats on continents which today are separated by thousands of kilometres of sea. As it would have been impossible for many of these species to cross such wide expanses of ocean, Wegener concluded that, in the past, the distance between the landmasses must have been negligible. Specific examples include the case of *Mesosaurus* – a small reptile which lived during Permian times. These animals are known to have existed only in freshwater habitats and could not have survived a journey across a wide ocean; yet the fossils of identical specimens have been found thousands of kilometres apart in matching habitats on continents which are widely separated (Figure 5.1). Similarly, Cambrian organisms known as *Archaeocatha* (a coral-like organism) now occur in rocks on different continents separated by entire oceans; yet coral cannot survive in deep, cold water. It is fossil evidence such as this that led Wegener to believe that a continental rift had occurred by the Cretaceous period, (because such shared fossil histories cease by that particular time).

2 **Palaeogeography** also suggested to Wegener that the continents were once joined. Rocks of similar age, type, formation and structure are found in both south-eastern Brazil and southern Africa. Wegener discovered that the coastlines of these two continents appear to match closely. More significantly, once the two halves had been re-joined, the palaeogeographic evidence fitted together, as Wegener put it 'like lines of print would match if you ripped a newspaper in half and then rejoined it along the tear' (Figure 5.1).

3 **Palaeoclimatology** (the study of climates in past times). Study of both sedimentary rocks and their fossils allows scientists to reconstruct past climates (and other environmental conditions). This was the evidence that most fascinated Wegener, for he was a meteorologist. He was intrigued that coal deposits occurred in North America, Britain, north-western Europe and Antarctica. Coal forms in warm (semi-tropical), wet conditions, but such conditions do not exist in any of these locations today. By the early twentieth century, scientists had also discovered overwhelming evidence of a major period of glaciation affecting the southern continents (including peninsular India) during the Carboniferous period (see Figure 5.2), yet these areas are tropical and sub-tropical today. Ice Ages are associated with the advance of polar ice and originate in polar regions. If areas which are now tropical have been glaciated in the past, we can confidently extrapolate that they must, at that time, have been nearer to the poles than they are today – or that the magnetic pole itself migrates. (Alternatively, of course, both of these possibilities may be true). Wegener did not elaborate upon these possibilities, but did support his hypothesis further with speculation about the distribution of rocks such as sandstone and limestone across western Europe; both rocks are tropical in origin, but are now widespread across temperate northern regions.

Wegener's greatest difficulty was that he could not support his theory with a mechanism which explained how his '**continental drift**' might come about. During a trip to Greenland he observed icebergs floating in the sea, and considered the possibility that the continents might 'float' upon underlying rock structures, but, whilst this captured the imagination of some scientists, it was deemed by many to be far too fanciful. Wegener's theory was, in fact, heavily attacked, not because of the lack of evidence but because his fellow academics, researchers and students could not begin to understand how the necessary movements might occur. Consequently, the basic ideas themselves together with Wegener's evidence were dismissed.

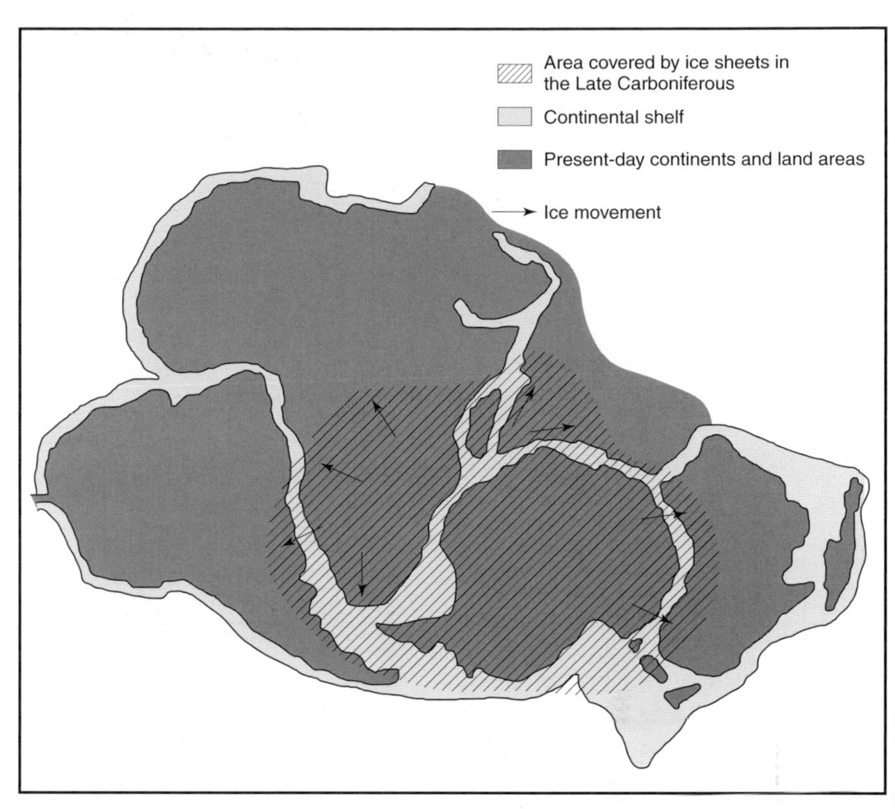

FIGURE 5.2 Palaeoclimatic evidence for continental drift

One of the difficulties in discovering the mechanics of continental movement lay in the fact that those scientists who were interested looked for answers on land; but the key to the solution actually lay deep on the ocean beds. It was not until the late 1940s that scientists turned their attentions to investigations of the sea/ocean beds. This interest had its origins not in geophysics – or even scientific curiosity – but rather in defensive operations following the Second World War, coupled with perceived threats during the escalating Cold War. Prior to the World War, it had been assumed that the ocean floors were flat, plain-like features, and because of this there had been little interest in the ocean floor. However, aerial reconnaissance of the oceans during the Second World War began to suggest that these preconceptions were wildly inaccurate and created renewed interest amongst the scientific community. New advances in radar and sonar facilitated attempts to map the ocean floor. The first major discovery was that ocean deeps do not abut the continental landmass. Instead, the continents are surrounded by a zone of shallow water (about 200 m deep) which extends for several hundred kilometres outward from the coast. At this point, the water deepens relatively quickly to reach depths of several thousand metres (Figure 5.3). This area of shallow, coastal water is known as the **continental shelf.** For those scientists who were still intrigued by Wegener's theories, this was a major advance; they discovered that if they defined

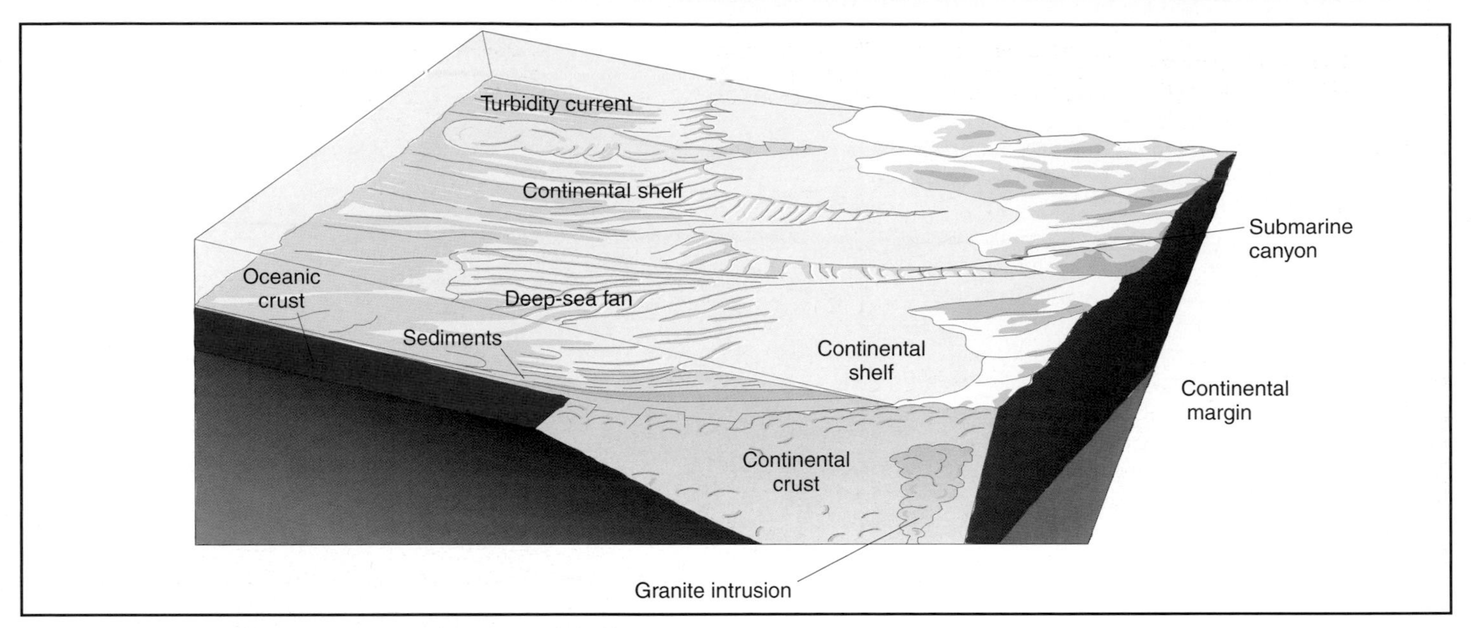

FIGURE 5.3 Cross-section through a typical continental shelf

the continents by the edge of their continental shelves instead of by the present coastline, many of the continents fit back together again even better than was previously thought.

In 1948, an American called Ewing discovered that there were marine mountain ranges which in some places extended the whole length of the ocean bed. Furthermore, Ewing discovered that these mountain ranges were made of volcanic rock and, even more importantly, were of very recent origin. This blew away existing ideas about the oceans, their topography, structure, geology and age. This time, although the scientific community could not explain what had been discovered, the evidence itself was accepted in its own right.

During the 1950s, there were three further major advances. One was the development of, and rapid advances in, diving equipment (especially mini-submarines able to operate at great depths) which would eventually allow first-hand observations of the sea floor. Secondly, there were important discoveries in the field of palaeomagnetism (the study of fossil magnetism) and finally there were significant advances in the accuracy of calculating the age of rocks and other sediments (primarily by radio-carbon dating).

In the field of geophysics, it was discovered that, periodically, Earth's magnetic field has been reversed. It is now believed that such reversals have occurred on more than 150 occasions during the last 76 million years. New evidence showed that, at various times during Earth's history, magnetic north had actually been located in those areas which we currently consider to be 'south'. Earth does not suddenly and unexpectedly 'flip' over – polarity merely reverses and magnetic north relocates in the opposite alignment. Studies of basaltic lavas also showed that minerals within the cooling rock align themselves in the direction of the magnetic pole as it exists at that time. Therefore, such rocks carry a permanent magnetic 'fingerprint' which allows us to establish where the magnetic pole was at the time they cooled. Basaltic rocks are now known to make up much of the ocean floor; consequently, the ocean floors contain information which allows us to correlate the palaeomagnetism in rocks at one site with that of rocks found at another site. During the 1950s, these basaltic ocean floor rocks were dated, and it was discovered that none of them are more than 200 million years old. This astounded the scientific community – which had assumed that all of the Earth's surface was of approximately the same age. Rocks 3850 million years old are found on land – how could the seabed be so much younger? Further discoveries, which followed on from this work, revealed that the sea floor is not of a constant age. Instead, it is considerably younger along the mid-ocean mountain ranges and oldest along the continental margins at either side (Figure 5.4).

The 1960s brought a revolution in thinking and understanding both about the ocean floors generally and continental movement specifically. This revolution was begun in 1960 by an American named Hess. Hess discovered that sea floor temperatures were significantly higher along the mid-ocean ridge crests than elsewhere on the ocean bed; he also suggested that such higher temperatures would cause rocks to expand and this expansion would both 'push up' the sea floor and lower the density of the rocks. He went on to postulate that mid-ocean ridges were possibly locations where hotter and less dense parts of Earth's deep interior rose to the surface simply because they 'obeyed' the basic Laws of Physics. Hess also observed that mid-oceanic ridges are inevitably free from sedimentary deposits, but that sediment coverage becomes thicker proportionally with decreasing distance from the coast. Whilst he considered that this may be merely the result of

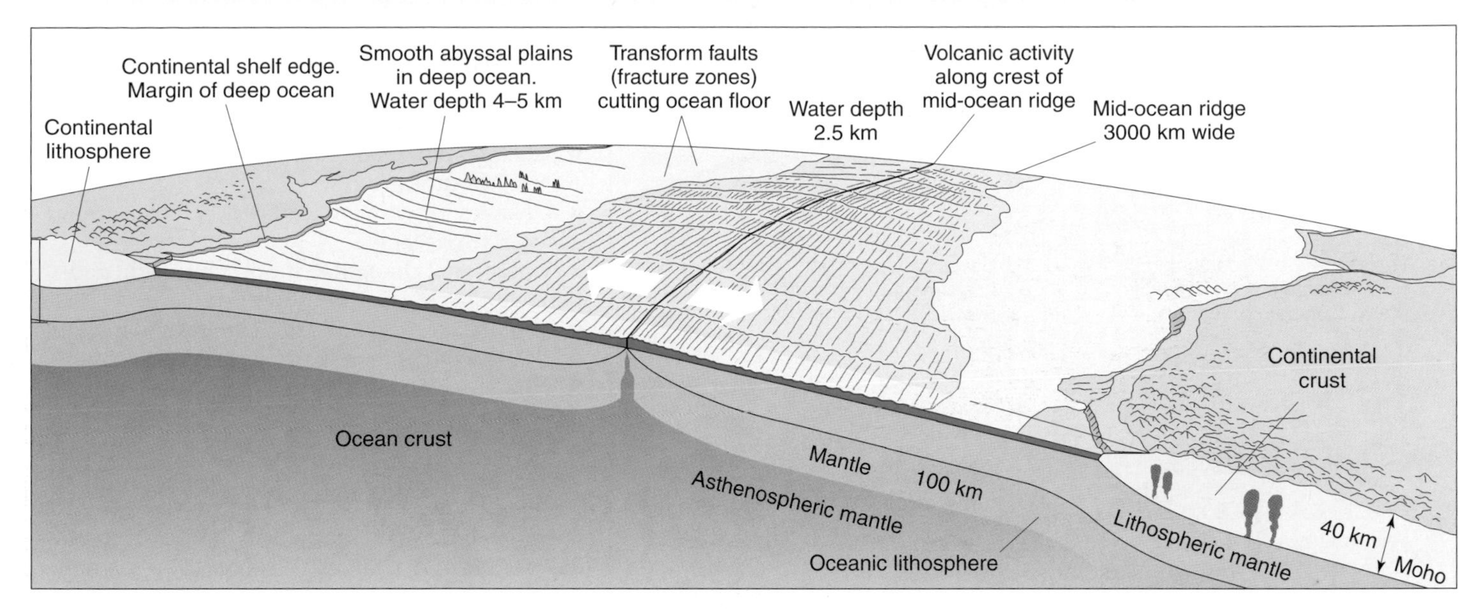

FIGURE 5.4 The topography of the sea floor

distance from the sediment-producing continents, he also hypothesised that there was a correlation with the age of the ocean floor; ocean floor which was older had increased sedimentary cover, whilst that which was younger has little or no sedimentary coverage. Hess introduced a new concept to the scientific community – **sea floor spreading**. By 1962, Hess's work had led him to suggest that the Atlantic Ocean may be opening or 'spreading' at a rate of about 5 cm/year.

Hess's work was largely substantiated during the latter part of the 1960s by the first underwater photographs of active submarine ridges, and also by palaeomagnetic investigations which revealed that magnetic reversals in the basalts of the Atlantic are almost symmetrical on both sides of the Mid-Atlantic Ridge and are of identical age on either side (Figure 5.5).

Even during the 1960s, and in the face of what now seems to have been overwhelming evidence collated from a wide range of disparate scientific fields, many scientists remained totally opposed to the concept of 'moving continents'. However, the latter half of the 1960s saw the development of computers capable of simulating and extrapolating features and events which occur in reality. They were limited only by the creativity of their programmers, not by pre-existing belief systems, and this opened the way for new and exciting developments in geology and geophysics. As new material was collected about ocean/continental margins scientists finally had sufficient data (and a means of processing large and complex datasets) to begin to 'piece together' the complexities of continental movements, and move from an elementary idea of continental drift to a new model of **plate tectonics**. Such computer programmes also allowed researchers to produce a 'best-fit' representation of Pangaea. The 'mysteries' today concern the existence/positions of continental landforms prior to Pangaea. When and how did land masses actually come into existence?

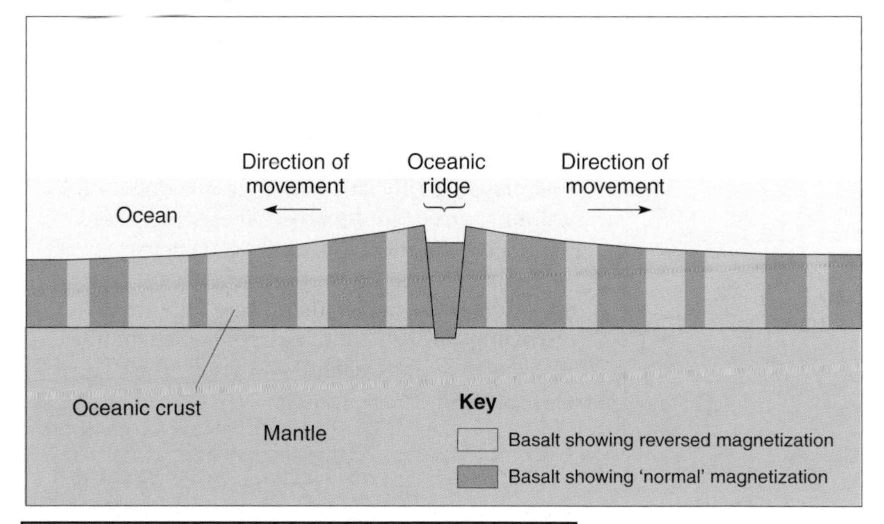

FIGURE 5.5 Palaeomagnetism in the Atlantic Ocean

ACTIVITIES

1 Research either:
a) Greater detail about the evidence for continental movement as introduced in this section.
or:
b) Further, additional, evidence for continental movement.

2 Choose one of the following essay-type questions to research and answer:
a) Describe and explain, with named examples, the mechanics of plate movement as it is currently understood and show how this explains the movement of the continents across time.
b) Does existing evidence, coupled with the theory of crustal movement, adequately explain the apparent movement of the continents? Support your answer with relevant illustrations and named examples.

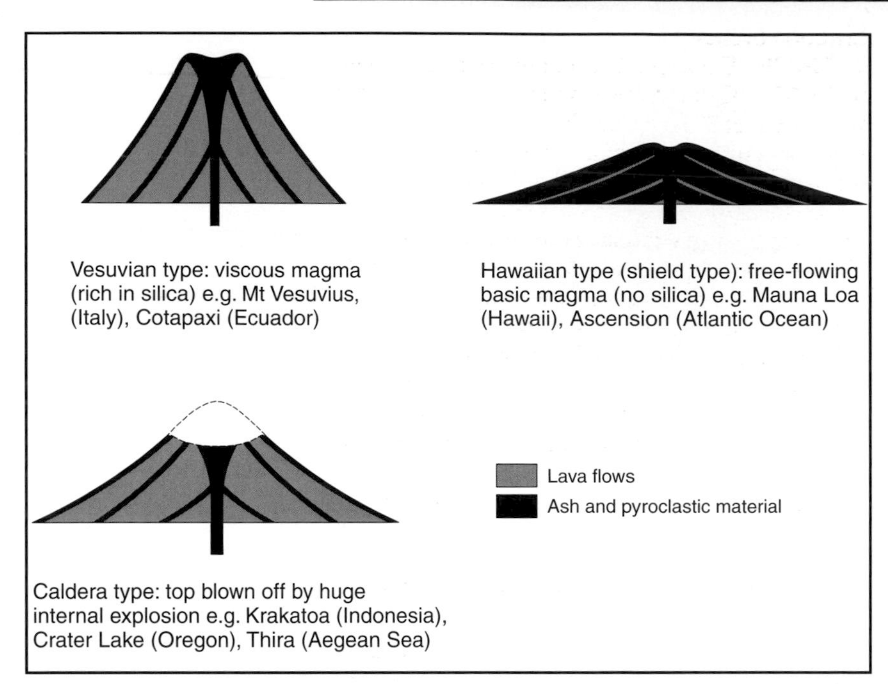

FIGURE 5.6 Types of volcano

Study 5.1 Volcano characteristics and eruption prediction – Southern Italy

The prediction of volcanic eruptions is not a precise science. It is, however, a very important research activity, because most eruptions occur with very little warning and it takes a considerable amount of time to evacuate large, densely populated areas. For example, 60 000 farmers and their families were only just rescued in time from the valleys of Mount Pinatubo in the Philippines, which erupted in 1991. The 1980 eruption of Mount St Helens had been predicted only 24 hours previously, by the observation of ground movements and seismic recordings associated with deep earthquake tremors. An estimated one million people have lost their lives over the last two millennia due to volcanic activity and many of these casualties have occurred in relatively few eruption events.

Eruption predictions may be based on very short-term evidence, some examples of which are listed in the next column:

- Seismic evidence of underground shock waves triggered by rock fracturing caused by large-scale movements of molten rock material.
- Gas emissions through fractured mountainsides.
- Changes in temperature and levels of ground water and surface lakes.
- Changes in sediment load within underground and surface streams.
- Unseasonal changes in the accumulation and depletion rates of ice and snow.
- Changes to electrical, gravitational and magnetic fields caused by the movement of molten rocks.
- Landslides triggered by subterranean movements.

Eruptions may also be predicted by studying the timings and nature of past volcanic events. Although it is possible to classify volcanoes, as shown in Figure 5.6, they are all unique structures and their eruptive materials may also have a range of different characteristics (Figure 5.7).

Italy's three major volcanoes (whose key features and eruptive events are shown in Figure 5.8) provide excellent examples of variations in eruption patterns, even though they all lie within a relatively small surface area and are located within the same subduction zone (Figure 5.9). Contrasting tectonic environments incorporating volcanoes are shown in Figure 5.10.

A recent evacuation plan for the Vesuvius area assumes that any major eruption of its volcano can be predicted at least two weeks ahead of the event. It also assumes that about 700 000 inhabitants can be evacuated within one week of the evacuation order being given. A proposed alternative plan called 'Vesuvius 2000' utilises computer modelling as its chief eruption predictor and assumes that a much shorter period of notice is likely to be available.

FIGURE 5.7 Volcanic lava comparisons

Comparison characteristic	Characteristics of andesitic (acid) lavas	Characteristics of basaltic (basic) lavas
Cooling rate	Rapid	Slower
Eruption pattern	Infrequent but violent	More frequent but less violent
Gas content	High	Low
Lava temperature	800 °C	1200 °C
Lava viscosity	High	Low
Material ejected	Ash, gases, lava, rocks and steam	Lava and steam
Plate margin-type locations	Destructive	Constructive
Silica content	Above 52 per cent	Below 52 per cent
Slope gradients on surface features produced when lava solidifies	Steep gradients	Gentle gradients
Example	Mount St. Helens, USA	Mauna Loa, Hawaii

Etna (3350 m), on the island of Sicily, has a long history of frequent volcanic eruptions. Its upper 400 m section is a strato-volcano consisting of several coalesced vents. Much of its surface is covered by solidified lava flows.

Eruption events

- 7 June 2000: Plume of ash 2.5 km high was so dense that it resembled a solar eclipse. City of Catania (population 1 million) 30 km away from summit – showered with black volcanic dust. Tremors created new crevices on the mountain's flanks.
- 2 June 2000: Lava and a column of ash ejected high into the air. Lava flowed 625 m down the slopes of the volcano.
- 27 April 2000: Fountain of incandescent lava ejected.
A plume of ash drifted into the path of an aeroplane, forcing it to land prematurely.
- 1 July 1998: Many volcanic bombs up to 1.5 m in diameter ejected; fell later on the eastern side of the summit cone.

Stromboli (900 m) forms one of the Aeolian Islands. It is 2 km in diameter, and rises 3000 m above the floor of the surrounding seabed. It is one of the most active of all volcanoes, having erupted almost continuously for at least 2000 years. Most Stromboli eruptions consist of small gas explosions which hurl incandescent blobs of molten lava above the crater rim. Several explosions occur every hour. Violent eruptions are rare, although some exceptions have resulted in fatalities.

Eruption events

- 1919: Bombs weighing up to 50 tonnes killed four people.
- 1930: Three people killed by **pyroclastic flows**. A fourth person scalded to death in the sea close to where flows entered the water.

Vesuvius (1281 m) is on the coast of mainland Italy, only a short distance from densely populated areas including the major seaport of Naples. Its caldera formed about 34 000 years ago, but the current Vesuvius cone is only 17 000 years old and formed after the collapse of Somma Caldera. Vesuvius is a polygenic volcano, which means that it was formed by a succession of eruptions rather than a single cataclysmic event. It has proved to be the most deadly of all three Italian volcanoes in terms of both fatalities and damage caused. It has exhibited a variable eruptive pattern in that its plinian eruptions have taken place every few thousand years, whilst its more regular sub-plinian explosive eruptions take place every few hundred years.

Eruption events

- 5960 BC: Cataclysmic event (no details available).
- 3580 BC: Cataclysmic event (no details available).
- 24 August 79 AD: Cataclysmic event. First volcanic eruption in history to be witnessed and recorded in detail (by Pliny the Younger, after whom this type of volcano has been named). Preceded by an earthquake on 5 February 62. Plinian-type eruption, which made the previously fertile slopes agriculturally unproductive for several centuries. 4 km^3 of ash ejected in 19 hours. Ejected column of ash 32 km into the atmosphere. Killed 2000 people. 3 m deep layer of tephra buried Pompeii, where excavations in sixteenth and twentieth centuries revealed molds of animals and humans.
  Herculaneum buried under 23 m-deep deposits of ash deposited by a pyroclastic flow of ash, lapilli and scoriae.
- 202, 472, 512, 685, 993, 1036, 1139, 1306, 1500: (minor)
- 1631: Cataclysmic event. Sub-plinian eruption. Killed 10 000 people. Destroyed many villages at foot of mountain. Lava flows reached coast.
- 1794: Cataclysmic event. Strombolian/effusive eruption. Destroyed Turre del Greco village.
- 1906: Cataclysmic event. Caused widespread damage to Ottaviano and San Guiseppe Vesuviano villages.
- 1913–1929: Long period of intense activity.
- 1933–1944: Long period of intense activity.
1933 eruption preceded by minor earthquake.
250 million cubic metres of lava flowed out of the volcano during this phase, which ended with frequent landslides of the crater walls.
Crater now 300 m deep.
- 18 March–19 April 1944: Effusive strombolian eruption. Destroyed two-thirds of buildings in San Sebastian al Vesuvio.
- 1983–85: 80 km^2 in the Bay of Naples area was uplifted by 1.8 m, damaging homes, the harbour and the tourist industry. 36 000 people had to be relocated.

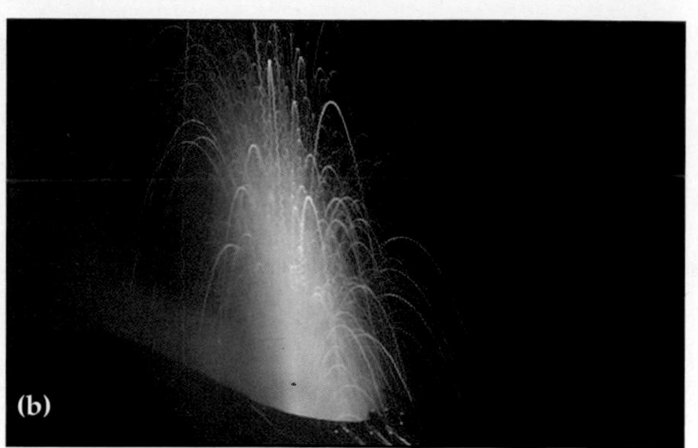

FIGURE 5.8 Italian volcanoes; a) Etna, b) Stromboli, c) Vesuvius

FIGURE 5.9 Vulcanicity in southern Italy

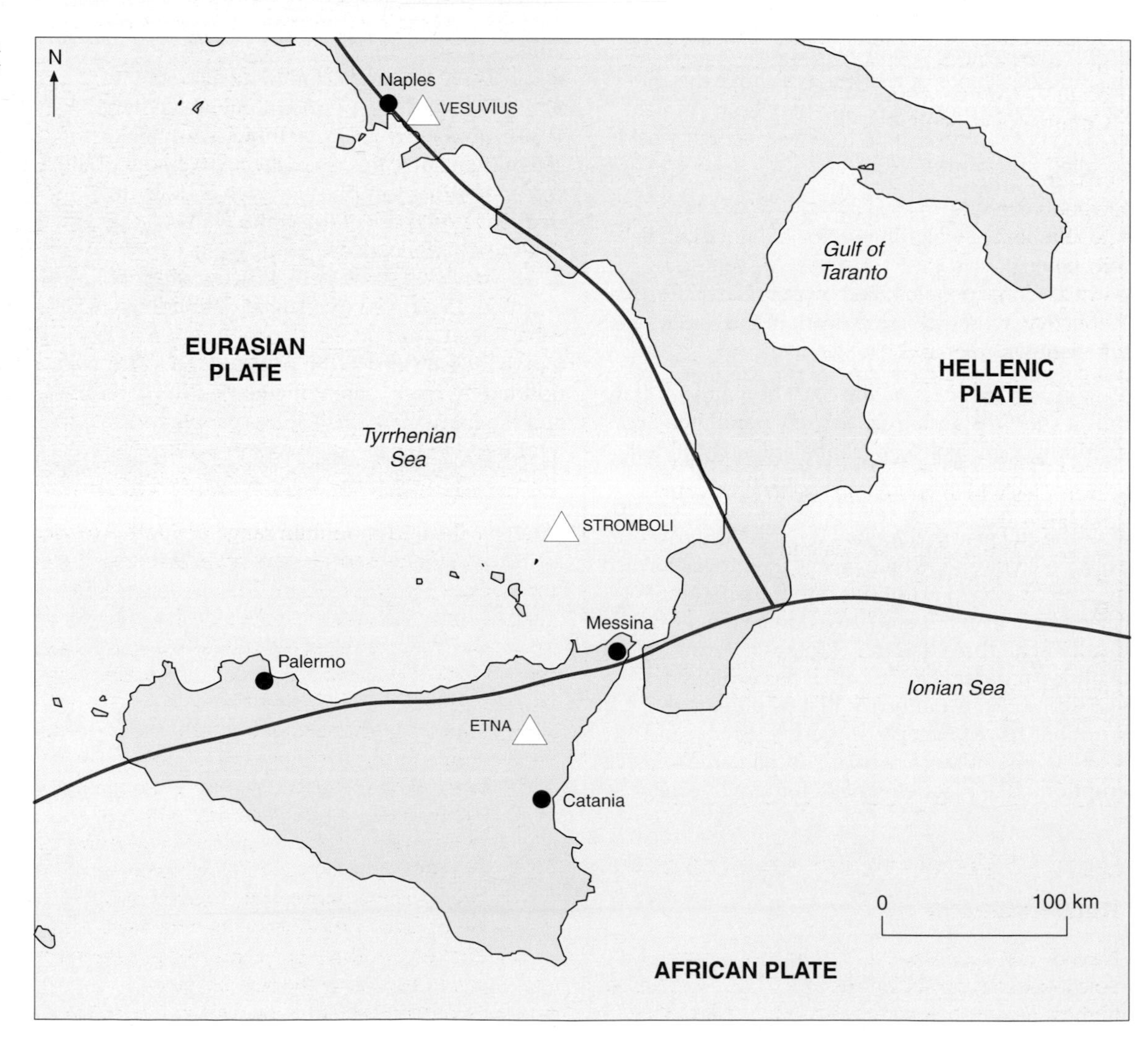

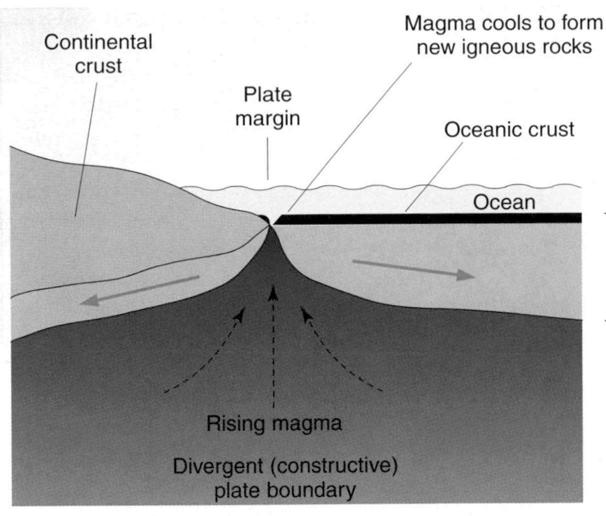

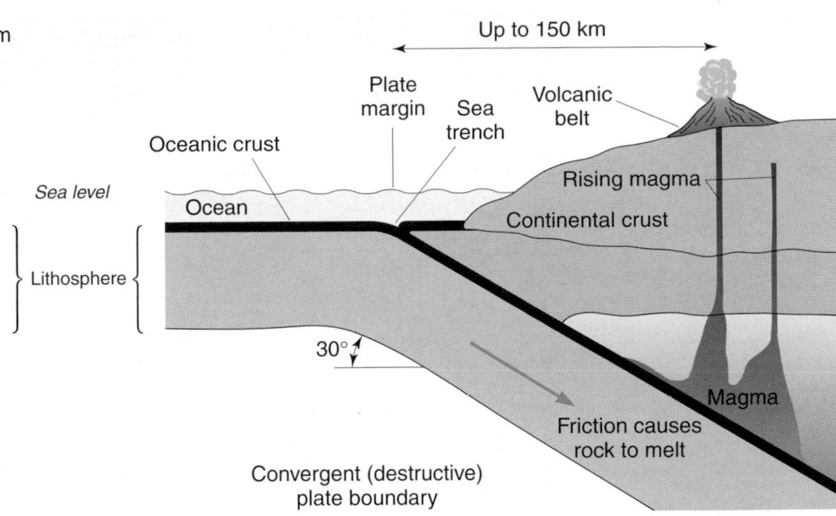

FIGURE 5.10 The two types of volcanic plate boundary

ACTIVITIES

1 Use a table (probably requiring no more than four columns) to compare the key features of Italy's three volcanic peaks.

2 a) Plot Vesuvius' listed eruptive events in the form of a time-line diagram, utilising appropriate ways of differentiating between the most dramatic and less dramatic events.

b) Computer simulations suggest that the most traumatic eruption events of Vesuvius appear to occur at approximately 500 year intervals. Comment on the likely accuracy of this assertion.

c) Assuming that this computer simulation is reasonably accurate, make a prediction as to when Vesuvius' next cataclysmic event is likely to take place.

3 Give reasons why the current Vesuvius evacuation plan is unlikely to prove sufficiently effective.

4 Use Internet sites to investigate examples of volcanic calderas (apart from Vesuvius), the events which created them and the effects which they have had on global climatic characteristics and ecosystems. Some of the better-known examples which you could use to start your research are:

- Lake Taupo, in New Zealand
- Krakatoa Island, in Indonesia
- Santorini Island (also known as Thira), in Greece.

5 Employing a range of methods of your own choice, utilise the wide range of information in Figures 5.6–5.10 to indicate the diversity of the environmental and eruptive characteristics of volcanoes.

Study 5.2 Volcanic lahars – Nevado del Ruiz, Colombia

Nevado del Ruiz is one of a cluster of active volcanoes in central Colombia. At 5321 m, it is the highest peak in this most northerly section of the Andes – the fold-mountain range of South America – and snowfields and glaciers cover its summit area throughout the year (Figure 5.11). In spite of its impressive height, Nevado del Ruiz is relatively easily to climb, and its accessibility to major urban centres makes it one of the most popular climbing and skiing localities in Colombia. Its fertile lower slopes support highly profitable coffee farms which have stimulated the development of many towns and villages. Volcanic ash breaks down into soil very quickly, releasing the minerals which were formed underground at very high temperatures as vital plant nutrients such as nitrogen and potassium. In some particularly active areas such as the island of Java in Indonesia, the volcanic soils are so rich that two or three crops may be grown every year without depleting the soil, and rural population densities are, therefore, amongst the

FIGURE 5.11 The summit of Nevado del Ruiz

FIGURE 5.12 Lahar on slopes of Nevado del Ruiz

highest in the world. Some of the farming communities in neighbouring Papua New Guinea perform special tribal rituals to stimulate fresh falls of the highly-prized ash.

This study is based on the eruption of Nevado del Ruiz which took place on 13 November 1985. Although this eruptive explosion was relatively modest, it did occur after nightfall, when the majority of the area's population was at its most vulnerable. 22 000 people died and a further 6000 were rendered homeless within a very short period after the initial eruption. This event proved to be the twentieth century's second most deadly eruption event, the 1902 Mont Pelee eruption (in Martinique) having resulted in 29 000 fatalities.

The 1985 eruption was by no means unique in the history of Nevado del Ruiz, as similar events had occurred previously, in 1595 and 1845, when summit eruptions also ejected hot ash which melted its surrounding glaciers to create **lahar** mudflows. The volcano remained continually active for several years after the 1985 event, culminating in particularly violent eruptions in 1991 and 1992.

The volcano had started to become restless in 1984, but there was a distinct lack of urgency amongst the local villagers and farmers to move to safer areas (due to fears that their properties would be looted during their absence). At first, there was no significant lava flow, but increasing emissions of gas, steam, rocks, ash and sulphur clouds combined to melt some sections of the summit's permanent cap of ice and snow. This meltwater combined with ancient layers of volcanic dust to produce a mudflow on 8 November (Figure 5.12). The lahar consisting of this deadly combination was 20 m deep in parts and followed the River Languillian valley over half way to Armero (population 22 000 and 45 km from the summit).

During the night of the 13 November, a much larger eruption took place and increasing quantities of magma in the volcano's upper pipe warmed the peak sufficiently to melt the remaining sections of the ice cap. Prolonged heavy rain then added to the volume of surface water to increase the mudflow's depth to 40 m at particular points and increase its speed to 50 km/hr. All bridges, houses and vehicles in its path were swept away, their debris adding to the lethal mass of mobile material. One huge boulder with a volume of 208 m^3 was carried a distance of 300 m.

This particular lahar moved with great speed. It began to engulf Armero at 11 pm and deposited 8 m of mud during the rest of that night. The Colombian Government declared the area a military zone and appealed for immediate international help. The International Red Cross and United Nations Disaster Relief Organisation acted as co-ordinators and flew in medical supplies, tents, digging equipment and personnel. Twenty-five Colombian Air Force helicopters as well as many British and French aircraft ferried survivors to safety, but were seriously hindered by clouds of volcanic ash which threatened to block their engine intakes. Similar difficulties were reported by many of the road vehicles involved in the rescue operation. Bloated human and animal corpses began decomposing where they had come to rest in the grey mud, prompting concern about outbreaks of yellow fever, gastro-enteritis and respiratory infections.

A detailed investigation into the Nevado del Ruiz disaster was subsequently co-ordinated by 'Centro Regional de Sismologia para America del Sur' (CERISIS), an organisation based in the Peruvian capital, Lima. CERISIS had been created to collate and interpret data on all active South American 'hot spots'. CERESIS used this processed data and the expertise it had acquired to assist South American countries to develop policies which are both effective and realistic for their level of economic development. The conclusions with respect to the 1985 Nevado del Ruiz disaster were that there had been:

- A shortage of two-way radios for communication.
- A previous lack of systematic volcano monitoring.
- An absence of effective policies for disaster prevention and preparedness.
- Insufficient co-ordination among the various organisations responsible for hazard assessment and emergency planning.
- A widespread failure to devise evacuation plans and lines of communication.
- An unwise over-reliance on warning systems alone.

CERESIS then submitted the following recommendations for future action:

- To produce a video aimed at educating communities at the greatest risk from volcano-related hazards.
- To arrange an international workshop at Bogota on the theme of earthquake response.
- To identify ways to share expertise and resources on a regional basis.
- To identify strategies for reducing the number of casualties in future disasters.
- To identify ways in which civil defence teams could be used more effectively.

- To produce maps identifying volcanoes likely to pose the greatest risk.
- To decentralise administrative responsibility for emergency planning and response.
- To establish more effective links between the mass media and the scientific community.
- To emphasise the importance of ongoing geological and historical analysis of high-risk areas.
- To stress the need for sufficient quantities of the most appropriate types of equipment needed to sustain the continuous monitoring of high-risk volcanoes, with international organisations being encouraged to provide such support.
- To undertake further research into the methods by which a lahar might be stopped or re-routed, e.g. building mudflow catchment basins, diversion channels, tunnels and concrete obstructions.
- To recognise that the best preventative method is to establish warning systems. These would include the use of seismometers able to track a lahar's passage down a valley as well as rainfall collection gauges to monitor changes in precipitation.

ACTIVITIES

1 Devise a flow-diagram to trace the chief causes of, and stages in, the Nevada del Ruiz lahar disaster described in this study.
2 List separately all the 'conclusions' and 'recommendations' of CERISIS with respect to the 1985 disaster – both lists to be in the order of importance that *you* attach to them.

Study 5.3 Volcanic features – British Isles

Although it is approximately 35 million years since the last period of major volcanic activity in the British Isles, its landscapes still exhibit many types of relict volcanic features. These may be usefully sub-divided into the following categories:

- Batholiths, stocks and **bosses**.
- Ring dykes, dykes and sills.
- Extinct volcanoes and volcanic plugs.
- Lava plateaux.
- Igneous rocks e.g. basalt, dolerite, gabbro and granite.
- Metamorphic rocks e.g. gneiss, schist and slate.

The table in Figure 5.13 lists selected examples of the main types of feature within the British Isles and provides details of their locations and implications for local populations. The photographs in Figure 5.14 illustrate some of the most important of these relict features.

Location	Type of volcanic feature/material	Relevance to local community/economy
Dartmoor	Batholith/granite	Rugged open countryside granted National Park status. Used for military exercises
Castle Rock, Edinburgh	Volcanic plug/basalt	Steep-sided defensive site of Edinburgh Castle
Giant's Causeway, County Antrim, Northern Ireland	Lava plateau/basalt	Tourist attraction
Helvellyn, Cumbria	Part of an eroded volcano/tuft, rhyolite and andesite	Rugged, mountainous landscape; steep crags ideal for rock climbing
Lismore Island, Argyll and Bute	Dolerite dykes	Natural field boundaries. Shelter for houses
Shap, Cumbria	Granite	Quarrying of building materials
Whin Sill, Northumberland	Sill/dolerite	Foundation for part of Roman wall. Initiated development of High Force Waterfall on River Tees. Formed Farne Islands at eastern extremity

FIGURE 5.13 Volcanic landforms in the British Isles

FIGURE 5.14 a) Granite batholith, Dartmoor b) Volcanic plug, Castle Rock, Edinburgh

FIGURE 5.14 *(cont.)*
c) Basalt lava columns, Giant's Causeway, County Antrim
d) Eroded volcano, Helvellyn
e) Dolerite sill, Northumberland
f) Lismore 'dyke'

ACTIVITIES

1 a) Use the information in Figure 5.13 to locate each of its listed features on an outline map of the British Isles.
b) Comment on any distribution patterns which are evident from your completed map.
2 Summarise the relevance of the relict features included in this study to people living in the British Isles. You should consider the following factors in your response to this activity:
- population distribution
- economic potential
- recreational opportunities.

3 Studies 5.1–5.3 have introduced a number of ways in which past and present volcanic activities have created opportunities for human enterprise. For this activity, the members of your class should form sub-groups, and each of these should investigate one area of opportunity in a place or places outside the British Isles. You should then present your findings to the assembled group. It will be necessary for you to conduct some research work. Suggested areas for research include:
- energy generation
- mineral and rock exploitation
- recreation and leisure opportunities
- soil formation and agriculture.

Study 5.4 Earthquake intensity assessment – California

The assessment of earthquake strength and impact is not an easy task. One of the first serious attempts at earthquake assessment took place in Italy towards the end of the eighteenth century, when over 1000 earthquakes were placed in four best-fit categories described as slight, moderate, strong and very strong. During the twentieth century, two scientists developed accurate and reliable assessment systems. Interestingly, they chose very different criteria by which their assessments should be based.

In 1902, the Italian seismologist Guiseppe Mercalli devised a system based on earthquake intensity. Mercalli's original scale has since undergone several refinements to produce more regular intervals between the steps and to up-date the criteria by which observations are made. The result of these refinements is called the Modified Mercalli Scale, which has twelve grades of intensity identified by the Roman numerals I–XII.

In 1935, an American physics graduate, Charles Richter, adopted a quite different approach, which was based on the magnitude of an earthquake. Shortly after being appointed in charge of the Seismological Laboratory at Caltech in southern California, Richter realised that much of his time was being wasted trying to provide local newspaper reporters with accurate assessments of earthquake incidents within the Los Angeles area. In order to save himself time in future, Richter devised a revolutionary system based on data provided by his laboratory's network of increasingly reliable seismometers (Figure 5.15).

Because the assessment systems developed by Mercalli and Richter are based on very different criteria, it has never been easy to devise a truly reliable comparison between them. Figure 5.16 represents one attempt to achieve such a comparison; it also indicates how frequently California has experienced earthquakes at each level of intensity.

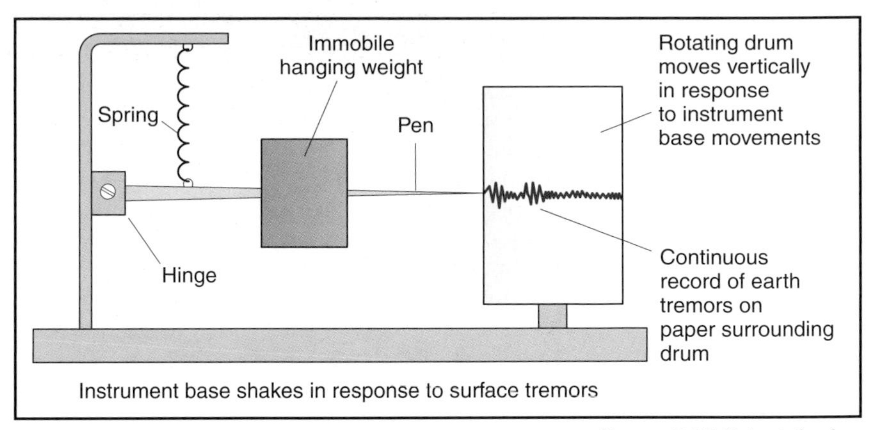

FIGURE 5.15 Principles by which seismometers function

Earthquake event scale descriptors	Mercalli Scale	Richter Scale	Number of earthquake events at Richter Scale ranges 6.0–6.4, 6.5–6.9, 7.0–7.4, 7.5–7.9, 8.0–8.4 during each of the following 50 year periods		
			1846–1895	1896–1945	1946–1995
Instrumental: detected only by seismographs	I	2			
Feeble: felt only by sensitive people	II	3			
Slight: like the vibrations due to a passing light lorry	III				
Moderate: like the passing of a heavy road vehicle; rocking of loose objects, including standing cars	IV	4			
Rather strong: felt by most people; church bells ring	V	5			
Strong: most people frightened; windows broken; dishes fall out of cupboards	VI				
Very strong: general alarm; walls crack; plaster falls	VII				
Destructive: car drivers find it difficult to steer; masonry cracked; chimneys fall	VIII	6	29	34	40
Ruinous: general panic; ground cracks appear and pipes break open	IX		11	12	18
Disastrous: ground cracks badly; many buildings destroyed; landslides on steep slopes	X	7	2	9	6
Very disastrous: most buildings and bridges destroyed; all services (railways, pipes and cable) out of action; great landslides and floods; dams badly damaged	XI		1	1	0
Catastrophic: total destruction; objects thrown into air; ground rises and falls in waves; cracks open and close	XII	8	1	1	0

FIGURE 5.16 A comparison of the Mercalli and Richter Scales

ACTIVITIES

1 Make your own copy of Figure 5.16 but without the information on Californian earthquakes.
2 a) Use appropriate graphical techniques to display the Californian earthquake frequency/intensity information in Figure 5.16.
b) Comment on any trends which are evident from your completed graphs.

Study 5.5 Tsumanis – Papua New Guinea

Tsunamis are almost certainly the most devastating of all the effects of earthquakes above Richter Scale 5.5. Such waves are usually at their most devastating when crossing inshore waters, particularly in and around enclosed coastal features such as bays and harbours. Tsunamis can be caused by non-seismic jolting events such as nuclear explosions, meteorite impacts and major landslides, but the most impressive examples have always been triggered either by earthquakes or the eruption of submarine volcanoes.

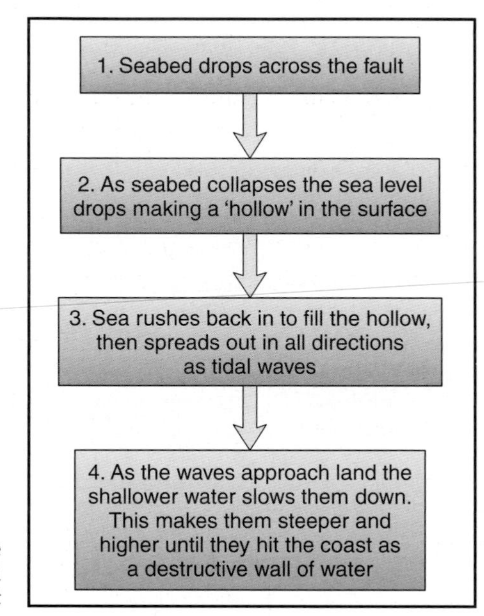

FIGURE 5.17 The sequence of a typical tsunami event

FIGURE 5.18 Tsunami events resulting in over 20m waves

Year	Earthquake location		Earthquake magnitude (Richter Scale)	Wave height (m)	Number of fatalities
	Lat	Long			
1611	39 °N	143 °E	8.0	25	5000
1746	12 °S	77 °W	8.0	24	4800
1771	24 °N	124 °E	7.4	85	13 486
1854	34 °N	138 °E	8.3	28	1000
1868	19 °N	155 °W	7.5	20	116
1871	2 °N	125 °E	N/A	25	277
1877	20 °S	70 °W	8.3	21	512
1883	17 °S	105 °E	N/A	35	36 500
1896	40 °N	144 °E	7.6	38	26 360
1933	39 °N	145 °E	8.3	29	3000
1944	34 °N	137 °E	8.0	20	998
1946	53 °N	164 °W	7.3	35	165
1960	40 °S	75 °W	8.6	25	1260
1964	61 °N	148 °W	8.5	70	123
1992	9 °N	122 °E	7.5	26	15 000
1993	43 °N	139 °E	7.7	32	330

Tsunamis should not be referred to as 'tidal waves' because they are totally unrelated to the Sun and Moon's gravitational forces which produce the familiar (and predictable) pattern of twice-daily high and low tides. Neither should tsunamis be confused with 'wind waves' as these are caused by local wind conditions and do not involve the mass-movement of seawater from surface to bottom.

In mid-ocean, tsunamis can be gently heaving swells capable of making even the largest ships rise and fall, but without endangering them in any way. Such tsunami waves may be no higher than the desk in your classroom, but they often travel great distances, and at astonishing speed, sometimes traversing entire oceans at over 1000 km/hr. When approaching a coastline, their waves slow down due to increasing friction with the sea floor. At the same time, their height increases very sharply, in extreme cases exceeding 65 m. Tsunamis rarely occur in isolation and it is usual for them to form a series of waves 15 minutes to one hour apart. Figure 5.17 shows a typical sequence of events from earthquake to tsunami. Figure 5.18 lists some of the world's most devastating tsunami events and their consequences in terms of loss of life.

ACTIVITIES

1 a) On a world map, plot the named locations of the tsunami events listed in Figure 5.18, adding the total number of fatalities given for each event.
b) Comment on the distribution of these events, with reference to the location of plate boundaries and volcanic hot spots.
c) Does there appear to be any significant trend in the number of fatalities due to the tsunamis included on your map? Offer plausible explanations for the presence (or absence) of any such trend.

Study 5.6 Major earthquake events in MEDCs – California, 1989

The earthquake which shook the San Francisco Bay area on 17 October 1989 was the most severe tectonic event to hit California since the disastrous earthquake of 1906. The 1989 earthquake resulted from a 40 km movement of the San Andreas Fault in the Santa Cruz region, 100 km south-east of San Francisco, resulting in fissures up to 8 m wide. Tremors were detected as far away as Los Angeles (700 km to the south) and damage occurred within a 1500 km radius (Figure 5.19).

The 1906 event was much more powerful than that of 1989 (8.3 on the Richter Scale as opposed to 7.1) and killed far more people (700 as opposed to only 62 deaths). In 1906 a total of 28 188 buildings were destroyed, many of them timber-built houses. Even the sturdy, stone-built San Fransisco City Hall was ruined on that occasion. The 1906 earthquake did, however, serve as a stern warning for the future and much was done to make the key buildings and infrastructure networks increasingly earthquake-proof.

An outstanding example of what had been achieved during the 83 intervening years was the survival of the Golden Gate Bridge during the 1989 earthquake, which had been strengthened to withstand shocks up to 8.5 on the Richter Scale. San Fransisco's many modern skyscraper office and apartment blocks had also been designed in accordance with much more stringent building regulations and they survived the 1989 earthquake remarkably well. An Earthquake Preparedness Budget had been established and by 1988 was set at $19 per person throughout the Bay Area. The chief features and consequences of the 1989 earthquake are listed below:

- It occurred at 1704 hours on 17 October 1989, i.e. during the build-up of traffic at the start of the evening rush hour.
- Its initial shock lasted 15 seconds and was originally estimated at 6.9 on the Richter Scale, but this was later revised to 7.1. A major aftershock of 5.2 struck 37 minutes later.
- 62 people were killed, 3757 injured and over 12 000 rendered homeless.
- Many bridges and viaducts suffered major damage throughout the San Francisco conurbation. The most serious road disruption occurred on a 1.6 km length of the twin-deck Nimitz Highway section of Interstate 880. Its upper deck carrying traffic into the heart of San Francisco collapsed to within only 45 cm above the lower deck being used by Oakland-bound vehicles (Figure 5.20). Cars, delivery trucks and even fully-loaded lorries were crushed between the two layers of road. 250 people were killed instantly. The Nimitz Freeway had been strengthened prior to the earthquake, but obviously not sufficiently effectively as its upright supports failed to withstand the whiplash action caused by the shock. California State Governor Deukmejian appeared surprised by what had happened, as he had been led to believe that all above-ground sections of highway within the Bay Area had been strengthened adequately.
- A 9 m-long section of the Bay Bridge collapsed, further disrupting communications and plunging a number of cars down the section incline (Figure 5.21).
- Road traffic throughout the area was seriously disrupted, the most notable incident being a 50-car pile-up at a major road junction in the San Jose district.
- At least 27 serious fires broke out, many of them the result of fractured gas mains. The most serious blaze was in the Marina District, where a series of apartment blocks sank into a lagoon filled with bay mud. The fire and gas explosions along San Francisco's northern shoreline were so intense that fire-fighting teams had to be continually sprayed with water. The smell of leaking gas was widespread and the public were cautioned against smoking and lighting open fires.
- Over one million people lost electrical power for

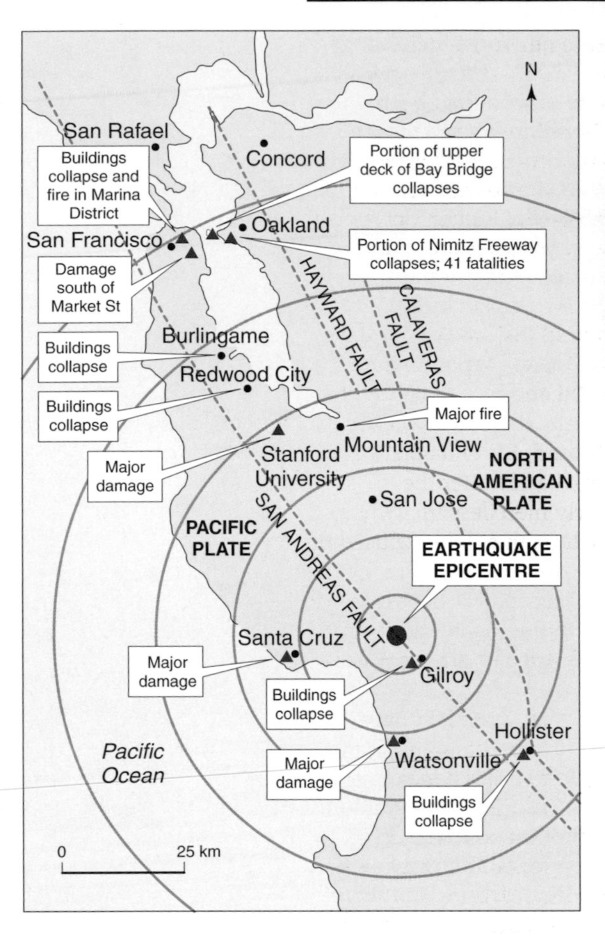

FIGURE 5.19 Effects of the 1989 earthquake in California

FIGURE 5.20 Earthquake damage to Highway 880

FIGURE 5.21 Earthquake damage to Bay Bridge

up to three days. Emergency telephone cables failed due to a fire in the main telephone equipment room.

■ Waste water and raw sewage filled the basement of the War Memorial and Veterans' Building.

■ Widespread looting was reported, including gangs wielding baseball bats to break into shops in the downtown area. District Attorney Arle Smith was moved to declare that anyone caught looting would be prosecuted and would not be released on bail whilst awaiting trial.

■ A 1.3 m high tsunami was triggered in Monterey Bay by a huge undersea landslide. The sea level at Santa Cruz dropped suddenly by 1 m.

■ The cost of earthquake-related damage was calculated at £6 billion, a huge financial burden, even to one of the richest states within the world's most affluent country.

A U-2 'spy-plane' from Beale Air Force Base flew reconnaissance missions over the San Francisco area, and its aerial photographs were later used to pinpoint structural weaknesses in major buildings. The US Navy helicopter carrier *PELELIU* provided temporary accommodation for 300 people who had lost their homes.

The most devastated areas were toured by Vice-President Dan Quayle, and President Bush declared seven districts to be a disaster zone. He pledged a $3.45 billion earthquake relief package for the Bay Area.

ACTIVITIES

1 Using detailed information as necessary, summarise the ways in which California's earthquake preparations appeared to have been:

a) effective

b) inadequate.

Study 5.7 Major earthquake events in LEDCs – Turkey, 1999

At 3.02 am local time on 17 August 1999 an earthquake of strength 7.4 on the Richter Scale shook north-western Turkey for approximately 45 seconds. It resulted in 18 000 deaths, 35 000 serious injuries, 60 000 dwellings being destroyed and over 200 000 people rendered homeless. Its **epicentre** was pinpointed at Izmit (population 500 000), an important industrial town 105 km to the east of Istanbul (Figure 5.22). Other towns which were seriously affected include Avcilar, Bolu, Bursa, Eskisehir and Golcuk. Widespread damage was also caused to some suburbs of Istanbul and shockwaves were felt as far away as Ankara (the Turkish capital city) 405 km away. The most seriously affected region covered almost 8000 km^2. Numerous aftershocks were recorded – some of them above 5.0 on the Richter Scale. This earthquake took place at the eastern end of the North Anatolian Fault – one of the most tectonically active faults in the world – whose effects along part of its length are very clearly described by an aerial observer in the newspaper extract in Figure 5.23.

The earthquake's immediate impact on the human population was considerable. The majority of the most serious casualties were residents of accommodation blocks, which had been hastily built to house migrants from much poorer, rural districts seeking better employment prospects (Figure 5.25). Many of these blocks were constructed of inferior materials such as beach sand instead of higher quality, but more expensive, quarried sands. The steel reinforcing rods most frequently used were so thin that they corroded very quickly and offered little additional structural strength. It is hardly surprising, therefore, that these blocks proved extremely vulnerable during an earthquake of such magnitude.

Unfortunately, corruption amongst both building contractors and local officials was widespread; it was commonplace for planning officials to overlook even the most serious failures to meet the government's theoretically-strict building regulations. Although the reaction of newspaper reporters immediately after the earthquake was to be utterly merciless in condemning these contractors, they later focused on the failure of government departments and regional planning authorities to monitor the implementation of building regulations. The reporters also accused both government and local officials of regarding their public appointments chiefly as personal money-spinners rather than opportunities to serve the communities which had voted them into office. Some of the rural-urban migrants were so poor that they had no alternative but to build their own shacks known locally as 'gecekonous', and their homes offered even less protection than the poorly-constructed accommodation blocks.

The earthquake victims in Golcuk were mainly naval personnel, whose barracks were swept away

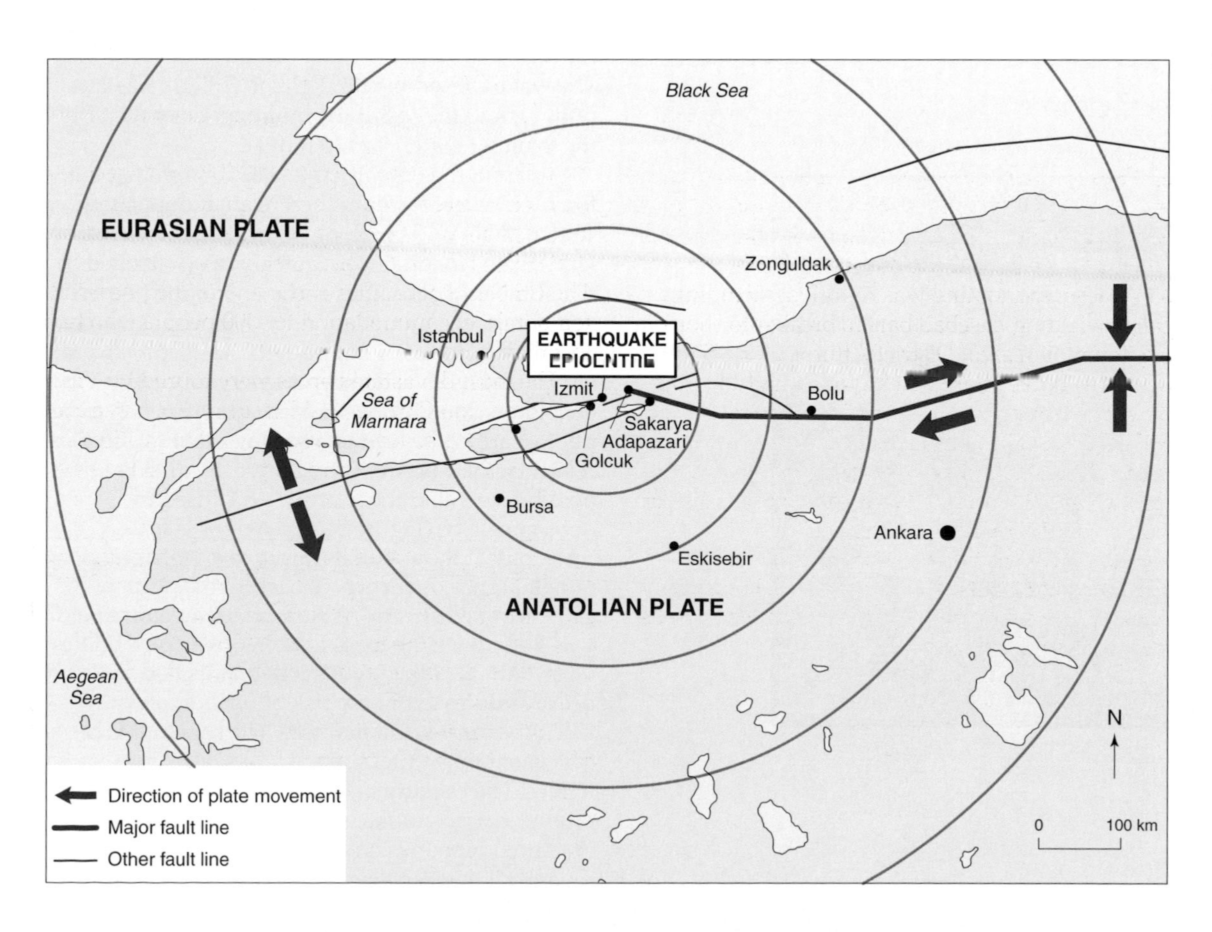

FIGURE 5.22 The location of the Turkish earthquake, 1999

Flying over the trail of death

The morning sky suddenly grew dark and a cold wind began to buffet the helicopter from side to side. The cause of this menace was a ball of fire still burning in Turkey's largest petrol refinery, which was surrounded by a black ball of smoke. A plane dived into the thick cloud to disperse its load of firefighting chemicals.

In the air, it is possible to follow almost all four kilometres of the fault. It begins near the small town of Arifiye, where an overpass once crossed the main Ankara highway. The fault wiggles across that very spot. An intercity coach crashed into the falling bridge and a dozen passengers died.

The fault then weaves its way towards an isolated housing estate. One five-storey building was still tottering but another seven had been turned to dust.

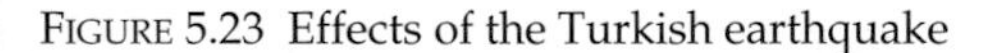

At Lake Sapanca it disappears, appearing to reappear on the other side and makes its way towards the sea. This is in fact a separate segment of the fault. The action of the two segments of the fault moving in opposite directions was responsible for the formation of the lake in the first place – "a classic example of a pull-apart structure," according to Aykut Barka, a geologist at Istanbul Technical University.

Professor Barka has documented this particular segment of the fault since 1991. It is a cruel cause for satisfaction to him that his model is being proved accurate by the evidence freshly etched into the ground. "The error looks to be about 230 metres on either side," he says. This means that the dimensions of the human tragedy *could* have been contained. "This earthquake should serve as a warning to Istanbul," Professor Barka said.

It is difficult, however, to see how Istanbul can take a lesson to heart in a city of more than ten million where much of the construction is of no better quality than that near Izmit.

Daily Telegraph, 20 August 1999

FIGURE 5.23 Effects of the Turkish earthquake

FIGURE 5.24 Rapid burial to reduce the risk of spreading disease

FIGURE 5.25 A damaged apartment block in the Marmora region

by a tsunami which had been triggered by the earthquake. Golcuk's victims included a Vice-Admiral and many other senior, highly qualified officers whose premature deaths would prove to be a serious loss to the Turkish Navy. A British Royal Navy Lieutenant-Commander visiting the Golcuk base at that time was amongst the fatalities. The onset of the tsunami at Golcuk was described by an eye witness as follows: 'The sea retreated at least 300 m . . . then there was a short pause . . . then this huge wave, which got bigger and faster as it came towards me, swept over everything and everyone. The wave chased me up flight after flight of stairs until I got to the top of the building. Only then did the foaming water start to retreat.'

At Izmit, the casualty rate was so high that the town's medical facilities there were immediately stretched beyond breaking point. The wards in Izmit State Hospital were quickly overwhelmed by the number of casualties and many of the injured had to be treated outside, in the hospital car park. Electricity and water supplies to the hospital were cut off and its own damaged emergency electricity generators could not be started. Fatalities had to be photographed instead of awaiting identification by relatives; their bodies were buried as quickly as possible, to avoid the health risks caused by decomposition (Figure 5.24). There was little ceremony at these burials, just a few words of remembrance over corpses hastily rolled up in carpets or bed sheets. At Adapazari, an entire field was dug up for the mass burial of over 1000 bodies. Over 100 000 of the Izmit's inhabitants fled to nearby hills to avoid the risk of further injury.

At the Tupras Oil Refinery, workers had to be evacuated when fuel storage tanks burst into flames. The refinery blaze lasted three days and its plume of acrid smoke stretched 40 km out to sea. Fractured pipelines created an offshore oilslick several kilometres long. This refinery was insured for losses up to £800 million, but only 10 per cent of Turkey's total assessed earthquake damage was covered by insurance, prompting the Prime Minister to create a special 'earthquake tax' to raise the money required to repair the damage.

At Degirmendere, a 7 m high tsunami created havoc along the whole length of the sea front, quickly followed by another smaller wave. A further 200 unusually large waves continued to pound the coast for several minutes. Fishing boats in the harbour were snapped like twigs or carried up to 50 m inshore. A passenger ferry was ripped from its offshore moorings and dumped in the middle of an amusement park. A huge refrigerator used by ice cream sellers on the promenade was lifted by the wave and deposited on the second floor balcony of a block of flats on the seafront.

The earthquake could not have come at a worse time for Turkey for a number of reasons. The country was just beginning to reduce its economically damaging high rate of inflation. One indication of Turkey's improving financial position was its increasing GDP per head of population relative to other countries.

Offers of international aid were immediate and generous. The IMF (International Monetary Fund) pledged £200 million; the countries of the EU (European Union) proved far more generous than Turkey's traditional Islamic neighbours in the Middle East and this led to a gradual shift of allegiance amongst Turkey's upper classes to those countries in the EU which had offered help without hesitation or thought of repayment. At key stages of the relief operation, over 2000 overseas volunteers were active in Turkey, including a team of highly skilled members of a British-based rescue team which remains on constant standby to respond to such emergencies anywhere in the world.

Most significant of all such offers of help was the airlift of ambulances and medical supplies funded by Greece (Turkey's traditional enemy) and humanitarian aid from Cyprus, with whom Greece had no political relations whatsoever. This help touched the hearts of many Turks and when the Greek capital itself suffered an earthquake shortly afterwards, the Turkish authorities had no hesitation in making similar offers of assistance.

The recovery work was long and arduous. The rescue teams had to operate in searing heat with daytime temperatures well over 30 °C (Figure 5.26). Rescue workers had to be immunised against cholera and dysentery and chlorine purification tablets were issued to them to ensure that their water was safe to drink. Heaps of rotting refuse lining the streets added to the health risk. Turkish children began to develop skin rashes – the result of constant heat and an almost total lack of washing facilities. Families in other parts of Turkey offered to shelter orphaned children for as long as necessary. Four days after the earthquake, vomiting and diarrhoea became an increasing problem – especially amongst children and the elderly. On the fifth day, the Turkish Prime Minister ordered rescue teams to cease searching for casualties and concentrate instead on bulldozing the rubble remains of devastated buildings. On 23 August, torrential rain made further rescue work virtually impossible and added to the misery of the survivors. Totally exhausted and dispirited, thousands of people left the cities in a mass exodus to seek rest and safety in the nearby countryside. Tent encampments were erected in public parks to shelter the homeless, but these were only made of thin material and proved totally inadequate by the onset of winter. Newspaper reports on 23 December reported that thousands of people were still living in these tents and beginning to suffer from hypothermia in temperatures below −8 °C (Figure 5.27).

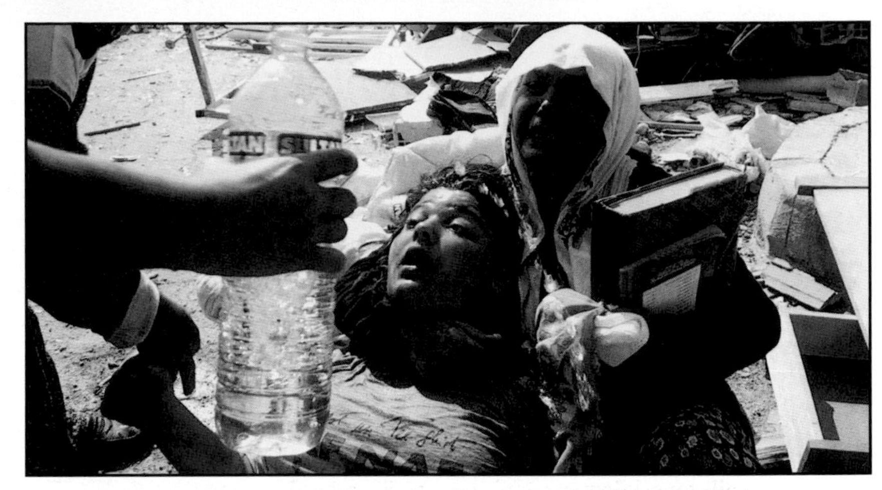

FIGURE 5.26 A dehydrated woman rescued from a collapsed home

FIGURE 5.27 Earthquake survivors battle against the cold

The resilience and stamina of the ordinary Turkish people proved to be quite remarkable. The WHO (World Health Organisation) maintains that 85–95 per cent of earthquake victims are likely to recover only if rescued within 48 hours, but a number of trapped people were recovered after 120 hours. Ultimately, a boy only five years old was rescued after surviving no fewer than 146 hours in the ruins of a collapsed building.

Later the same month, another serious earthquake (7.2 on the Richter Scale) shook the town

Murderers!
Headline in Hurriyet (Turkish Daily Newspaper).

It was all sand – just sand!
Words spoken by a woman who had just lost her home and thrown a piece of its concrete rubble onto the ground.

This was a disaster waiting to happen . . . I am ashamed of my country.
Maureen Freely, Novelist.

The blame for this earthquake is not nature's alone. The corrupt politicians who take bribes to allow these builders to make these wretched projects are guilty for all the deaths. Why did we let them build? Some of these buildings went up in only one month. I hope these people rot in hell.
Gunur Bilgin, earthquake survivor who lost her mother and four brothers in a collapsed building in which 123 other people died.

I implore all pilgrims to abandon their plans to visit the holy Islamic sites in Mecca and Medina for the time being and to donate their travel money instead to charities helping the earthquake victims. You will receive many more blessings from Allah if you do so.
Mehmet Nuri Yilmaz, Religious Leader in Turkey.

This wasn't a natural disaster; it was a human disaster.
Tseneji Rikitake, Director-General, Japan Earthquake Prediction Study.

This city has been built in the wrong place – across tectonic faultlines and on fragile soil, providing weak foundations.
Idris Kurtkaya, Deputy Governor of Adapazari.

On the road from Golcuk to Izmit yesterday, I smelt the death zone before I reached it. A thick, sweet stench hangs over the town. It fills your nostrils, making it impossible even to drink water or chew chewing gum without feeling sick.
Payne, Head of the British relief team "Rapid UK", which assisted with the post-earthquake rescue operations.

There is a risk of toxins being released during the oil refinery fire; these toxins could be washed out of the atmosphere by rain and onto the land. I strongly advise people living near to the oil refinery to leave their homes.
Osman Durmus, Turkish Health Minister.

The long-established Turkish tradition of a strong, centralised style of government is now being challenged. It is time to change. In western democracies, the state serves the people; at present, in Turkey, the people serve the state and this is wrong.
Ilnur Cevik, Turkish Daily News.

No government could have coped with the scale of the disaster which hit Turkey last week.' *and* 'The government promises to supply thousands of prefabricated houses to solve the housing crisis before the onset of winter. I also pledge the construction of entirely new settlements outside the earthquake zone with a year.
Buent Ecevit, Prime Minister of Turkey.

FIGURE 5.28 Quotations regarding the 1999 Turkish earthquake

of Duzce in Western Turkey, 185 km from Istanbul. It caused a further 362 deaths, seriously injured 2660 people and disrupted the main Istanbul–Ankara highway. On this occasion, government assistance was much speedier and the Turkish Army was observed to be very active in helping the general public.

The quotations in Figures 5.28 provide additional insights into the impact of the 17 August earthquake on the Turkish people and the depth of their feelings about the issues involved. Figure 5.29 outlines some positive and negative effects of earthquakes.

ACTIVITIES

1 Give a wide-ranging and detailed account of the many reasons for the severity of the 17 August 1999 earthquake in Turkey.

Positive effects

■ **International relations** Major earthquakes provide opportunities for countries which have been traditionally hostile to each other to improve their relationships. Beginning to co-operate on humanitarian and other 'safe' issues has already led to greater political stability in a number of regions which have experienced hatred and/or armed conflict.

■ **Research potential** Such events advance our knowledge of global natural processes. They also enable us to develop more effective protection against future disasters.

Adverse effects

■ **Vermin** Rats (which are carriers of many diseases such as The Plague) are highly sensitive to earthquakes and flock to the surface at the first sign of minor tremors.

■ **Fracture of underground systems** These include: broken sewers, which also spread diseases such as cholera; burst water mains, which disrupt supplies of safe drinking water and hinder the work of fire-fighting teams; split electricity cables, which make the night-time work of emergency teams extremely difficult. They also cause the failure of lifts, trapping people inside them and on the upper floors of high-rise buildings; fractured gas mains, which cause explosions when in contact with bare electricity cables; cut telephone cables, which hinder the exchange of up-to-date information and the passing of instructions to emergency services.

■ **Power stations disabled** Electricity is vital to all aspects of both normal life and the work of the emergency services. Food stocks are lost when refrigerators and freezers are disconnected. Damaged nuclear power stations may become a radiation risk.

■ **Burst reservoir dams** Flood valleys and low-lying coastal areas.

■ **TV/radio stations and newspaper offices destroyed** The general public cannot be warned of danger, informed of new developments or reassured by community leaders.

■ **Emergency facilities** Hospitals, fire stations and ambulance services cease to operate effectively.

■ **Looting** Food shops are likely to be looted by hungry people. Police are less able to monitor criminal activity, making banks more vulnerable to theft and shops stocking attractive goods such as alcohol and electronic goods more vulnerable to burglary.

■ **Disruption of transport infrastructure** Underground train services are immobilised, trapping commuters and preventing survivors from escaping the worst affected areas; fractured road surfaces and railway lines also prevent survivors from escaping to safer areas and delay essential food, water and medical supplies; bridges are especially vulnerable to earthquakes.

■ **Shattered buildings** These trap and injure occupants. Shards of glass from broken windows cause injuries, especially to areas of exposed skin. Many people are rendered homeless and without shelter, with the old and infirm at greatest risk from exposure to extremes of temperature.

■ **Schools/colleges destroyed** Education of school pupils and college students is disrupted, with serious long-term implications for the national economy.

■ **Aftershocks** Often occur after the main earthquake. These increase stress and panic amongst the population and heighten the risk of heart failure.

■ **Landslides** Often demolish buildings and block roads.

■ **Disruption to public gatherings** Earthquakes during the day or early evening may panic and cause injury to large numbers of people attending sports events and social gatherings.

■ **Damage to factories, offices and warehouses** Disruption to businesses leads to job losses and financial ruin for investors who have bought shares in companies.

■ **Financial cost** The cost of making new and existing structures earthquake-resistant is very high. The cost of replacing damaged buildings and services will increase the national debt and insurance companies may collapse due to the size of compensation claims.

■ **Monitoring** Countries such as Japan which are at constant risk from earthquakes have to maintain costly monitoring and research programmes.

■ **Tourism** Sudden, adverse publicity results in holiday cancellations and reduces the long-term attractiveness of popular holiday destinations.

■ **Pollution** Punctured garage fuel tanks and factory waste tanks are fire and health hazards.

FIGURE 5.29 Positive and negative effects of earthquakes

Study 5.8 Earthquake tremors – British Isles

The British Isles are in a comparatively 'quiet' area in terms of earthquake activity, simply because the region does not lie close to any major plate boundaries. It would, however, be wrong to regard this area as completely dormant. In fact, earthquake tremors do occur with surprising frequency within many regions of the British Isles. Figure 5.30 lists nine such tremors for the month of December 1999 alone, the majority of them fortunately below 3.0 on the Richter Scale, the level at which trees sway and items of furniture might be expected to fall over. Between 300 and 400 earthquakes are recorded in England, Scotland and Wales every year by the British Geological Survey's (BGS) network of over 140 seismic monitoring stations. In a typical year, 40 such events will exceed 2.0, about 20 of them being powerful enough to be detectable by local residents. BGS monitoring equipment is sufficiently sensitive to be able to differentiate between true earthquakes and the wide range of non-natural tremor sources such as quarry blasts, sonic booms and landslides.

Date of event	Location	Magnitude (Richter Scale)
3 December	Jura, Argyll and Bute	1.8
3 December	Jura, Argyll and Bute	2.2
5 December	Burnley, Lancashire	2.3
7 December	Anglesey, Gwynedd	0.4
8 December	Torridon, Highland Region	0.3
14 December	Altrincham, Greater Manchester	2.8
20 December	Gilmanscleuch, Scottish Borders	1.4
24 December	Papplewick, Nottinghamshire	1.3
26 December	Mount's Bay, Cornwall	1.1

FIGURE 5.30 Tremors in the British Isles, December 1999

The largest accurately-recorded British earthquake occurred in 1931 in the North Sea and is usually referred to as the Dogger Bank Earthquake. Its epicentre was some distance from the nearest inhabited coastlines and the damage caused was correspondingly minor for an earthquake measuring 6.1 on the Richter Scale. The largest onshore British earthquake took place in the Lleyn Peninsula of north-west Wales in 1984 and was followed by an unusually long sequence of aftershocks. Two less powerful earthquakes took place at the same location in 1992 and February 1994, with strengths of 3.5 and 2.9 respectively. In common with other earthquakes at or very close to 5.4, the 1984 shock was able to dislodge chimney pots, already-weakened sections of walls and crack underground water drains and sewerage pipes. Indoor items such as flower pots, ornaments, crockery and upright pieces of furniture were toppled to the floor and beds caused additional distress by sliding from one bedroom wall to another! Whilst it is possible that such earthquakes may have hastened the deaths of some elderly people already in a delicate state of health, the only fatalities proved to have been a direct consequence of earthquake action occurred in 1580 – when an unusually severe earthquake affected the Dover Straits and the adjacent land areas of south-east England.

The British public have become much more familiar with earthquake-related activity on the global rather than a local scale. This is due to the intensity of media coverage of the more dramatic overseas events and the relatively negligible consequences of British earthquakes. Nevertheless, greatly increased earthquake monitoring has been undertaken in recent years, to ensure that sensitive industrial installations such as nuclear power stations, radioactive waste disposal sites, petrochemical plants and networks of underground oil and natural gas pipelines are located in the least tectonically-vulnerable locations.

Research suggests that the British Isles might normally be expected to experience:

- One earthquake of at least 3.7 annually.
- One earthquake of at least 4.7 once every decade.
- One earthquake of at least 5.6 once every century.

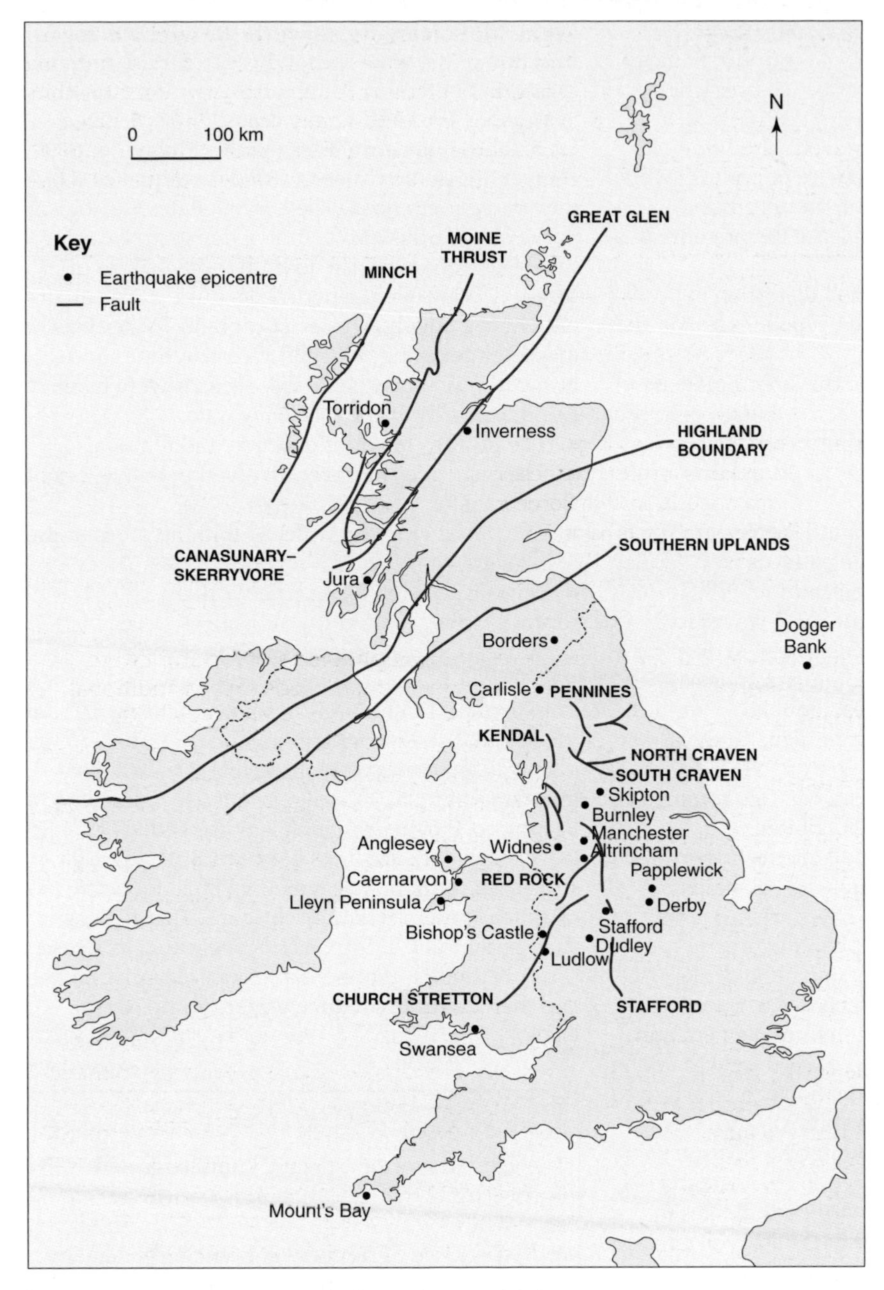

FIGURE 5.31 Earthquakes and faults in the British Isles

ACTIVITIES

1 List a wide range of problems which might be experienced in the British Isles close to earthquake epicentres; be prepared to add to the problems included in the text.
2 Describe the general distribution of British earthquakes as shown in Figure 5.31. You should refer to specific country/regional/county names as far as possible and locate the affected areas by using compass directions (e.g. north-west Scotland) whenever appropriate.
3 Study the two distribution patterns shown by Figure 5.31 and outline how closely, if at all, the earthquake distribution pattern matches the major fault line pattern.

Study 5.9 Earthquake prediction – Japan and the USA

Most people who live along or near to plate boundaries are acutely aware of the potential earthquake hazard which they face. Their responses to this threat are very wide-ranging and tend to be linked to their ability to take some form of positive action either as individuals or as members of a community. Experience of recent events suggests that any measures which can be taken to minimise the impact of earthquakes represent a very sound investment in both economic and humanitarian terms.

Many inhabitants of the world's poorest countries have little option but to adopt a philosophical or fatalistic approach. They are certainly not oblivious to danger but respond to it by following the example set by their parents and neighbours – which is usually to accept the inevitable, but hope and pray that the local area will not be affected. The absence of personal savings, the lack of family transport and consistently poor literacy rates are three key factors which limit people's ability to make effective preparations or respond speedily in the event of an emergency. Providing a basic education to young children is a priority for most LEDCs, but actually achieving higher standards is often a slow process.

It is, however, becoming increasingly apparent to countries and individuals alike that quite modest investments (often achievable only with governmental help or international aid) can make a significant difference to domestic safety. Figure 5.32 illustrates some recommended features for low cost earthquake-resistant housing and it is now standard practice for relief agencies to supply copies of such plans to communities engaged in disaster-recovery. These plans are usually printed on weather-proof cloth so that they can be used by many people and under the most difficult of conditions.

MEDCs (and those LEDCs whose financial situation is improving) are able to take much more effective measures than those discussed above. The examples in Figure 5.33 taken from California and Japan illustrate some of the options made possible by technological advances. Most of these illustrated measures are a direct response to the current trend for people to spend most of their time working, socialising and sleeping within high-rise buildings in densely populated areas. Another major concern which has accelerated research into earthquake hazards is the safety of earthquake-sensitive installations such as nuclear power stations and oil refineries. The frequency of earthquakes in Japan has led its capital city, Tokyo, to hold annual drills on 1 September (traditionally referred to as 'Disaster Day', in memory of the Great Kanto Earthquake which struck Tokyo in 1923).

Earthquake prediction is a combination of common-sense and technological support. Some of the most common and more effective prediction techniques are listed below:

- Monitoring water levels in wells and boreholes – these reflect the changing capacity of porous layers to hold liquids due to changing underground stresses which compress or release the pressure on their water-holding pores.
- Monitoring any unexplained distortion of surface features such as fences, pipelines, roads and railway lines.
- Use of satellites to monitor the lateral movement of surface features.
- Monitoring changes in ground level such as uplift or subsidence and water levels in lakes and reservoirs.
- Monitoring changes in ground surface gradients.
- Comparing current seismic patterns with similar patterns in the past, and the seismic activities which were induced by them; the ratio of P-waves to S-waves (Figure 5.34) has been observed to drop for a period of time then suddenly return to normal immediately before a major earthquake.
- Monitoring changes in local magnetic and gravitational fields.
- Observing animal behaviour e.g. rats coming to the surface and fish becoming agitated.

Our knowledge of fault movements (Figure 5.35) – the primary cause of earthquakes – remains imprecise, but is improving. There are numerous theories which attempt to explain the precise mechanics of fault movement, but all of these theories agree on one fundamental truth; that 'lubrication' of any kind aids their movement and hence increases the likelihood of earthquakes taking place. Within fault lines, lubrication can be achieved by softer, more mobile solids such as sand, clay or rock fragments, as well as genuine liquids such as water or oil. Study 1.10 of this book has already noted the increased tendency for earthquakes to occur in regions where major dam-reservoir projects have recently been completed – with specific reference to the Aswan High Dam on the Nile.

The 'liquids in faults' theory is borne out by observations of significant water content in faults exposed by erosion or human excavation. Further evidence became available in the 1960s, following an erratic but increasing pattern of seismic activity near to Denver, in the American state of Colorado – a region not previously noted for such activity. In the early 1960s, a series of earthquakes occurred near Denver. Between April 1962 and September 1963, local seismological stations registered more than 700 tremors having magnitudes of up to 4.2. A sharp decline in seismic activity took place in 1964, followed by a more frequent series of earthquakes during the next year. Retrospective investigation into this curious, irregular pattern of activity revealed that the US army had been injecting contaminated water from weapons production at its Rocky Mountains arsenal north-east of Denver. Injection of the water began in March 1962 and ceased in September for about a year. It resumed in September 1964 and finally ceased in September 1965, following complaints made by inhabitants of Denver about the water-induced earthquakes which they had experienced. The United States Geological Survey later utilised the knowledge gained during the 'Denver Experience' to undertake a series of carefully controlled experiments during 1969 at an oilfield in Rangely, western Colorado. Water was injected in existing oil wells at certain times and pumped out at others. Precise records were taken simultaneously of the following data:

- The timings of water injection and removal.
- The quantities of water involved in both types of water transfer.
- The pressure of the water within the pores of the rock layers.
- Seismic activity, using a network of seismographs.

Subsequent analysis revealed very high correlations between changes in pore water pressure and fluctuations in seismic activity. It follows that stresses along fault lines can be manipulated – maybe to the levels at which the more traumatic earthquake events are unlikely to take place. Further research is now needed to discover the precise links between water-transfer and stress-reduction. It may then be judged safe to use such strategies in densely-populated regions such as central California. One alternative is to use controlled underground explosions designed to halt and then gradually reduce underground stress levels.

ACTIVITIES

1 With reference to specific knowledge and events, argue the validity of the following statements:

a) 'Experience of recent events suggests that any measures which can be taken to minimise the impact of earthquakes represent a sound investment in both economic and humanitarian terms'.

b) 'Earthquake prediction can now be described as a reasonably precise science'.

c) 'Earthquake events tend to be easier to predict than volcanic eruptions'.

Study 5.10 Earthquake precautions – The USA

It is impossible to devise safety precautions guaranteed to protect people against all the dangers posed by serious earthquake and tsunami events. Earthquake precautions have, however, become

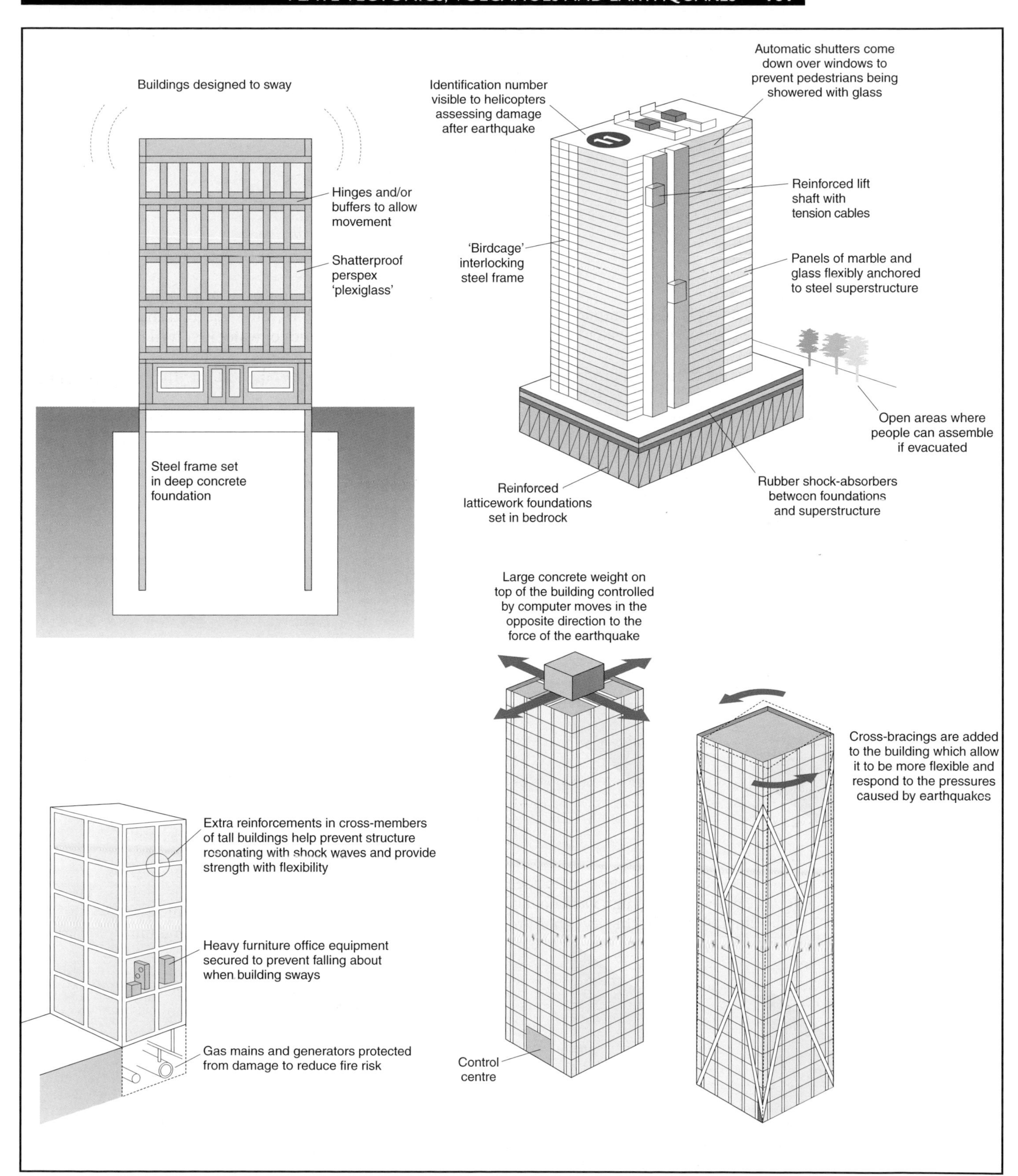

FIGURE 5.32 Earthquake resistant structures in MEDCs

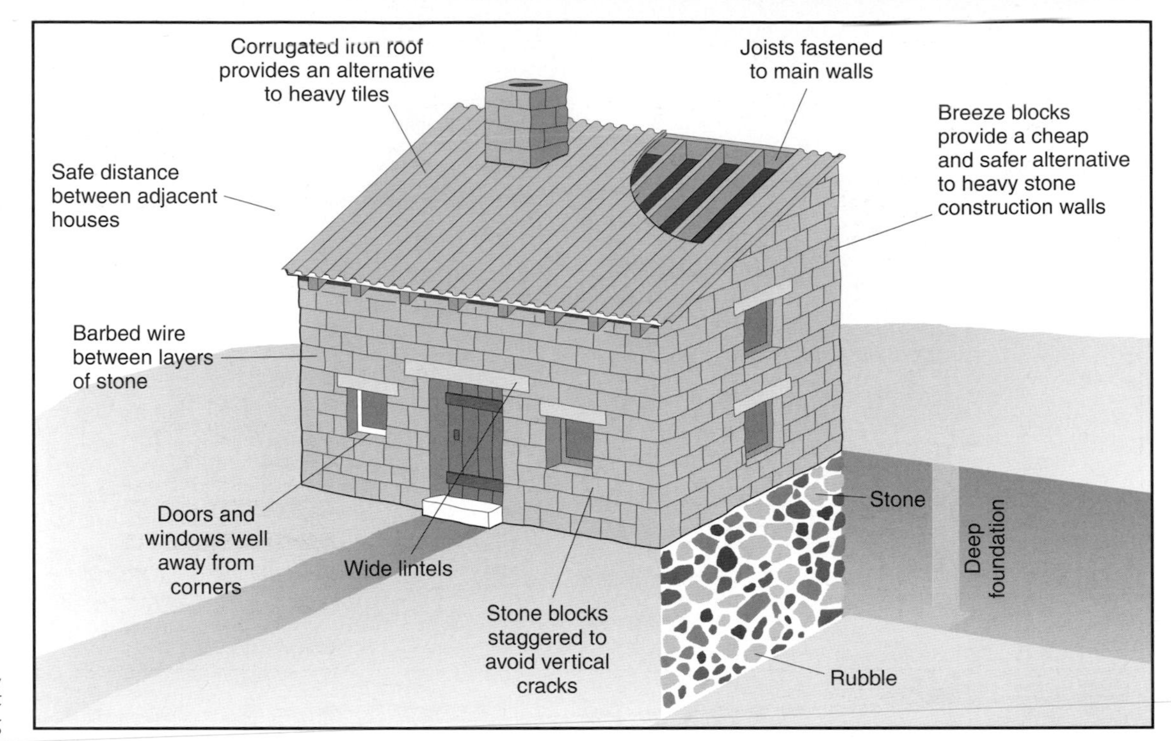

FIGURE 5.33 Low cost, earthquake resistant housing

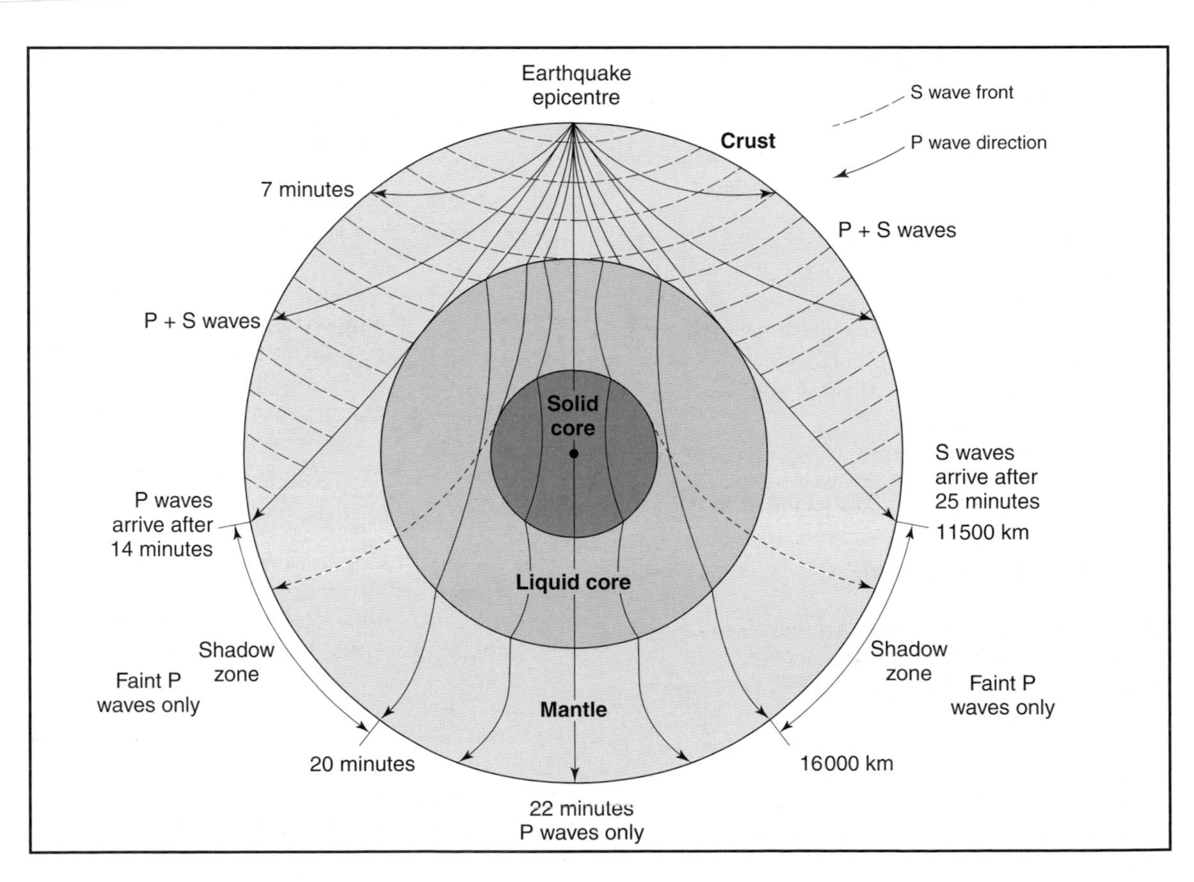

FIGURE 5.34 Shock waves triggered by earthquakes

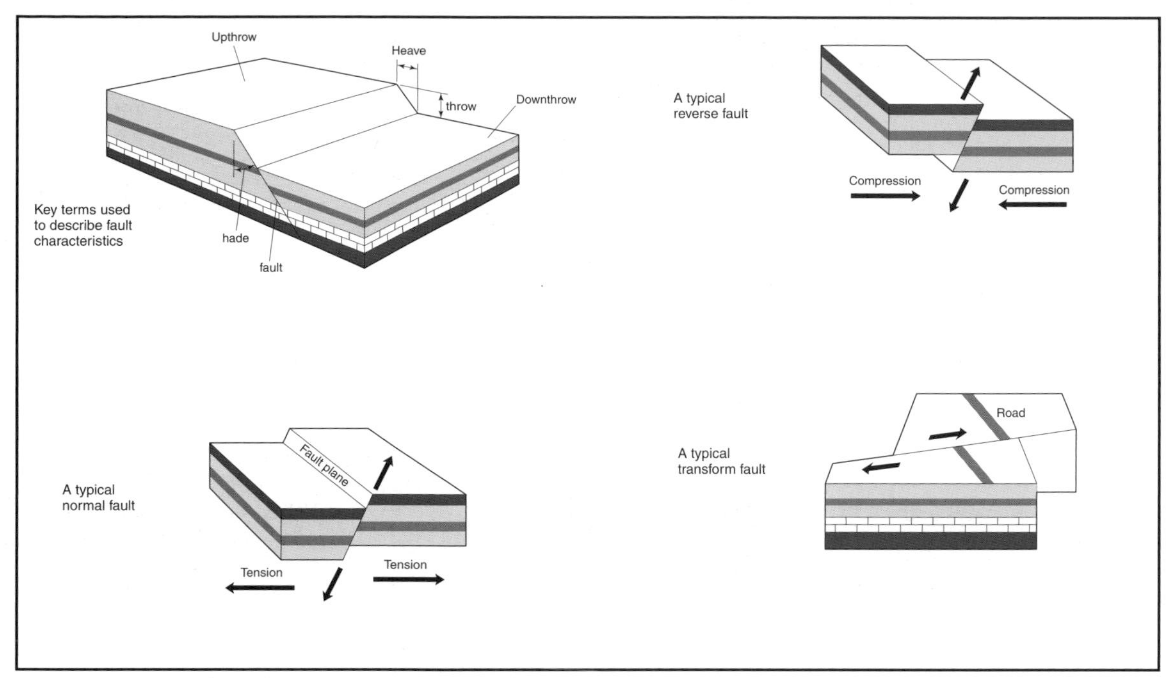

FIGURE 5.35 Crustal movements associated with faults

increasingly necessary, and much more difficult to put into effect, as people continue to be attracted to life in densely populated communities. Many of these are in lowland areas, where surface materials are so soft they vibrate easily or in coastal areas, which are vulnerable to tsunamis.

The following list of precautions is a summary of the recommendations currently issued by the USA's Federal Emergency Management Agency. Most of these precautions are basic common-sense and it may well be that their most useful function is to boost morale amongst populations living under the constant strain of seismic threat.

- Discover whether your house and place of work are in places at risk from natural hazards. If they are in a coastal area, find out your height above sea level, the distance you are from the nearest stretch of coastline and the locations and elevations of any nearby higher areas.
- Become familiar with the standard warning signs used to alert people to an increased threat.
- Be alert to any underground rumbling sounds or sub-surface tremors.
- When near to the shore, respond immediately to any sudden drop in sea level, because this almost certainly means that a tsunami will occur later.
- Devise a family evacuation plan which will enable you to escape from your local area and which includes at least one alternative escape route in case main roads are blocked.
- Make sure that all members of your family know how, where and when to turn off gas, electricity and water supplies to your house.
- Agree an emergency plan which will bring together members of your family in a safe meeting place.
- Ask a relative or friend who lives in a different area to act as a 'family contact' and make sure that everyone knows that person's telephone number. It is often easier to make long distance phone calls than trying to contact people within the same disaster zone as yourselves.
- Stay out of – and well away from – any damaged buildings.
- During an emergency, always test the quality of water before drinking it. Keeping ice cubes in your fridge/freezer will provide a source of emergency drinking water. Filling-up your bath as soon as an emergency is suspected will provide you with a longer term water supply.
- During an emergency, check for damage to electrical systems such as bare wires and dislodged wall sockets; always check for gas leaks before lighting matches or fires; avoid using your toilet if you suspect that sewage pipes are damaged.
- Maintain a stock of 'disaster supplies' which can

be reached at very short notice. This should include the following items:
- torch plus extra batteries
- toilet rolls
- portable, battery-operated radio plus extra batteries
- first aid kit and manual
- emergency food and water (rotate stocks of both at regular intervals)
- non-electric can opener
- range of basic medicines, pain-killers and other items which you would normally buy at a chemist
- cash and credit cards
- sturdy shoes and protective dry clothing.

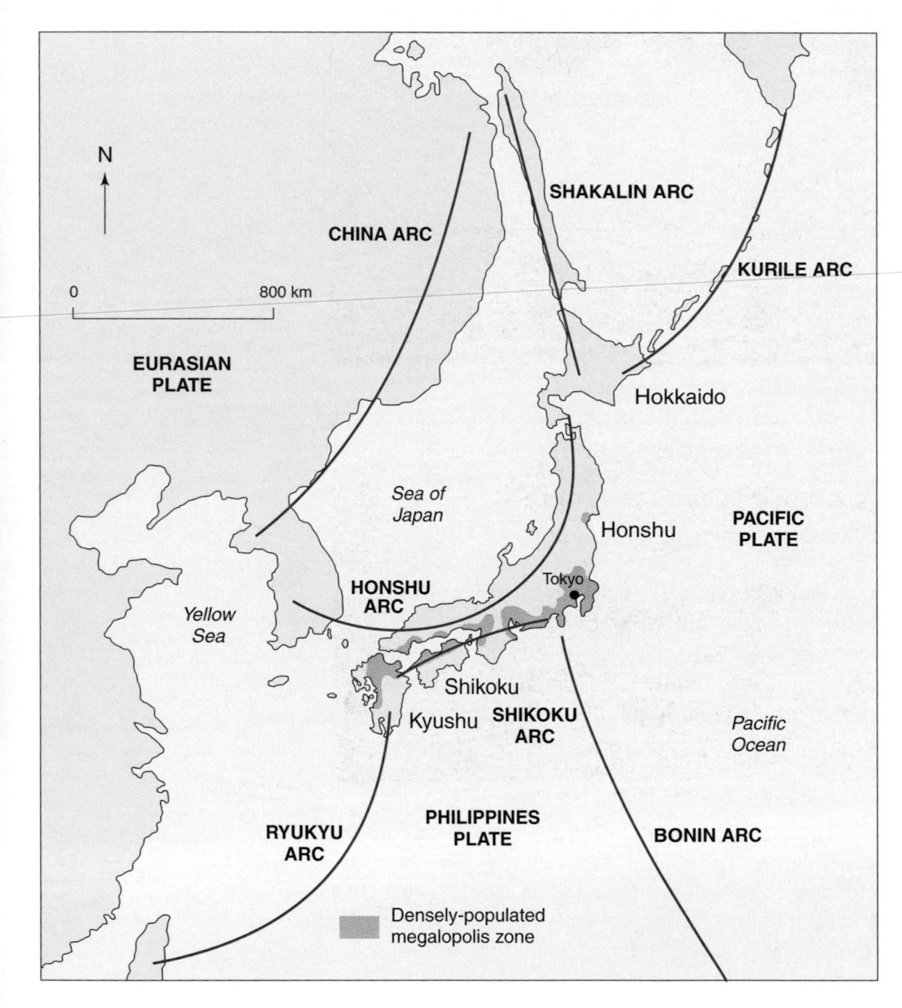

FIGURE 5.36 The Japanese Megalopolis

ACTIVITIES

1 a) Consider whether any important precautions have been left out of the Federal Emergency Management Agency's list.
b) Decide which precautions are suitable for including in a 'short-list' of no more than six priority items which could be given to children aged 8–11 to learn.

Study 5.11 Multiple hazards in the Japanese Megalopolis

This last section is somewhat different from all the previous case studies in a number of important ways:
- It provides an introduction to more than one type of physical phenomenon within its chosen region.
- It refers to hazards which are induced by the human population, as well as those which occur naturally.
- It invites you to identify possible 'links' between both of these groups of phenomena.
- It expects you, not the book, to be the chief provider of detailed information about these phenomena.
- The activities provide preparation for the synoptic type of question, which all examination Awarding Bodies are required to set as part of the final A2 assessment. The chief purpose of such questions is to test your ability to select information from a range of topics – often chosen from both the physical and human domains – and to use such information either to solve problems or to examine broadly-based issues. The ability to do this is important for advanced level as well as degree course students of any subject. By their very nature, synoptic questions do tend to be more open-ended, which means that they allow considerable flexibility in the choice of subject matter. Further opportunities to practice this type of question are given in the last set of questions in the Revision Section which follows.

The subject chosen for this unit is the Japanese Megalopolis (Figure 5.36). **Megalopolis** is the term reserved for the largest urban concentrations in the world, the others including the Ruhr in Germany and the USA's Atlantic and Pacific coastlines. The

Year	Location of hazard	Chief cause	Number of fatalities
1923	Tokyo	Earthquake (Richter 8.3)	143 000
1933	Sanrico Coast	Tsunami (30 m high)	4000
1948	Fukai	Earthquake (Richter 7.1)	4000
1959	Honshu	Typhoon Vera	5100
1959–85	Minamata	Mercury waste poisoning	1100
1991	Kyushu	Typhoone Mireille	62
1993	Hokkaido	Tsunami	250
1995	Kobe	Earthquake (Richter 7.2)	5000

FIGURE 5.37 Japanese hazards

FIGURE 5.38 a) Kobe earthquake, 1955; b) Minamata disease victim; c) Tsunami; d) Industrial air pollution; e) Landslide; f) River flooding; g) Typhoon Mireille, 1991; h) *Diamon Grace* oil spillage, 1997

Japanese megalopolis extends the full length of the southern and central sections of its Pacific Ocean coastline; it also includes the shores of the Inland Sea, between the islands of Kyushu, Shikoku and the mainland, Honshu. Like its American counterparts, this megalopolis is a closely-connected series of dense population concentrations – some of them the size of major global conurbations such as London and Paris. Tokyo, whose population is expected to exceed 25 million by 2015, is the 'primate' (by far the largest) conurbation within the megalopolis as well as being the national capital and one of its chief maritime, industrial and commercial centres.

There are many reasons why people tend to congregate in coastal locations. The Japanese are certainly the most extreme example of this, simply because they have little alternative but to occupy the 15 per cent of the total land area which is sufficiently low and flat to farm and build on. The existing available coastal land has been the subject of such intense demand that Japanese industrialists and port developers have resorted to creating 'reclaimed' land to satisfy their expansion plans (see study 2.10). Another important factor is Japan's almost total dependence on imported raw materials such as petroleum, coal, and the mineral ores needed by highly-efficient industries which have dominated world trade for almost half a century.

The Japanese have worked extremely hard to achieve their rapid industrial success but, until quite recently, have allowed any concerns for the state of the natural environmental to be totally eclipsed by their determination to become the world's most competitive producer of mass-produced high-tech goods. The result of this miscalculation has been a series of pollution incidents such as the outbreak of Minamata Disease in the 1950s, which proved to be so serious that it shattered the nation's previous complacency towards environmental issues. In addition to such hazards induced by human behaviour, the lowlands of Japan are also extremely vulnerable to a number of natural phenomena such as earthquakes, which are capable of causing widespread devastation. Study 5.9 examined some strategies adopted by Japan to minimise the potential effects of this particular type of hazard.

Figure 5.37 lists the most important types of hazards faced by the Japanese Megalopolis and its surrounding areas during the nineteenth century and Figure 5.38 illustrates such hazards.

ACTIVITIES

1 a) Investigate, in considerable detail, the Minamata Disease disaster and at least two examples of the natural types of hazard listed in Figure 5.37. Reference books and the Internet are likely to be your most fruitful resources when doing this. (A very important part of your research task is to discover how the Japanese people responded to each event.)

b) Write a summary of the relevant dates, circumstances and impacts which your research has revealed for each hazard which you have researched for a) above.

2 Now attempt to draw some valid generalised conclusions from your research summaries. In particular, you should make reasoned judgements about how the Japanese have regarded hazard events in the past, then decide whether these traditional attitudes have changed significantly over time. You should, of course, be able to justify your decisions by quoting relevant information from your earlier work for Activity 1.

Revision and synoptic activities

ACTIVITIES BASED ON O.S. MAPWORK SKILLS

Activities 1–4 are based on the O.S. map extract (Figure 5.39) on page 168 and the three photographs in Figures 5.40–5.42.

1 a) In which general compass directions would you be walking if you were to walk directly (i.e. take a straight-line route) each time from Loch an t-Siob to:
- the hut circles on the northern shore of Lowlandman's Bay?
- the summit of Beinn Chaolais?
- Leargybreck, on the A846?
- Lochan Gleann Astaile?
- Loch an Aircill?

b) What is the direct distance (in km) between the summits of Beinn an Oir and Corra Bheinn?
c) What is the total distance (in km) of the shoreline around Lowlandman's Bay from the headland at Sròn Gharbh to the most southerly tip of Rubh'an Lèim?
d) Write down the four-figure grid references of the squares which include the following features:
- Knockrome village
- the summit of Beinn Shiantairdh
- the western shore of Jura, which includes a natural arch and a stretch of raised beach
- Loch an Oir
- The largest area of woodland by the side of the A846.

e) Write down the six-figure grid reference positions of the following features:
- the milestone in grid square 5472
- the confluence of the Corran River and Allt an t-Sagairt
- the bridge by which the A846 passes over the Corran River
- the highest spot height by the side of the A846
- the more southerly of the two hut circles at the northern end of Lowlandman's Bay.

2 a) Of which type of surface drainage does the Corran River provide an excellent example?
b) Outline the chief reasons for your decision in a) above.
c) Suggest reasons which might help to explain the existence of a large lake in grid square 5173.
d) In what ways are the features of the Corran River system different in grid squares 5373 and 5471?
e) What are the key processes operating on the river system in each of the two grid squares above; how might these processes change its features in the foreseeable future?

3 The western and eastern coasts of Jura are very different from each other. Quoting evidence from the map extract and the two photographs in Figures 5.41 and 5.42:
a) describe some of the more significant differences between the two coastlines.
b) outline the chief reasons for the differences you have just described; when doing this, you should refer to local factors as well as basic physical processes.

4 The 'Paps of Jura' forms a cluster of mountains whose relief has been greatly influenced by rock-shattering as well as glaciation.
a) Explain why and how rock-shattering takes place.
b) Identify one type of feature in Figure 5.40 which has been caused by widespread frost-shattering.
c) Also with the help of Figure 5.40, identify the frost-shattering/glaciation features at and around each of the following grid reference locations: grid square 5176, 485728, 495758 and 507750.
d) Southern Jura is one of the most sparsely-populated districts within the Hebridean Islands of north-west Scotland. Account for this situation, with reference to both highland and lowland relief factors.

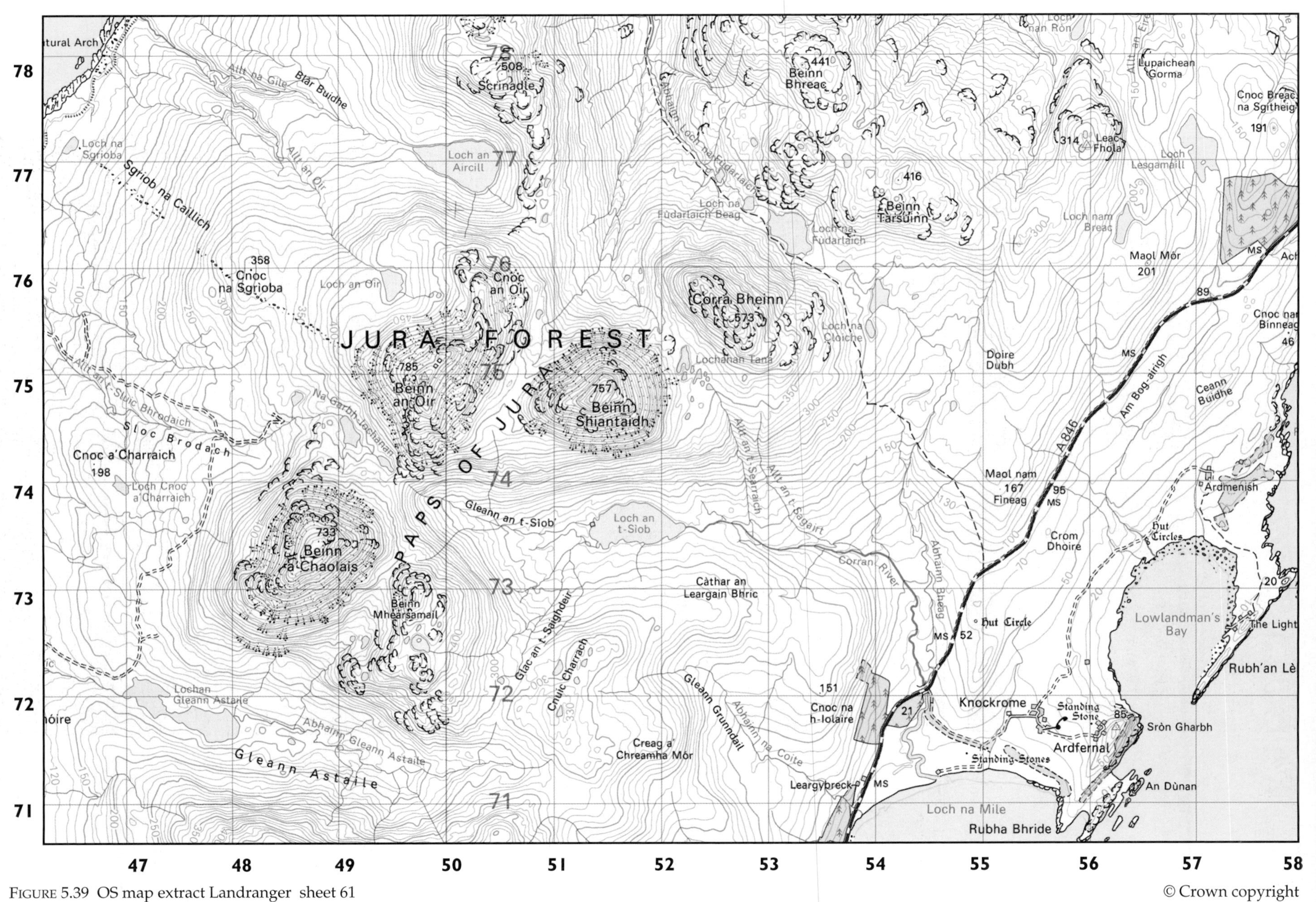

FIGURE 5.39 OS map extract Landranger sheet 61

FIGURE 5.40 Oblique-aerial view of Paps of Jura

FIGURE 5.41 Oblique-aerial view of the south-western coast of Islay

FIGURE 5.42 Oblique-aerial view of the south-eastern coast of Jura

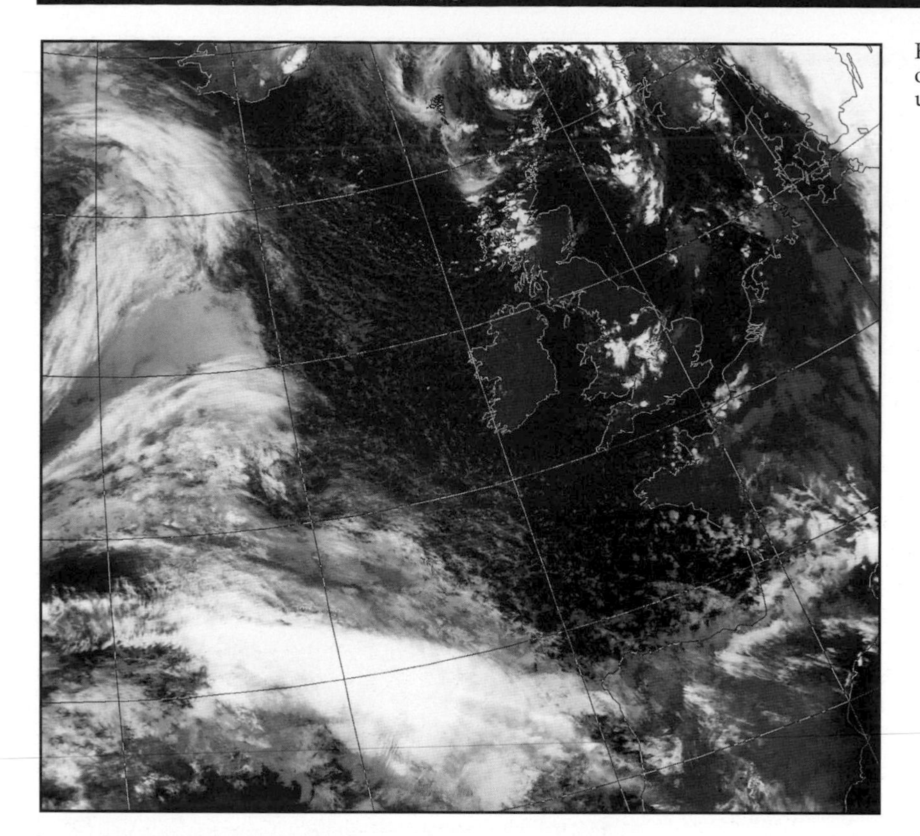

FIGURE 5.43 Satellite map of a weather system for use with Activity 11

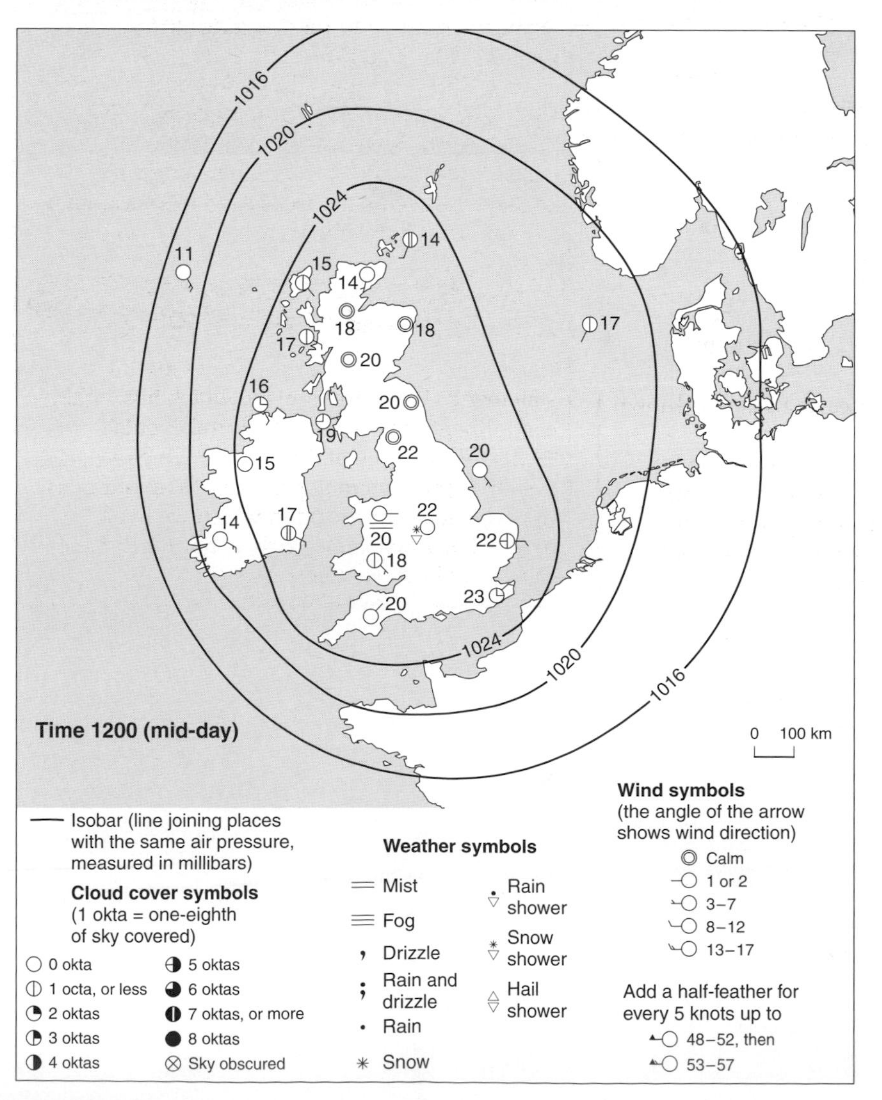

FIGURE 5.44 Synoptic chart of a second weather system for use with Activity 11

ACTIVITIES BASED ON SPECIFIC TOPICS

5 a) Describe ways in which 'physical' processes may lead to management problems on the floodplain of a large river.
b) Outline at least three possible river management strategies. For each strategy:
- explain how it is used to manage the river
- suggest how it might affect river channel processes and/or the river ecosystem.

6 With reference to a range of named examples, outline the more important advantages and disadvantages of building dams as part of a river management strategy.

7 Water management schemes aimed at ensuring adequate fresh water supplies tend to be motivated largely by political rather than environmental and 'quality of life' issues. Evaluate the likely validity of this statement, with reference to examples from both LEDCs and MEDCs.

8 For a flood event which you have studied:
a) Draw a hydrograph showing both precipitation input and river discharge response patterns.
b) Outline the major factors which contributed to that particular flood event.
c) Describe the immediate and longer-term effects of that flood event.

9 a) Explain how coastal sand dunes are formed.
b) Draw a fully-labelled transect diagram through a typical sand dune succession from beach to dune grassland/heathland.
c) Suggest reasons why species diversity tends to be variable across such transects.
d) In what ways are coastal sand dune environments under increasing threat from human activity; refer to named examples when responding to this question.

10 Explain how long term changes in relative sea level affect both coastal ecosystems and human activities in coastal regions.

11 a) Identify the two air pressure systems shown in Figures 5.43 and 5.44.
b) Identify, in some detail, at least five major differences between these two systems and the weather patterns which invariably accompany them as they migrate across the British Isles and the mainland of Western Europe.

12 a) What is a volcano?
b) Where, why and how do volcanoes tend to occur?
c) Describe a volcanic event which has occurred within the last 25 years, then assess its impact on the local environment.

13 a) Describe briefly how intrusive landforms occur.
b) Outline some of the economic benefits of such features.
c) For three different types of intrusive volcanic features:
- name one noteworthy example
- draw a labelled sketch of the feature showing clearly its mode of origin.

14 a) How does the changing overhead position of the Sun influence the temperature and rainfall patterns in different parts of the world?
b) Describe a range of constraints that tropical climates place on agricultural activity.
c) Explain how certain forms of natural vegetation have adapted to increase their chances of survival in tropical regions.

15 With reference to a range of named locations, make a judgement as to whether the legacy of past glaciation has proved to be more advantageous or problematical for the human population of the British Isles/western European region.

16 Antarctica is the most comprehensively 'protected' wilderness area on Earth. Outline the nature of its special environmental status, then argue the case either for or against the continuation of its protective status.

17 The 1999 earthquake in Turkey was described by one observer in the following terms: 'This wasn't a natural disaster; it was a human catastrophe'. Use your own knowledge of earthquake events to discuss and then evaluate this observer's statement.

ACTIVITIES BASED ON THE SYNOPTIC DIMENSION

18 Debate and then evaluate the following statement: 'There are limits to which humankind can influence the course of natural events'.

19 a) Outline some of the ways in which a systems approach may prove useful in our quest to understand the world in which we live.
b) In what ways might such an approach be capable of improvement?

Glossary

abiotic component non-living part of an ecosystem, biome or aquatic life zone
abrasion erosion caused by the movement of wind, water or ice *and* any transported material (e.g. sand/boulders)
acid rain precipitation that has absorbed sufficient sulphur or nitrogen to have a corrosive effect on rocks such as Carboniferous limestone
adaptation genetic characteristic that assists an organism's survival in response to its environment
adiabatic lapse rate rate at which air temperature changes with increasing/decreasing height above sea level
anticyclone area of high atmospheric pressure usually associated with dry, calm weather conditions
aquatic life zone combined marine/freshwater section of the ecosphere/biosphere
aquifer porous rock capable of storing liquids below the surface of the ground
arch opening through a headland formed when two caves are eroded backwards until they meet
arête knife-edged ridge between two adjacent cirques
artesian well borehole drilled into an aquifer
aspect direction towards which a slope/other feature faces
autotroph 'self-feeder'; alternative term for 'producer'
avalanche rapid down-slope movement of surface material, especially ice and snow
barrage dam constructed across an estuary or bay
barrier island island formed by a bar; separated from the mainland by a lagoon
bar deposit of sand/shingle formed across a bay/estuary
basalt fine-grained igneous rock formed when lava cooled
batholith large dome-shaped intrusion of igneous rock formed when magma cooled within Earth's crust
beach nourishment (restoration) strategies designed to counter the removal of beach sand by natural processes
bedding plane line of weakness between strata (rock layers) within a sedimentary rock
bifurcation ratio relationship between the number of streams of a particular order of magnitude within a drainage basin and those of the next highest order
biodiversity the number and variety of living organisms within a habitat
biomass total global quantity of organic material
biome naturally-occurring community of plant and animal life within a given terrestrial (land-based) area
biosphere (ecosphere) zone of Earth in which all forms of life occur
biota all the animals and plants that inhabit a given area
biotic component part of an ecosystem, biome or aquatic life zone that is living
biotic factor living element that contributes to any process or change
black box system system in which the inputs/outputs are identifiable but its processes are not known/understood
bog waterlogged area thickly covered by partially-decomposed vegetation
boreal (coniferous) forest type of forest occurring in latitudes above 55°
boss mass of intruded igneous rock; generally circular in shape with steep contact zones
boulder clay unsorted glacial material deposited by glaciers/ice sheets
browser animal that feeds by nibbling leaves and twigs
Carboniferous limestone massively-bedded, hard limestone deposited during the Carboniferous Period
carnivore animal that feeds on other animal life
cascading system related set of sub-systems through which energy and/or mass may pass
catchment area (drainage basin) region drained by a river and all its tributaries
cerrado Brazilian species-rich woodland savanna
chalk fine-grained, pure white limestone formed during the Upper Cretaceous Period
char low-lying island created by deposition within the Ganges Delta
chelation occurs when compounds bond to metallic ions especially aluminium, iron and magnesium; allows relatively insoluble minerals to become soluble/be released into soil
china clay (kaoline/kaolinite) deposits produced as the result of decomposition of the feldspar mineral within granite
cirque/corrie/cwm large, steep-backed glaciated hollow
climax community/vegetation final stage of succession; reached when a community is in equilibrium with its abiotic environment
clint block limestone pavement feature; individual block of limestone separated by grikes (grykes)
closed system system that allows the input or output of energy but not matter
community combined populations of species that live and interact in an area or environment at a specified time
community development (ecological succession) process by which plant and animal species are replaced by other (often more complex) communities over a period of time
conglomerate sedimentary rock formed when rounded fragments (often pebbles) are cemented together by finer-grained materials
consumer organism that cannot synthesise the nutrients it needs and so obtains these by feeding on producers (plants) or other consumers (animals)
continental drift theory suggesting that continents were formed by the break-up of a single land mass (*see* **plate tectonics**)

continental shelf area of shallow sea surrounding a land mass; water depth rarely exceeds 200m

control system type of model describing a natural system modified by human activity

convection(al) rainfall precipitation that results from upward convection in a mass of moist air; cooling causes condensation and further cooling results in rainfall

cool temperate maritime climate climatic pattern associated with the ocean fringes of mid-latitude locations; annual precipitation totals and temperature patterns are both influenced by the nearness of the ocean

cyclone see **hurricane**

decomposer organism that lives off dead organisms/the waste of living organisms

deforestation removal of trees from a forested area – without the intention of replacing them

delta area at the mouth of a river where the rate of deposition greatly exceeds the rate of erosion

depression area of low atmospheric pressure associated with variable; often wet weather conditions

desertification the change of grassland or marginal agricultural land into a true desert-like environment

detrital food web food web in which waste material (detritus) is the major food source

detritivore consumer that feeds on detritus

differential erosion erosion occurring at different rates due to variations in the resistance of adjacent rock areas

drainage density ratio between the combined lengths of all the streams within a drainage basin and its total area

drought occurs when an area receives significantly below-average precipitation

drumlin small hill composed of deposited glacial till

dry weight weight of any organism once all water/moisture has been removed; used to calculate biomass (as water has no nutrient/energy value)

dynamic equilibrium state of changing balance between components within a system/sub-system

early succession species low-growing/ground-hugging plants able to colonise harsh environments

ecology study of the interactions of living organisms both with each other and their abiotic environment

ecological optimum condition narrow range of ambient temperatures that actively facilitates the growth of a specific plant species

ecological succession process by which plant and animal species are replaced by different (often more complex) communities over a long period of time

ecosystem community of different species interacting with each other and their abiotic environment

ecotone transition from one ecosystem to another

eco-tourism tourism focussed on an appreciation and enjoyment of the natural environment

edaphic factor any element relating to the soil

effective moisture net soil moisture available for vegetation to use as and when required

eluviation downward movement of material through a soil profile as a result of water movement

energy flow pyramid way of illustrating/representing energy lost as it flows through the trophic levels of a food chain/web

ephemeral stream short-lived stream occurring only after heavy rainfall and/or when underlying permeable rocks become saturated

epicentre point on the Earth's surface directly above the focus of an earthquake

eutrophication physical, biological and chemical changes that occur within a body of water as a result of nutrient enrichment

fault fracture in a rock along which movement may occur

feedback response by a system to changes in its inputs

fetch maximum distance a wave may travel from its point of inception to where it breaks on the shore

fiord long, narrow, steep-sided 'U'-shaped glaciated valley flooded by the sea

flash flood occurs as a result of a temporary but very substantial increase in channel discharge

flood plain broad valley floor liable to flooding when a river overflows its banks

flow rate of movement of energy/matter through a system

focus point within Earth's crust where an earthquake originates

food chain flow of energy/nutrients through organisms at various trophic levels

food web complex inter-relationship between food chains

freshwater life zone aquatic life system occurring where the salt content of the water is less than 1%

frost-shattering (freeze–thaw action) weathering caused by alternate freezing/thawing of water in cracks on exposed rock surfaces

GDP (Gross Domestic Product) total value of all the goods purchased and services provided within a country

gelic podsol variety of podsol soil that develops where permafrost occurs within 20 cm of the surface

gleyed podsol variety of podsol soil that is result of long-term water logging of upper soil horizons due to the formation of hard-pan

glacier 'tongue' of downward-moving ice within a valley

global warming increase in Earth's overall temperatures; believed to be linked to the greenhouse effect

Gondwanaland ancient southern continent formed when Pangaea split into two parts; includes today's continents of Africa, India, Australia and Antarctica – together with some parts of South America

gorge deep, steep-sided river valley; usually caused by the up-stream retreat of a waterfall

granite course/large-grained igneous rock formed when magma cooled deep within Earth's crust

grazer animal that eats grasses and herbs

grazing food web food web in which energy flows from plants to herbivores (grazers) before reaching carnivores and decomposers (see **detrital foodweb**)

greenhouse effect warming of the earth's atmosphere due to the burning of fossil fuels

grike (gryke) joint in limestone widened and deepened by weathering

gross primary productivity rate at which an ecosystem's producers capture/store energy or biomass

groundwater water that percolates through soil/rock to become stored in natural underground reservoirs (aquifers)

growing season season of the year when vegetation growth is possible due to favourable rainfall and/or temperature conditions

groyne beach breakwater built to reduce longshore drift
gryke see **grike**
habitat place where an organism/population lives
halophyte plant that can survive in saline conditions
halosere plant succession evolving under saline conditions
hard-pan hardened, cement-like, impervious layer within the soil
herbivore animal that feeds on plants
heterotroph 'other-feeder'; alternative term for 'consumer'
hierarchical system representation of the world (or an environment) in which sub-systems are arranged hierarchically so that a system at one level becomes a component within a higher-order system
high-energy coastline stretch of coast along which erosional processes are dominant
hurricane strong tropical cyclonic (low air pressure) system formed over large ocean areas with temperatures above 27°C occurring between latitudes 5° N and S
hydraulic action erosion caused solely by the force of moving water
hydraulic quarrying use of high-pressure water jets to dislodge kaolinite (china clay)
hydrograph graph displaying river/stream discharge across time
(global) hydrological cycle world-wide system of water collection, purification, movement and distribution that re-cycles Earth's water supply
hydrosere plant succession occurring under submerged or waterlogged conditions
hydrosphere total volume of Earth's water store; as water, snow/ice and water vapour
ice age periods when glaciers/ice sheets covered large of land; warmer phases within ice ages called inter-glacials
ice budget balance between the annual accumulation of ice (input) and the loss of ice due to ablation (output – due mainly to melting and evaporation, avalanches and iceberg calving)
ice cap *either* an alternative name for an ice sheet *or* a small(er) mass of permanent ice and snow
ice sheet mass of permanent ice and snow that covers an extensive area of land and/or sea
illuviation deposition in the lower soil horizon of material removed from the upper soil horizon by the process of eluviation; hardpan may be created by this process
input introduction of energy or matter into a system
interference competition occurs when one species limits or prevents another species gaining access to a resource
interlocking spurs sloping ridges of higher land that project alternately from the sides of a river valley
isolated system any system that has no inputs or outputs of either energy or matter
isostatic change adjustment to the balance believed to exist between the Earth's highland and lowland areas
joint near-vertical line of weakness in sedimentary rock
kaolin (kaolinite) see **china clay**
lagoon shallow area of water in the sheltered zone behind a coastal spit, bar or barrier island
lahar flow of volcanic mud formed when water mixes with volcanic ash; highly destructive to objects in its path
Laurasia the more northerly of the two ancient continents into which Earth's land masses were originally split (North America and Eurasia)
leaching process by which dissolved material (e.g. plant nutrients) is removed from topsoil layers by the percolation of water through the soil
levée river bank created when sediment is deposited during periods of flood
limestone pavement exposed, level Carboniferous limestone area composed of alternate clint blocks and grikes
lithosphere the crust and the upper mantle combined; marked at its lower boundary by the Mohorovicic Discontinuity (Moho Line)
littoral cell length of coastline that is relatively self-contained so far as the movement of sand and shingle is concerned and where interruption to such movement should not have a significant effect on adjacent cells
llanos savanna or grassland area of the Orinoco Basin/Guiana Highlands in South America
longshore drift transportation of beach material along a coast due to wave action
low-sun period season of the year during which the sun is at its lowest point in the overhead sky at mid-day
marine (saltwater) life zone marine 'biome' associated with salt-water environments (>1% salinity)
mature plant community stable, undisturbed community based upon one or more long-lived plant species with the ability to sustain and replace itself
meander bend in a stream or river
megalopolis extensive urban area created by the sprawling together of a series of neighbouring conurbations
Mercalli Scale (Modified) measurement of earthquake intensity
mitigation banking American system whereby new wetlands are created and later bought by companies wishing to develop existing wetland sites; its aim is to maintain and possibly increase the overall area of established wetland sites
model representation (at a small-scale) of one or more aspects of the real world
monocot vegetation adaptation in which the vascular bundles of each plant are scattered throughout the stem; scorching of the trunk by fire does not kill the plant
monsoon climate climate type typified by the reversal of seasonal wind direction; generally found within the tropics; associated with heavy rainfall during one season contrasted with dry/drought conditions during the rest of the year
moraine mound of material deposited by melting ice; may be ground, lateral, medial or terminal moraine
morphological system simple model identifying the components of a system and the links between components
mudflat accumulation of mud deposits in sheltered waters; when colonised by vegetation, forms salt marsh
mutualistic relationship occurs when different species interact in ways that benefit all the participating species in some way
negative feedback occurs when a system as a whole

responds to an input by lessening the effect of the change(s) brought about by that input
net primary productivity (NPP) rate at which all the plants in a system produce useful energy
nunatak isolated peak standing above an ice sheet
nutrient flow movement of nutrients in chemical form through the abiotic environment – into organisms – and their return to the abiotic environment
Old Red Sandstone sandstone deposited during the Devonian Period
omnivore animal able to feed on both animals and plants
open system system that allows both energy and matter to enter and leave it
organism any individual animal or plant
orographic (relief) rainfall rainfall caused when moist air is forced to rise because of mountainous areas in its path
output mass, energy or change that occurs when an input passes though a system
outwash plain area of alluvial deposits created by deposition by meltwater streams
overgrazing result of the depletion of vegetation by the grazing of excessive numbers of animals; a major cause of soil erosion
palaeobiology study of living organisms throughout geological history using evidence in sedimentary rocks and fossils
palaeoclimatology study of climate conditions during geological history
palaeoecology study of ecological factors determining life during geological history
palaeogeography study leading to the reconstruction of the geography of the Earth's surface during geological history
Pangaea original super-continent believed to have existed before its division into Gondwanaland and Laurasia
perennial vegetation plant species that lives for more than two years
permafrost permanently frozen ground
photosynthesis process by which plants obtain food by converting energy from sunlight
pillar (column) cave feature formed when a stalactite joins with a stalagmite beneath it to form a single unit
pioneer species the first ground-hugging plants that begin the colonisation process on a newly-exposed surface – often mosses and lichen; the first stage of ecological succession
plagioclimax succession ecological succession that has been shaped or influenced by human intervention
plate tectonics theory suggesting that sea floor spreading 'moves' continental land masses across Earth's surface and creates new crustal material at mid-oceanic ridges; this process is balanced by subduction of existing crust at destructive plate margins
plunge pool pool at bottom of waterfall resulting from a combination of cavitation, eddying, hydraulic action and pothole erosion
podsol soil characteristic of coniferous (boreal) forests; leaching removes minerals and clay from the top horizons; these are then deposited by illuviation to form hard pan in the lower soil horizons
pollution the existence of energy or substances that cause undesirable change in the natural environment
positive feedback occurs when a system as a whole responds to input(s) by increasing the effect of the change(s) brought about by that input
precautionary principle protection measures introduced on the assumption that an environment is threatened; evidence or proof of ecological threat/damage need not exist
precipitation collective term for hail, rain, sleet and snow
predator organism that feeds by capturing and devouring organisms of another species
prevailing wind wind that blows most frequently from a particular direction, e.g. the prevailing south-westerly winds that the British Isles experience for most of the year
primary consumer organism that feeds on plants (ie a herbivore) or other producers
primary producer organism that uses solar or chemical energy to manufacture nutrients from inorganic compounds available in the environment
primary succession development of communities in an area that was previously bare and that had never before supported a community of organisms (see **secondary succession**)
prisere sequence of events over a period of time that produce a mature community from a barren expanse of land or water
process series of actions, developments or changes
process-response system system formed when morphological and cascading systems are joined in order to model how changes can be both inputs and outputs to a system/sub-system
psammosere ecological succession that develops on sand – especially sand dunes
pyramidal peak formed when three or more corries develop on a single mountain causing a sharp, steep-sided horn with arêtes radiating from the peak
pyroclast material blown high into the atmosphere during a volcanic eruption; particles vary in size from millimetres to metres in diameter
pyroclastic flow movement of pyroclastic debris from the source of a volcanic eruption
pyrophytic plants plants that have adaptations allowing them to withstand fire; often fire-resistant bark
quality of life overall standard of living of the human population in an area – based on such factors as the quality of housing, diet and medical provision
raised beach landform created when land rises relative to sea level, causing former beaches to rise; often backed by a relict cliff line with fossilized caves, stacks, etc.
re-afforestration the planting of trees on recently deforested sites
reclaimed land additional coastal land created by human activity/intervention
relatively stable community (see **mature plant community**) stable, uninterrupted community having the ability to sustain and replace itself
resource-partitioning process of dividing-up a scarce resource in an ecosystem so that different species requiring it have access at different times, in different ways or at different sites
resurgent stream underground stream forced to re-surface upon meeting a layer of impermeable rock

ria 'V'-shaped river valley flooded due to rising sea level
ribbon lake long, narrow lake within a glaciated valley
Richter Scale logarithmic scale used to measure earthquake strength
risk assessment means of evaluating the amount of damage an area may experience as a result of a disaster and/or the likelihood of a disaster actually happening
roche moûtonné small rocky outcrop on a valley floor formed by glacial erosion; its slope facing the approaching glacier is smooth and gentle due to abrasion whereas the opposite slope is steep and rough due to freeze-thaw action
salinity measure of a body of water's salt content
saltmarsh flat, sheltered coastal area that originally developed as tidal mudflats but is now vegetated
salt pan hollow that once contained water but now only contains salt deposits left when the water evaporated
sand dune pile of sand deposited then sculpted by wind
savanna tropical biome usually dominated by grasses; most common in areas subject to seasonal drought; may also be a plagio-climax response in some areas
scavenger organism that feeds off dead organisms
scree slope steep hill/mountainside covered by piles of angular rock fragments resulting from freeze-thaw action on exposed, higher sections of slope
sea floor spreading movement of crustal plates away from constructive plate margins resulting in new crust being formed by up welling magma
secondary consumer organism that feeds only on primary consumers
secondary succession development of a community in an area of land from which the previous vegetation cover, but not the soil, has been removed
sedimentary cycle passage of minerals/nutrients within sedimentary rocks through erosion, transportation and deposition processes; consolidation and uplift eventually expose the 'new' rock to erosion once again
sere particular kind of plant succession
slate metamorphic rock produced from fine-grained sedimentary material; well-marked cleavage lines allow it to be split into thin, parallel sections
soil horizon distinct, well-developed horizontal layer within a soil profile – distinguished by colour, texture and/or chemical composition
Spearman's Rank Correlation Coefficient statistical tool producing a numerical value indicating the degree to which there is a relationship between two sets of data
species group of organisms resembling one another in appearance, behaviour, chemical processes and genetic structure
spit long, low and often curved feature formed due to longshore drift and the deposition of beach materials such as sand and shingle
Site of Special Scientific Interest (SSSI) UK site designated as having especially rare/valuable biological, geological or physiographical features requiring protection
stack tall, narrow coastal feature that results from the erosion and collapse of the arch that previously linked it to the headland
store part of system/sub-system able to hold energy and/or mass
storm surge rapid rise in sea level, resulting in significantly higher than usual tidal level
stream ordering a way of ranking a length of stream/river within its own drainage basin
striation groove etched in rock surface; resulting from abrasion by rock fragments frozen within a moving body of ice
sub-system small part/section of a larger system; with identifiable inputs and outputs of its own
succession species species of different (more complex) plants or animals that replace pre-existing species
sustainability the ability of a population or system to survive over time
swallow hole (sink hole) hollow – usually in a Carboniferous limestone area – through which a surface stream enters the ground
system model that shows how parts of the 'real' world fit together or function
taiga see **boreal forest**
tertiary consumer animal that feeds on carnivorous species
tombolo beach or bar that joins an 'island' to the mainland
topography the pattern of surface features within an area (especially land height and slope gradients)
tor hill of exposed rock (usually granite)
transhumance seasonal movement of livestock between grazing lands near human settlement and distant pastures; the pasture land may be at higher altitudes or great distance from the home settlement
transitional zone (zone of transition) area that demonstrates a mix of use/species as the usage/biome to one side gives way to a different one on the other side
tree-line biome boundary beyond which trees cannot survive
trophic level 'feeding level' of an organism within an ecosystem; is an indication of the number of transfers between an organism and its original energy source
trophic (energy-flow) pyramid way of representing energy flow/loss between succeeding trophic levels
tropical storm cyclonic weather system developing within tropical latitudes; associated with cumulus cloud, heavy rainfall and strong winds
tsunami devastating sea wave – formed as a result of an earthquake above 5.5 on the Richter Scale; may also follow submarine volcanic activity/coastal landslides/the calving of large coastal ice masses
tundra northern continental plains experiencing cool summers and intensely cold winters; permafrost results in poor surface drainage; wetlands are common in summer
typhoon hurricane in the western Pacific Ocean region
urbanisation process by which there is an increase in the percentage (of the total population) living in towns and cities
washlands coastal area whose flooding is considered acceptable in order to protect more valuable areas further inland
water table level beneath which rocks are saturated
wave-cut notch zone of undercutting at the base of a cliff where erosive action by the sea is greatest
wave-cut platform gently-sloping shelf of rock caused by coastal erosion

wetland area of bund that frequently becomes waterlogged
wilderness area region unchanged by human activity/intervention; term originally referred to such areas in North America, but is now increasingly used to refer to similar regions worldwide
wind-chill cooling effect of air movement
xerophytic vegetation plants able to grow in dry conditions
xerosere plant succession occurring under dry conditions

Index